KIAS Springer Series in Mathematics

Volume 3

The KIAS Springer Series in Mathematics publishes original content in the form of high level research monographs, lecture notes, proceedings and contributed volumes as well as advanced textbooks in English language only, in any field of Pure and Applied Mathematics. The books in the Series are connected to the research activities carried out by the Korea Institute for Advanced Study (KIAS), and will discuss recent results and analyze new trends in mathematics and its applications. The Series is aimed at providing useful reference material to academics and researchers at an international level.

Sung-Soo Byun · Peter J. Forrester

Progress on the Study of the Ginibre Ensembles

 Springer

Sung-Soo Byun
Department of Mathematical Sciences,
Research Institute of Mathematics
Seoul National University
Seoul, Korea (Republic of)

Peter J. Forrester
School of Mathematics and Statistics
University of Melbourne
Melbourne, VIC, Australia

ISSN 2731-5142 ISSN 2731-5150 (electronic)
KIAS Springer Series in Mathematics
ISBN 978-981-97-5172-3 ISBN 978-981-97-5173-0 (eBook)
https://doi.org/10.1007/978-981-97-5173-0

Mathematics Subject Classification: 60B20, 15B52, 33C45, 60-02, 60K35, 60G55, 15A18, 82B05

This Springer imprint is published by the registered company Springer Nature Singapore Pte Ltd.
The registered company address is: 152 Beach Road, #21-01/04 Gateway East, Singapore 189721, Singapore

If disposing of this product, please recycle the paper.

Preface

By their very definition, the Ginibre ensembles are a class of non-Hermitian random matrix ensembles. At the same time, the statistical states formed by the eigenvalues in the complex plane permit interpretations in terms of the equilibrium statistical mechanics of particular two-dimensional Coulomb gases. Ginibre's work dates back to 1965 [293]. Come the decade of the 2020s and its lead-up, more and more researchers in mathematics and theoretical physics have been drawn to the field of non-Hermitian quantum mechanics on one front, and that of higher dimensional point processes relating to long-range potentials on another. With the Ginibre ensembles impacting on both of these broad themes, their relevance is on the increase. It is moreover the case that research on these themes has contributed to the development of the theory and applications of Ginibre ensembles.

With these circumstances in mind, the present project has been embarked on with the aim of giving an extended account of the Ginibre ensembles. The ongoing research programs of both co-authors have contributed to the original literature on this topic. In fact the senior co-author, Forrester, was introduced to the Ginibre ensembles GinUE (complex entries Gaussian matrices) and GinSE (quaternion entries Gaussian matrices) through their Coulomb gas analogies during his M.Sc. year in 1982. His supervisor, E. R. Smith, had just worked out the details of a solvable two-dimensional one-component plasma with Neumann boundary conditions [504]. This related to calculations reported in the first edition of Mehta's book [435] for the eigenvalue correlation functions of GinSE. The work of Smith was in turn inspired by the work of Jancovici, who in the year before had modified methods of study of the correlation functions for GinUE to obtain a solvable model of the two-dimensional one-component plasma with hard wall boundary conditions [337]. The research of co-author Byun on topics relating to the Ginibre ensembles began in 2017, at the commencement of his Ph.D. His supervisor, N.-G. Kang, had worked on developing the probabilistic and complex analytic articulation of the conformal field theory and its applications [352], which included the theory of Ward's equations for finite Boltzmann–Gibbs ensembles [61]. At the same time, as a participating doctoral student of the Bielefeld–Seoul graduate exchange program, International Research

Training Group 2235, Byun was mentored by G. Akemann, who has contributed vastly to the study of exactly solvable non-Hermitian random matrix models [13].

While the combined research of the co-authors relating the Ginibre ensembles provides a base for the present work, it is the sum effort of many researchers over several decades that has revealed the extent of knowledge as it now stands. We have been most happy immersing ourselves in many of these works, both to further inspire our own current research programs, and to assimilate the content for use in properly showcasing the extent of the theory and applications.

Accounts prior to the year 2010 of the Ginibre ensembles with emphasis similar to our own can be found in [237, Chap. 15, with several proofs], [362, some proofs sketched], with the latter overlapping mainly with Chap. 2. To make the presentation as self-contained as possible, this material is also part of the present work, albeit with some reordering and additional context. And when practical from the viewpoint of the space required, proofs are presented of a number of the results. Progress during the period from 2010 to the present has been rapid, as can be gleaned by inspection of the publication dates in the References section. While on the one hand our work consolidates the state of knowledge relating to the Ginibre ensembles up to the end of 2022, with their ever-increasing relevance, it is expected that subsequent years will see new applications emerge, which will in turn motivate a new generation of theoretical studies. We put forward the present work primarily to facilitate such a future, whereby foundations already laid can be built upon.

Seoul, Korea (Republic of)   Sung-Soo Byun
Melbourne, Australia   Peter J. Forrester
August 2023

Acknowledgements

This research is part of the program of study supported by the Australian Research Council Discovery Project grant DP210102887. SB was partially supported by the National Research Foundation of Korea grant NRF-2019R1A5A1028324, Samsung Science and Technology Foundation grant SSTF-BA1401-51, and KIAS Individual via the Center for Mathematical Challenges at Korea Institute for Advanced Study grant SP083201. The authors gratefully acknowledge Gernot Akemann, Christophe Charlier, Yan Fyodorov, Grégory Schehr, and Aron Wennman for helpful feedback on the first draft of this work. Furthermore, the support of the KIAS Springer Series in Mathematics editors, and the effort made by the referees in their reviews, is most appreciated and has been of much value.

Contents

Chapter 1
Introduction

The Ginibre ensembles are non-Hermitian random matrices with real, complex or quaternion Gaussian entries. For reasons to be discussed below, these are denoted by GinOE, GinUE and GinSE respectively. With the quaternion entries of the latter identified as the 2×2 complex block structure

$$\begin{bmatrix} z & w \\ -\bar{w} & \bar{z} \end{bmatrix},$$
(1.1)

members of each of these ensembles are readily realised; Fig. 1.1 gives example plots of the corresponding eigenvalue distribution in the complex plane. Ginibre introduced these non-Hermitian random matrix ensembles in 1965 [293]. Although with a self-described motivation of mathematical curiosity [31, Sect. 2.2, quoting correspondence with Ginibre], the line of investigation follows on from ideas and methods put forward by Dyson a few years earlier.

Thus in a fundamental paper on random matrix theory published in 1962, Dyson [205] isolated three ensembles of Hermitian matrices, and three ensembles of unitary matrices. This was done by seeking the minimal requirement of a quantum Hamiltonian H, respectively evolution operator U, to exhibit a time reversal symmetry or not, and then imposing a probability measure. First, for a time reversal operator T—defined in general by the requirement that it be anti-unitary—it was first shown that there are two possibilities, either $T^2 = \mathbb{I}$ or $T^2 = -\mathbb{I}$, with the latter requiring that the Hilbert space be even-dimensional. For T to commute with H or U it was then shown that in the case $T^2 = \mathbb{I}$ both H and U should be invariant under the transpose operation. In the other possible case, that $T^2 = -\mathbb{I}$, an invariance under the so-called quaternion dual $M \mapsto Z_{2N} M^T Z_{2N}^{-1}$ was deduced. Here Z_{2N} is the $2N \times 2N$ anti-symmetric tridiagonal matrix with entries all -1 in the leading upper triangular diagonal, and all 1 in the leading lower triangular diagonal, and moreover in this case it was shown that T has the realisation $T = Z_{2N} K$ where K corresponds to complex conjugation and $2N$ is the dimension of the Hilbert space. In the case of a quantum Hamiltonian H, it was then shown that $[H, T] = 0$ with $T^2 = \mathbb{I}$ implies a basis can be chosen so that the elements are real, while with $T^2 = -\mathbb{I}$ it implies that

© The Author(s) 2025
S.-S. Byun and P. J. Forrester, *Progress on the Study of the Ginibre Ensembles*, KIAS Springer Series in Mathematics 3, https://doi.org/10.1007/978-981-97-5173-0_1

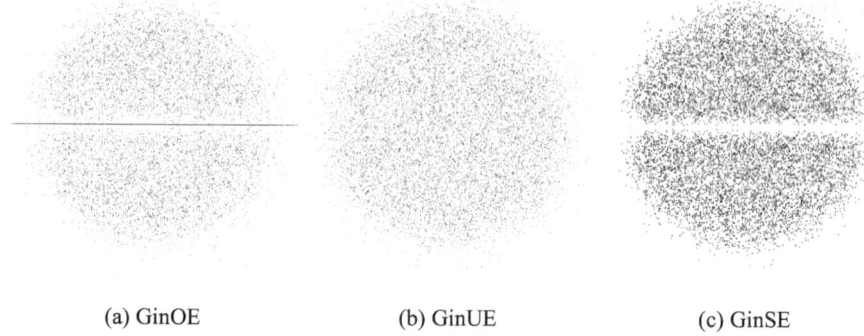

(a) GinOE   (b) GinUE   (c) GinSE

Fig. 1.1 Eigenvalues of 500 Ginibre matrices of size 20

the elements can be chosen to have a 2×2 block structure (1.1), which as already remarked can be identified with a member of the (real) quaternion number field. Hence for quantum Hamiltonians, Dyson was led to the requirement that matrices in his sought ensemble theory should have real entries, or have a 2×2 block structure corresponding to the quaternion number field in the presence of time reversal symmetry, or to be complex without a time reversal symmetry. This requirement is in addition to the matrices being Hermitian and thus having real eigenvalues and a matrix of eigenvectors which can be chosen to be unitary.

The work [205] specifies the eigenvalue probability density function (PDF) for an ensemble of quantum Hamiltonians H modelled as random matrices, and chosen from a Gaussian distribution on the elements proportional to

$$\exp(-\beta \operatorname{Tr} H^2/2). \tag{1.2}$$

The scaling factor β is chosen for convenience, and takes on the value of the number of independent parts of the corresponding number field—thus $\beta = 1$ for real entries (time reversal symmetry with $T^2 = \mathbb{I}$), $\beta = 2$ for complex entries (no time reversal symmetry), and $\beta = 4$ for quaternion entries (time reversal symmetry with $T^2 = -\mathbb{I}$). With this specification, β is referred to as the Dyson index. A detail is that Hermitian matrices commuting with the quaternion dual must have doubly degenerate eigenvalues (Kramer's degeneracy), with the convention in (1.2) that the trace operation relates to the independent eigenvalues only. The result of Dyson, known earlier in the case $\beta = 1$ by Wigner (see the Introduction section of the book edited by Porter [463] for references and moreover reprints of the original works) is that the eigenvalue PDF is given by [205, Eq. (146)]

$$\prod_{l=1}^{N} e^{-\beta \lambda_l^2/2} \prod_{1 \leq j < k \leq N} |\lambda_k - \lambda_j|^\beta, \tag{1.3}$$

up to proportionality. If instead of the distribution on elements being chosen as (1.2), a weighting

$$\exp(-\beta \operatorname{Tr} V(H)/2) \tag{1.4}$$

for some real-valued function $V(\lambda)$ is chosen instead, the modification of (1.2) is that it now reads

$$\prod_{l=1}^{N} e^{-\beta V(\lambda_l)/2} \prod_{1 \le j < k \le N} |\lambda_k - \lambda_j|^{\beta}. \tag{1.5}$$

Here it is being assumed that $V(\lambda)$ decays sufficiently fast at infinity for (1.5) to be normalisable. A weighting of the form (1.4) is said to specify an invariant ensemble, since it is invariant under conjugation by a unitary matrix $H \mapsto UHU^{\dagger}$ (and where too the elements of U are restricted to be real ($\beta = 1$) and quaternion ($\beta = 4$) so that H remains in the same ensemble). A weighting of the form (1.2) is said to specify a Gaussian ensemble.

Ginibre in his work [293] initiated a study of non-Hermitian Gaussian ensembles of matrices $G = [g_{ij}]_{i,j=1}^{N}$, with either real, complex or quaternion entries, by replacing (1.2) with the joint distribution on elements proportional to

$$\exp(-\beta \operatorname{Tr} G^{\dagger}G/2) = \prod_{i,j=1}^{N} \exp(-\beta|g_{ij}|^2/2). \tag{1.6}$$

Note that the second form in this expression shows that the real and imaginary parts (there are β such parts; e.g. in the quaternion case $\beta = 4$ there is one real and three imaginary parts) are all independent, identically distributed Gaussians. The concern of Ginibre was with the functional form of the eigenvalue PDF, and the implied eigenvalue statistics. The most obvious difference with the Hermitian case is that the eigenvalues are now in general complex. Also significant is the fact that the eigenvectors no longer form an orthonormal basis. Notwithstanding these differences, it was found in the complex case (referred to as the complex Ginibre ensemble, or alternatively as GinUE, where in the latter the U stands for unitary and refers to the bi-unitary invariance of (1.6) being unchanged by the mapping $G \mapsto UGV$ for U, V unitary matrices) that the eigenvalue PDF is proportional to

$$\prod_{l=1}^{N} e^{-|z_l|^2} \prod_{1 \le j < k \le N} |z_k - z_j|^2, \tag{1.7}$$

which is in direct correspondence with the Hermitian result (1.3) with $\beta = 2$, obtained essentially by replacing λ_j by z_j. We mention that the invariant non-Hermitian ensembles, known as the normal matrix models, with a general potential beyond the weight $|z|^2$ (i.e. the counterpart of (1.5)), have been extensively studied as well. See Chap. 5 for more details. For a classification of non-Hermitian random matrices in a similar spirit to Dyson's classification of Hermitian matrices explained above, we refer to Remark 6.3.

The eigenvalue PDF in the case of real and quaternion entries does not follow a direct correspondence with their Hermitian counterparts. First, in both these cases the eigenvalues come in complex conjugate pairs. Appreciating this point, the functional form of the eigenvalue PDF as found by Ginibre in the quaternion case can be obtained from (1.7) (not (1.3) with $\beta = 4$) by first replacing N by $2N$, then identifying z_{j+N} as $\bar{z}_j$ and ignoring terms which involve only $\{\bar{z}_j\}$ (which are thought of as part of the image system [245]) to obtain

$$\prod_{l=1}^{N} e^{-2|z_l|^2}|z_l - \bar{z}_l|^2 \prod_{1 \le j < k \le N} |z_k - z_j|^2 |z_k - \bar{z}_j|^2, \quad \text{Im } z_l > 0. \tag{1.8}$$

The real case is still more complicated. First, the eigenvalue PDF is not absolutely continuous. Rather, it decomposes into sectors depending on the number of real eigenvalues. Its precise functional form was not obtained in Ginibre's original work, with a further 25 years or so elapsing before this was achieved in a publication by Lehmann and Sommers [401], followed a few years later by an independent calculation of Edelman [211]. To present this, define the normalisation and the weight by

$$C_N^{\mathrm{g}} = \frac{1}{2^{N(N+1)/4} \prod_{l=1}^{N} \Gamma(l/2)}, \qquad \omega^{\mathrm{g}}(z) = e^{-|z|^2} e^{2y^2} \mathrm{erfc}(\sqrt{2}y) \tag{1.9}$$

respectively, where $z = x + iy$. Note that with $z = x$ and thus real, the weight simplifies to $\omega^{\mathrm{g}}(x) = e^{-x^2}$. The joint eigenvalue PDF for k real eigenvalues $\{\lambda_l\}_{l=1,\ldots,k}$ and the $(N-k)/2$ complex eigenvalues $\{x_j + iy_j\}_{j=1,\ldots,(N-k)/2}$ in the upper half plane (note that the remaining $(N-k)/2$ complex eigenvalues are the complex conjugate of these and so not independent) is then given by

$$C_N^{\mathrm{g}} \frac{2^{(N-k)/2}}{k!((N-k)/2)!} \prod_{s=1}^{k} (\omega^{\mathrm{g}}(\lambda_s))^{1/2} \prod_{j=1}^{(N-k)/2} \omega^{\mathrm{g}}(z_j)$$

$$\times \left| \Delta(\{\lambda_l\}_{l=1,\ldots,k} \cup \{x_j \pm iy_j\}_{j=1,\ldots,(N-k)/2}) \right|, \tag{1.10}$$

where $\Delta(\{z_p\}_{p=1,\ldots,m}) := \prod_{j<l}^{m}(z_l - z_j)$. Here $\lambda_l \in (-\infty, \infty)$ while $(x_j, y_j) \in \mathbb{R} \times \mathbb{R}_+$, $\mathbb{R}_+^2 := \{(x, y) \in \mathbb{R}^2 : y > 0\}$. In addition to the distinction in eigenvalue PDFs, we mention that GinUE forms a determinantal point process, while GinSE and GinOE form Pfaffian point processes. Further details can be found in Sects. 2.2, 7.5, and 10.4.

It would seem that the first occurrence of a Ginibre ensemble in applications (especially the real Ginibre ensemble or GinOE, where the "O" stands for orthogonal and refers to the bi-orthogonal invariance of the matrices) arose in the 1972 work of May [431] on the stability of complex ecological webs. Upon linearising about a fixed point, and the modelling of the fluctuations away from an attractor by a real Ginibre matrix G, May was led to the first-order linear differential equation system for the perturbed populations—an $n \times 1$ column vector $\mathbf{x}$—specified by

$$\frac{d}{dt}\mathbf{x} = (-\mathbb{I} + \alpha G)\mathbf{x}, \tag{1.11}$$

where α is a scalar parameter. The stability is then determined by the maximum of the real part of the spectrum of G, the precise determination of which has only recently become available in the literature [85, 161].

At the beginning of the 1980s the interpretation of (1.7), written in the Boltzmann factor form

$$e^{-\beta U(z_1,\ldots,z_N)}, \qquad U = \frac{1}{2}\sum_{j=1}^{N}|z_j|^2 - \sum_{1 \le j < k \le N} \log|z_k - z_j|, \quad \beta = 2, \tag{1.12}$$

as a model of charged particles, repelling pairwise via a logarithmic potential, and attracted to the origin in the plane via a harmonic potential, gained attention [41, 131]. This viewpoint was already prominent in the works of Dyson in the context of (1.3) (see too the even earlier work of Wigner [533] as reprinted in [463]), and was noted for (1.8) in Ginibre's original article [293]. In contrast to (1.3), in (1.12) the domain is two-dimensional, and the logarithmic potential is the solution of the corresponding Poisson equation (in one spatial dimension, the solution of the Poisson equation is proportional to $|x|$), so (1.12) corresponds to a type of Coulomb gas. Broad aspects of the latter have been the subject of the recent reviews [141, 403]. Also in the early 1980s, for all positive values of $\beta/2$ odd, (1.12) gained attention as the absolute value squared of Laughlin's trial wave function for the fractional quantum Hall effect [389].

Fast-forward 40 years, and there are now a multitude of applications which require knowledge of properties of Ginibre matrices, in particular their eigenvalues and eigenvectors. These topics will be further discussed in Chap. 6. This is due in no small part to a resurgence of interest in non-Hermitian quantum mechanics [67, 83]. Moreover, the progression of time has seen a much deeper understanding of the mathematical structures associated with the Ginibre matrices, and the theoretical progress has been considerable. It is the purpose of this book to give a presentation of a number of these advances, both in the theory and the applications.

Part I is on GinUE, and consists of four main themes. These form Chaps. 2 to 5: eigenvalue PDFs and correlation functions, fluctuation formulas, sum rules and asymptotic behaviours, and normal matrix models. There is also a sixth chapter entitled "Further theory and applications". Here the topics considered are the analogy between GinUE and the quantum many-body system for free fermions in the plane subject to a perpendicular magnetic field, the relevance of GinUE statistics to studies in quantum chaos, singular values and statistical properties, eigenvectors of GinUE matrices and non-Hermitian Wigner matrices. When excluding the Gaussian case, this latter class of ensembles has the distinct feature relative to the other matrix generalisations of GinUE considered in our presentation of not permitting explicit formulas at the level of the joint eigenvalue PDF.

Part II is on GinOE and GinSE. Of the themes considered in Part I, the one on eigenvalue probability density functions and correlation functions shows the most complete analogy between GinUE, GinOE and GinSE. This is notwithstanding the already-mentioned fact that the eigenvalues of GinUE form a determinantal point process, while those of GinOE and GinSE form a Pfaffian point process. Nor the fact that eigenvalue PDF of in the GOE is not absolutely continuous, but rather according to (1.10) divides into sectors depending on the number of real eigenvalues. We know in the theory of GinUE that there are elliptic and induced extensions, as well as one giving rise to a spherical ensemble, one coming from truncating Haar distributed unitary matrices, and a product ensemble of GinUE matrices or truncated Haar unitary matrices. These are distinguished by having an explicit formula for the joint eigenvalue PDF, with the correlation kernel relating to certain special functions, the properties of which allow for a detailed asymptotic analysis. We will see that each of these ensembles has a counterpart in the theory of GinOE and GinSE, which furthermore permit detailed asymptotic analysis making use of the same classes of special functions. To varying degrees of generality, fluctuation formulas, gap probabilities, sum rules, asymptotic expansion of the partition function and eigenvector statistics, discussed in Part I for GinUE, are again amenable to exact analysis.

Topics distinct from those seen in Part I show themselves. The fact that the eigenvalues form a Pfaffian point process is one reason for this. Thus associated with the Pfaffian structure is an underlying skew inner product, and associated skew polynomials. In GinOE theory, essential use is made of their form as a matrix average (7.18). But for GinSE, a relation with the corresponding GinUE orthogonal polynomials turns out to have a wider scope. Asymptotic analysis is more challenging for GinSE, with a technique based on differential equations found to be powerful. The topic of real eigenvalues is unique to GinOE. The corresponding statistics have features distinct to those exhibited in GinUE studies. They also provide for a number of applications, coming from areas as diverse as diffusion processes and persistence in statistical physics, topologically driven parametric energy level crossings for certain quantum dots, and equilibria counting for a system of random nonlinear differential equations.

Chapters 7–9 relate to GinOE, and Chaps. 10 and 11 to GinSE.

Part I
GinUE

Chapter 2
Eigenvalue PDFs and Correlations

A GinUE random matrix has all entries as independent standard complex Gaussians. The eigenvalue PDF can be explicitly calculated, revealing an analogy with the Boltzmann factor for a particular two-dimensional Coulomb gas. The functional form of the eigenvalue PDF leads to a determinantal structure for the general k-point correlation function. The corresponding correlation kernel admits bulk and edge scaling limits, while the density also admits a global scaling limit giving rise to an example of the circular law. Various generalisations of the GinUE permit analogous analysis. Those considered in the present chapter are the elliptic GinUE consisting of a linear combination of Hermitian and anti-Hermitian Gaussian random matrices; the induced GinUE which relates to a polar decomposition involving a rectangular generalisation of a GinUE matrix; the complex spherical ensemble of matrices $G_1^{-1}G_2$, with both G_1, G_2 GinUE matrices; the sub-block of a Haar unitary random matrix; and products of both GinUE matrices, and of truncated Haar unitary matrices.

2.1 Eigenvalue PDF

In the original paper of Ginibre [293], the diagonalisation formula $G = V \Lambda V^{-1}$, where Λ is the diagonal matrix of eigenvalues, and V is the matrix of corresponding eigenvectors which are unique up to normalisation, was used as the starting point to derive the eigenvalue PDF (1.7). The matrix V was then further decomposed $V = UTD$, where U is unitary, T is upper triangular with all diagonal elements equal to 1, and D is a diagonal matrix with real positive elements. As a consequence

$$\operatorname{Tr} G^\dagger G = \operatorname{Tr} \bar{\Lambda} B \Lambda B^{-1}, \qquad B = T^\dagger T. \tag{2.1}$$

This is independent of U and D, and B^{-1} is a simpler structure than V^{-1}, since it has determinant unity. It is necessary to integrate out the variables of B, which was

© The Author(s) 2025
S.-S. Byun and P. J. Forrester, *Progress on the Study of the Ginibre Ensembles*, KIAS
Springer Series in Mathematics 3, https://doi.org/10.1007/978-981-97-5173-0_2

done in N steps with each one consisting of integrating out over the last remaining row and column.

A more versatile (equally applicable to the GinOE, for example) method of derivation of (1.7) has since been found. It is due to Dyson, and first appeared in published form in [436, Appendix 35]. Here, instead of using the diagonalisation formula for G, the starting point is the Schur decomposition

$$G = UZU^\dagger. \tag{2.2}$$

Here U is a unitary matrix, unique up to the phase of each column, and Z is an upper triangular matrix with elements on the diagonal equal to the eigenvalues of G.

Proposition 2.1 *For GinUE matrices, specified by the distribution on elements proportional to (1.6) in the case $\beta = 2$ (complex elements), the eigenvalue PDF is equal to $1/C_N$ times (1.7), where upon relaxing the ordering constraint on the eigenvalues implied by (2.2), the normalisation constant C_N is specified by*

$$C_N = \pi^N \prod_{j=1}^{N} j!. \tag{2.3}$$

Proof We have from (2.2) that

$$\operatorname{Tr} G^\dagger G = \sum_{j=1}^{N} |z_j|^2 + \sum_{1 \le j < k \le N} |Z_{jk}|^2, \tag{2.4}$$

where $\{z_j\}$ denotes the diagonal elements of Z (which are the eigenvalues of G), and $\{Z_{jk}\}$ denotes the upper triangular elements. Note the simplification relative to (2.1).

After this brisk start, it is still quite a challenge to compute the Jacobian corresponding to (2.2). The strategy of [436, Appendix 35] is explained in more detail in [326], and repeated in [237, Proof of Proposition 15.1.1]. Here one begins by computing the matrix of differentials $U^\dagger dG\, U$, with the Jacobian corresponding to the (absolute value of) factor which results from the corresponding wedge product. For the latter task, proceeding in the order of the indices (j, k), with j decreasing from N to 1, and k increasing from 1 to N, gives the factor

$$\prod_{j<k} |z_j - z_k|^2. \tag{2.5}$$

However, the product of differentials so obtained is not immediately recognisable in the factorised form

$$\wedge_j dz_j^{\mathrm{r}} dz_j^{\mathrm{i}} \wedge (U^\dagger dU) \wedge_{j<k} dZ_{jk}^{\mathrm{r}} dZ_{jk}^{\mathrm{i}}, \tag{2.6}$$

where the superscripts indicate the real and imaginary part. Further arguing involving a count of the number of independent real variables associated with U, which implies some apparent differentials contribute zero to the wedge product, is required to make the simplification to this form. With (2.6) established, the integration over $\{Z_{jk}\}$ in (2.4) is immediate.

To deduce the normalisation (2.3) from this calculation requires first that the normalisation of (2.4) be included throughout the calculation, and second knowledge of the integration formula (see e.g. [194, Eq, (4.4)])

$$\int_U (U^\dagger dU) = \text{vol}\left(U(N)/(U(1))^N\right) = 2^{N(N-1)/2} \prod_{l=1}^{N-1} \frac{\pi^l}{\Gamma(l+1)}. \tag{2.7}$$

Finally, an extra factor of $N!$ is required in C_N to account for relaxing an ordering of the eigenvalues. □

2.2 Correlation Functions

With the joint eigenvalue PDF denoted by $p_N(z_1, \ldots, z_N)$, the k-point correlation function $\rho_{(k),N}(z_1, \ldots, z_k)$ is specified by

$$\rho_{(k),N}(z_1, \ldots, z_k) = N(N-1)\cdots(N-k+1) \int_{\mathbb{C}} d^2 z_{k+1} \cdots \int_{\mathbb{C}} d^2 z_N \, p_N(z_1, \ldots, z_N), \tag{2.8}$$

where, with $z := x + iy$, $d^2 z := dxdy$. In the simplest case $k = 1$ this corresponds to the eigenvalue density. Ginibre [293] showed that

$$\rho_{(k),N}(z_1, \ldots, z_k) = \det\left[K_N(z_j, z_l)\right]_{j,l=1}^k, \tag{2.9}$$

for a particular function $K_N(w, z)$, referred to as the correlation kernel. The structure (2.9) makes the eigenvalues of GinUE an example of a determinantal point process [103].

Proposition 2.2 *The kernel function in (2.9) is specified by*

$$K_N(w, z) = \frac{1}{\pi} e^{-(|w|^2+|z|^2)/2} \sum_{j=1}^N \frac{(w\bar{z})^{j-1}}{(j-1)!} = \frac{1}{\pi} e^{-(|w|^2+|z|^2)/2} e^{w\bar{z}} \frac{\Gamma(N; w\bar{z})}{\Gamma(N)}, \tag{2.10}$$

where $\Gamma(j; x) = \int_x^\infty t^{j-1} e^{-t} \, dt$ denotes the (upper) incomplete gamma function.

Proof The second equality follows from the first by the identity

$$\frac{\Gamma(N; x)}{\Gamma(N)} = e^{-x} \sum_{j=1}^{N} \frac{x^{j-1}}{(j-1)!}.$$

To deduce the determinantal structure (2.9) with K_N specified by the first equality in (2.10), the first step is to rewrite the product in (1.6) according to

$$\prod_{1 \le j < k \le N} |z_k - z_j|^2 = \prod_{1 \le j < k \le N} (z_k - z_j)(\bar{z}_k - \bar{z}_j), \qquad (2.11)$$

then to rewrite each of the product of differences on the right-hand side (RHS) as a Vandermonde determinant,

$$\prod_{1 \le j < k \le N} (z_k - z_j) = \det[z_j^{k-1}]_{j,k=1}^{N}; \qquad (2.12)$$

see e.g. [237, Exercises 1.9 q.1] for a derivation. Multiplying (2.12) by its conjugate and taking the transpose of the matrix on the RHS (which leaves the determinant unchanged) shows

$$\frac{1}{C_N} \prod_{l=1}^{N} e^{-|z_l|^2} \prod_{1 \le j < k \le N} |z_k - z_j|^2 = \det \left[K_N(z_j, z_k) \right]_{j,k=1}^{N}. \qquad (2.13)$$

The significance of the form (2.13) for purposes of computing the integrations as required by (2.8) are the reproducing and normalisation properties of K_N,

$$\int_{\mathbb{C}} K_N(w_1, z) K_N(z, w_2) \, d^2z = K_N(w_1, w_2), \qquad \int_{\mathbb{C}} K_N(z, z) \, d^2z = N. \quad (2.14)$$

Using these properties, a cofactor expansion along the bottom row can be used to show [208]

$$\int_{\mathbb{C}} \det[K_N(z_j, z_k)]_{j,k=1}^{m} \, d^2z_m = (-(m-1) + N) \det[K_N(z_j, z_k)]_{j,k=1}^{m-1}; \quad (2.15)$$

see also [237, Proof of Proposition 5.1.2]. Applying this inductively gives (2.9). □

According to (2.9) with $k = 1$ and (2.10), the eigenvalue density is given by the rotationally invariant functional form

$$\rho_{(1),N}(z) = \frac{1}{\pi} \frac{\Gamma(N; |z|^2)}{\Gamma(N)}. \qquad (2.16)$$

Ginibre [293] identified a sharp transition for $|z| \approx \sqrt{N}$ from the constant value $\frac{1}{\pi}$ for $|z|$ less than this critical value, to a value approaching zero for $|z|$ greater than this critical value. An equivalent statement is the limit law

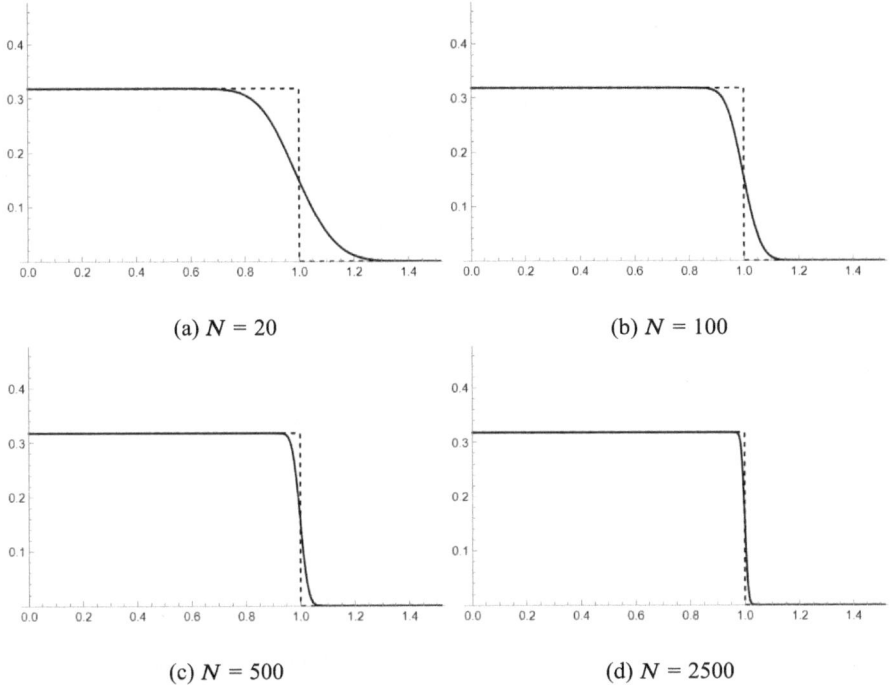

(a) $N = 20$ (b) $N = 100$

(c) $N = 500$ (d) $N = 2500$

Fig. 2.1   Graphs of $x \mapsto \rho_{(1),N}(\sqrt{N}x)$ (full line) and their comparison with $\frac{1}{\pi}\chi_{x<1}$ (dashed line)

$$\lim_{N \to \infty} \rho_{(1),N}(\sqrt{N}z) = \begin{cases} \dfrac{1}{\pi}, & |z| < 1, \\ 0, & |z| > 1. \end{cases} \tag{2.17}$$

(See Fig. 2.1.) This was the first example of what now is termed the circular law, which specifies the (global) scaled limiting eigenvalue density for a wide class of non-Hermitian random matrices, with identically and independently distributed elements to be constant inside a particular circle in the complex plane, and zero outside [68, 100, 294, 300, 302, 336, 514].

Remark 2.1   1. In the sense of probability theory, (2.17) is a statement in the mean, due to the ensemble average. The strong version of the circular law establishes that for large N the eigenvalues of a single GinUE matrix obey the circular law almost surely [100].

2. Define the numerical range of an $N \times N$ matrix X by $W(X) = \{\overline{X\mathbf{u}} \cdot \mathbf{u} \mid ||\mathbf{u}|| = 1\}$. By the variational characterisation of eigenvalues, for X Hermitian $W(X)$ must be contained in the interval of the real line $[\lambda_{\min}, \lambda_{\max}]$, and in fact is equal to this interval. It is proved in [408] that for X a global-scaled Ginibre matrix, and thus with eigenvalue density obeying the circular law (2.17), $W(X)$ converges to the centred disk in the complex plane of radius $\sqrt{2}$. Here, the annulus $\{z \in \mathbb{C} : 1 \leq |z| \leq \sqrt{2}\}$

of width $\sqrt{2} - 1$ can be interpreted as a quantification of the non-normality of the GinUE.

The correlation kernel (2.10), and thus the correlation functions (2.9), admit distinct scaling limits depending on the centring of the variables being in the bulk region, where the density is constant, or the edge region, where the density begins to decrease to zero. The first was specified in Ginibre's original paper [293], whereas the latter was not made explicit until some time later [252].

Proposition 2.3 *Let $K_N(w, z)$ be specified by (2.10). We have*

$$K_\infty^b(w, z) := \lim_{N \to \infty} K_N(w, z) = \frac{1}{\pi} e^{-(|w|^2 + |z|^2)/2} e^{w\bar{z}}, \qquad (2.18)$$

$$K_\infty^e(z_1, z_2) := \lim_{N \to \infty} K_N(-i\sqrt{N} + z_1, -i\sqrt{N} + z_2)$$

$$= e^{-(|z_1|^2 + |z_2|^2)/2} e^{z_1 \bar{z}_2} h\left(\frac{1}{2}(-iz_1 + i\bar{z}_2)\right), \qquad (2.19)$$

where

$$h(z) = \frac{1}{2\pi}\left(1 + \mathrm{erf}(\sqrt{2}z)\right).$$

Proof The limit (2.18) is immediate from (2.10), and the fact that for fixed $w\bar{z}$,

$$\lim_{N \to \infty} \frac{\Gamma(N; w\bar{z})}{\Gamma(N)} = 1.$$

The derivation of (2.19) relies on (the first term of) the asymptotic expansion [444]

$$\frac{\Gamma(N; N + \tau\sqrt{N})}{\Gamma(N)} = \frac{1}{2}(1 - \mathrm{erf}(\tau/\sqrt{2})) + \frac{1}{3\sqrt{2\pi N}} e^{-\tau^2/2}(\tau^2 - 1) + O\left(\frac{1}{N}\right). \qquad (2.20)$$

With the leading support of the eigenvalues the disk $|z| < \sqrt{N}$, it is natural to consider as a point in the bulk any $z_0 = s\sqrt{N} + it\sqrt{N}$ for some $|s|, |t| < 1$. Making use of the fact that $\Gamma(N; u)/\Gamma(N) \to 1$ for $|u|/\sqrt{N} \to c, c < 1$ as $N \to \infty$, as is consistent with (2.20), it follows that

$$\lim_{N \to \infty} K_N(z_0 + w, z_0 + z) = K_\infty^b(w, z)$$

independent of z_0; see also [104, Appendix C]. Similarly, for any $|\nu| = 1$,

$$\lim_{N \to \infty} K_N(\nu(\sqrt{N} + z_1), \nu(\sqrt{N} + z_2)) = e^{-(|z_1|^2 + |z_2|^2)/2} e^{z_1 \bar{z}_2} h\left(\frac{1}{2}(-z_1 - \bar{z}_2)\right) \qquad (2.21)$$

as calculated in [104, Appendix C with $s_k = \nu z_k$].

Generally, (2.9) gives for the appropriately scaled two-point correlation

$$\rho_{(2),\infty}(z_1, z_2) = \rho_{(1),\infty}(z_1)\rho_{(1),\infty}(z_2) - K_\infty(z_1, z_2)K_\infty(z_2, z_1).$$

For large separation of z_1 and z_2 the leading-order of the RHS is given by the first term which is the product of the densities. The second term $-K_\infty(z_1, z_2)K_\infty(z_2, z_1)$ must decay sufficiently rapidly for it to be square integrable, since the limiting form of the first integration formula in (2.14) remains valid,

$$\int_{\mathbb{C}} K_\infty(w_1, z) K_\infty(z, w_2)\, d^2z = K_\infty(w_1, w_2). \tag{2.22}$$

This can be verified from the results of Proposition 2.3 by the evaluation of appropriate Gaussian integrals. To separate off the product of densities, one defines the truncated (or connected) two-point correlation

$$\rho_{(2),\infty}^T(z_1, z_2) := \rho_{(2),\infty}(z_1, z_2) - \rho_{(1),\infty}(z_1)\rho_{(1),\infty}(z_2) = -K_\infty(z_1, z_2)K_\infty(z_2, z_1). \tag{2.23}$$

In particular, with bulk scaling, we read off from this and (2.18) that

$$\rho_{(2),\infty}^{b,T}(z_1, z_2) = -\frac{1}{\pi^2}e^{-|z_1-z_2|^2}, \tag{2.24}$$

which thus exhibits a Gaussian decay. With edge scaling, (2.23) and (2.19) give

$$\rho_{(2),\infty}^{e,T}(z_1, z_2) = -e^{-(x_1-x_2)^2-(y_1-y_2)^2}\left| h\left(\frac{1}{2}(y_1 + y_2 - i(x_1 - x_2))\right)\right|^2. \tag{2.25}$$

Use of the asymptotic expansion of the error function [450, Eq. (7.12.1)] gives to leading-order

$$\rho_{(2),\infty}^{e,T}(z_1, z_2) \underset{|z_1-z_2|\to\infty}{\sim} -\frac{1}{2\pi^3}\frac{e^{-2y_1^2-2y_2^2}}{(y_1 + y_2)^2 + (x_1 - x_2)^2}. \tag{2.26}$$

While this decays in all directions, parallel to the boundary of the leading-order density (i.e. in the x-direction) we see that the decay is algebraic, as an inverse square.

Remark 2.2 1. Generally for a matrix with eigenvalues $\{z_l\}_{l=1,...,N}$ the characteristic polynomial is given by $Q_N(z) := \prod_{l=1}^{N}(z - z_l)$. From the functional form (1.7) and the formula (2.8) we see that for GinUE

$$\rho_{k,N+k}(w_1, \ldots, w_k) = N(N-1)\cdots(N-k+1)e^{-\sum_{l=1}^{k}|w_l|^2}\prod_{1\le j_1 < j_2 \le k}|w_{j_2} - w_{j_1}|^2$$

$$\times \frac{C_N}{C_{N+k}}\left\langle \prod_{s=1}^{k}|Q_N(w_s)|^2\right\rangle_{p_N}, \tag{2.27}$$

where the average is with respect to the GinUE eigenvalue PDF $p_N(z_1, \ldots, z_N)$ of $\{z_l\}_{l=1,\ldots,N}$. In words, this says that for the $(N+k) \times (N+k)$ version of GinUE, the k-point correlation is up to multiplicative factors given by the average with respect to the $N \times N$ GinUE of a product of absolute value squared characteristic polynomials. As a generalisation of the average in (2.27) one can consider

$$I_k(u_1, \ldots, u_k; v_1, \ldots, v_k) := \left\langle \prod_{s=1}^{k} Q(u_s)\overline{Q(v_s)} \right\rangle_{p_N}.$$

This first arose in the work [449] where, using Grassmann calculus a $k \times k$ matrix integral evaluation was obtained. In the simplest case $k = 1$ the matrix integral reduces to an ordinary integral, giving

$$I_1(u, v) = \int_0^{\infty} e^{-s}(s + u\bar{v})^N \, ds;$$

see also [354]. Subsequently, it was noted in [40] that an orthogonal polynomial method underlying the proof of Proposition 2.2 gives

$$I_k(u_1, \ldots, u_k; v_1, \ldots, v_k) = \left(\prod_{l=0}^{k-1}(l+N)! \right) \frac{\det[K_N(u_{j_1}, v_{j_2})]_{j_1, j_2=1,\ldots,k}}{\prod_{1 \le j_1 < j_2 \le k}(u_{j_2} - u_{j_1})(\bar{v}_{j_2} - \bar{v}_{j_1})}.$$

2. It follows from (2.20) that the edge scaling of the eigenvalue density, in the coordinates of (2.19) with $z = x + iy$, has the large N expansion

$$\rho_{(1),N}^{e}(y) = \frac{1}{2\pi}\left(1 + \mathrm{erf}(\sqrt{2}y)\right) + \frac{1}{3\pi\sqrt{2\pi N}}e^{-2y^2}(y^2 - 1) + \mathrm{O}\left(\frac{1}{N}\right), \quad (2.28)$$

and thus in particular

$$\rho_{(1),\infty}^{e}(y) = \frac{1}{2\pi}\left(1 + \mathrm{erf}(\sqrt{2}y)\right), \quad (2.29)$$

where this latter expression is consistent with (2.19) upon setting $z_1 = z_2 = z$. See Fig. 2.2 for the graph of $\rho_{(1),\infty}^{e}$. There is interest in the functional form of the $1/\sqrt{N}$ correction term in our discussion of Sect. 4.2 below; see too [51, 397]. Integrating (2.29) over $y \in (-\infty, 0]$ and multiplying by $2\pi\sqrt{N}$ (the length of the bounding circle of the leading-order support) shows that to leading-order the expected number of eigenvalues with modulus greater than $\sqrt{N}$ is $\sqrt{N}/(2\pi)$ [293]. Note too that setting $y = 0$ in (2.29) gives $\rho_{(1),\infty}^{e}(0) = \frac{1}{2\pi}$, or equivalently $\lim_{N \to \infty} \rho_{(1),N}(\sqrt{N}) = \frac{1}{2\pi}$, which is exactly $\frac{1}{2}$ of the limiting value inside the unit circle as given by (2.17).

3. The two-dimensional classical Coulomb system interpretation (1.12) of the eigenvalue PDF (1.7) allows for (2.17) to be anticipated. For this, one scales $z_j \mapsto \sqrt{N}z_j$ and introduces a mean field energy functional

Fig. 2.2 Graph of $y \mapsto \rho^{e}_{(1),\infty}(y)$ in (2.29)

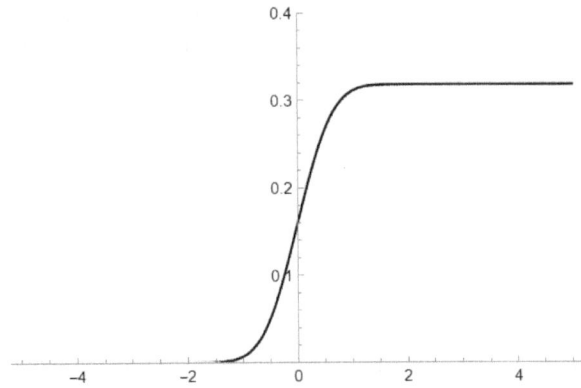

$$\frac{N}{2} \sum_{j=1}^{N} |z_j|^2 - \sum_{1 \le j < k \le N} \log |z_k - z_j|$$

$$\sim \frac{N}{2} \left(\int_{\Omega} \rho_{(1)}(z)|z|^2 \, d^2z - \int_{\Omega} d^2w \, \rho_{(1)}(w) \int_{\Omega} d^2z \, \rho_{(1)}(z) \log |z - w| \right), \quad (2.30)$$

with the hypothesis that $\rho_{(1)}(z)$ is chosen so that this functional is minimised and furthermore integrates over Ω to unity. Characterising the minimisation property by the vanishing of the functional upon variation with respect to $\rho_{(1)}(z)$ gives

$$|z|^2 - 2 \int_{\Omega} \rho_{(1)}(w) \log |z - w| \, d^2w = C, \quad (2.31)$$

where C is a constant, valid for $z \in \Omega$. Applying the Laplacian operation ∇_z^2 to this, using the standard fact

$$- \nabla_z^2 \log |z - w| = -2\pi \delta(z - w),$$

it follows that

$$\rho_{(1)}(w) = \frac{1}{\pi} \chi_{|w|<1}. \quad (2.32)$$

Here the restriction to $|w| < 1$ is implied by the rotational invariance, together with the minimisation and normalisation requirements. The notation χ_A denotes the indicator function of the condition A, taking on the value 1 when A is true, and 0 otherwise. The support Ω of $\rho_{(1)}(w)$ — which here is the unit disk — in such a Coulomb gas picture is typically referred to as the droplet (see e.g. [7, 322]). Moreover, this potential theoretic reasoning can rigorously be justified; see e.g. [141, Sect. 3.1] and references therein, as well as the discussion and references in the paragraph including (5.7) of Sect. 5 below.

4. Let X_0 have finite rank, and X be a GinUE matrix, scaled so that the leading support of the eigenvalues is the unit disk, and consider the sum $X_0 + X$. Assume too that the eigenvalues of X_0 are inside of the unit disk and near the boundary. In the limit $N \to \infty$ it has recently been shown that the edge correlation functions centred on the eigenvalues of X_0 form a determinantal point process with kernel involving generalisations of the error function [412].

2.3 Elliptic GinUE

In the 1980s, a one-parameter generalization of the non-Hermitian complex Gaussian matrices specifying GinUE was introduced [295, 507]. In this generalisation, varying the parameter allows for the Hermitian ensemble of Gaussian matrices known as the GUE (Gaussian unitary ensemble) to be obtained. A Hermitian matrix H from the GUE can be constructed from a scaled non-Hermitian GinUE matrix $\tilde{G} = (1/\sqrt{2})G$ according to

$$H = \frac{1}{2}\left(\tilde{G} + \tilde{G}^\dagger\right). \tag{2.33}$$

Proposition 2.4 *Let the parameters* $0 < \tau, v < 1$ *be related by* $\tau = (1 - v^2)/(1 + v^2)$. *For* H_1, H_2 *elements of the GUE, define*

$$J = \sqrt{1 + \tau}(H_1 + iv H_2). \tag{2.34}$$

The eigenvalue PDF of the ensemble of matrices $\{J\}$ *is given by*

$$\frac{1}{C_{N,\tau}} \exp\left(-\frac{1}{1-\tau^2}\sum_{j=1}^{N}\left(|z_j|^2 - \frac{\tau}{2}(z_j^2 + \bar{z}_j^2)\right)\right) \prod_{1 \le j < k \le N} |z_k - z_j|^2, \tag{2.35}$$

where $C_{N,\tau} = \prod^N (1 - \tau^2)^{N/2} \prod_{j=1}^{N} j!$.

Proof Following [275] and [237, Exercises 15.1 q.1], the starting point is to note that the joint element PDF of $\sqrt{1+\tau}H_1$ and $\sqrt{1+\tau}H_2$ is proportional to $\exp(-\frac{1}{1+\tau}\mathrm{Tr}(H_1^2 + H_2^2))$. The definition of J gives

$$\mathrm{Tr}\, H_1^2 = \frac{1}{2}\left(\mathrm{Tr}(JJ^\dagger) + \mathrm{Re}\,\mathrm{Tr}(J^2)\right), \qquad \mathrm{Tr}\, H_2^2 = \frac{1}{2v^2}\left(\mathrm{Tr}(JJ^\dagger) - \mathrm{Re}\,\mathrm{Tr}(J^2)\right).$$

As a consequence, it follows that the joint element PDF of J is proportional to

$$\exp\left(-\frac{1}{1-\tau^2}\mathrm{Tr}(JJ^\dagger - \tau\mathrm{Re}\,J^2)\right). \tag{2.36}$$

With this knowledge, the strategy of the proof of Proposition 2.1 leads to (2.35). $\square$

The correlations for (2.35) have, for an appropriate correlation kernel $K_N(w, z; \tau)$, the determinantal form (2.9). This involves the scaled monic Hermite polynomials

$$C_n(z) := \left(\frac{\tau}{2}\right)^{n/2} H_n\left(\frac{z}{\sqrt{2\tau}}\right), \qquad z \in \mathbb{C}. \tag{2.37}$$

Proposition 2.5 *In the notation specified above, we have*

$$K_N(w, z; \tau) = \frac{1}{\pi} \frac{1}{\sqrt{1 - \tau^2}}$$

$$\times \exp\left(-\frac{1}{2(1 - \tau^2)}\left(|w|^2 + |z|^2 - \tau(\operatorname{Re} w^2 + \operatorname{Re} z^2)\right)\right) \sum_{l=0}^{N-1} \frac{C_l(w) C_l(\bar{z})}{l!}.$$

$$\tag{2.38}$$

Proof Following the same procedure as used to begin the proof of Proposition 2.2, we modify (2.12) so that it reads

$$\prod_{1 \le j < k \le N} (z_k - z_j) = \det[p_{k-1}(z_j)]_{j,k=1}^N, \tag{2.39}$$

where $\{p_l\}$ is a set of monic polynomials, each p_l of degree l. Choosing these polynomials according to (2.37), with this choice being motivated by the orthogonality [193, 215]

$$\int_{-\infty}^{\infty} dx \int_{-\infty}^{\infty} dy\, e^{-x^2/(1+\tau)-y^2/(1-\tau)} C_m(z) C_n(\bar{z}) = \pi m! \sqrt{1 - \tau^2} \delta_{m,n},$$

the remaining working of the proof of Proposition 2.2 establishes the result. $\qquad\square$

Applying the Coulomb gas argument of Remark 2.2.2, with the scaling $z_l \mapsto \sqrt{N} z_l$, we conclude that within some domain Ω the density is constant, taking the value $\rho_{(1)}(w) = 1/(\pi(1 - \tau^2))$. This domain, or equivalently droplet, can be determined to be an ellipse with semi-axes $A = 1 + \tau$, $B = 1 - \tau$ and area equal to $\pi(1 - \tau^2)$. The shape can be verified directly, by showing that with Ω so specified, and $|z|^2$ in (2.31) replaced by

$$\frac{1}{1 - \tau^2}(|z|^2 - \tau \operatorname{Re} z^2), \tag{2.40}$$

the required minimisation equation is indeed satisfied [155], [237, Exercises 15.2 q.4]; see also [114, 193, 257, 397]. This droplet shape explains the terminology elliptic GinUE in relation to (2.35). For complex non-Hermitian matrices (2.34), now with the Hermitian random matrices H_1, H_2 constructed from (2.33) with general zero mean, unit standard deviation identically distributed entries that are not required to be Gaussian, a constant density in an ellipse was first deduced by Girko [295] upon additional assumptions. A complete proof has been given by Nguyen and O'Rourke [446].

As for the GinUE, as the boundary of the leading support of the elliptic GinUE is approached, there is a transition from a constant density to a density which decays to zero. Beginning with (2.38) in the case $w = z$, the analysis of the eigenvalue density in this edge regime is more complicated than for the deduction of (2.20). It was carried out by Lee and Riser [397]; see also [25, 441].

Proposition 2.6 *Consider the elliptic GinUE, with the eigenvalues scaled $z_j \mapsto \sqrt{N} z_j$ so that for N large the leading-order support is the ellipse Ω. Let z_0 be a point on the boundary of Ω. Denote the unit vector corresponding to the outer normal at this point by* **n**, *and the corresponding curvature by κ. We have*

$$\rho_{(1),N}\left(z_0 + \frac{(\alpha + i\beta)\mathbf{n}}{\sqrt{N}}\right) = \frac{1}{2\pi}\left(1 - \mathrm{erf}(\sqrt{2}\alpha)\right) + \frac{\kappa}{\pi\sqrt{2\pi N}}e^{-2\alpha^2}\left(\frac{\alpha^2 - 1}{3} - \beta^2\right) + \mathrm{O}\left(\frac{1}{N}\right).$$
$$(2.41)$$

Remark 2.3 1. The leading term in (2.41) is identical to that in (2.28) with the identification $y = -\alpha$. The universality of this functional form has been established for a wide class of normal matrix models (see Sect. 5 in relation to this class), and moreover extended to the edge-scaled correlation kernel (2.19) [61, 324]. This has recently been proved too for non-Hermitian random matrices constructed according to (2.33) with the elements of G identically distributed mean zero, finite variance random variables [159].
2. Consider the elliptic-shaped domain

$$\{z \in \mathbb{C} : 1 - \frac{1}{1 - \tau^2}(|z|^2 - \tau\,\mathrm{Re}\,z^2) > 0\}.$$

Let the function on the left-hand side (LHS) of the inequality be denoted by $1 - f(z)$. We see that $f(z)$ coincides with (2.40). It has been shown in [34, 443] that the polynomials $\{p_n(z)\}$ satisfying the orthogonality

$$\int_\Omega p_m(z)p_n(\bar{z})(1 - f(z))^\alpha \, d^2z \propto \delta_{m,n},$$

are simply related to the Gegenbauer polynomials $\{C_n^{(1+\alpha)}(z)\}$. After scaling, the result (2.37) can be reclaimed by taking the limit $\alpha \to \infty$.
3. We see from (2.34) that as $\tau \to 1^-$, J is proportional to a GUE matrix, and in particular its eigenvalues are then all real. It was found by Fyodorov, Khoruzhenko and Sommers [276] that setting $\tau = 1 - \pi^2\alpha^2/2N$ and scaling the eigenvalues $z_j \mapsto \pi z_j/N = \pi(x_j + iy_j)/N$ that

$$\lim_{N\to\infty} \frac{\pi^2}{N^2} K_N(\pi w/N, \pi z/N; \tau)\Big|_{\tau=1-\pi^2\alpha^2/2N}$$

$$= \sqrt{\frac{2}{\pi\alpha^2}} \exp\left(-\frac{y_1^2 + y_2^2}{\alpha^2}\right) \frac{1}{2\pi} \int_{-\pi}^{\pi} \exp\left(-\frac{\alpha^2 u^2}{2} + iu(w - \bar{z})\right) du.$$

$$(2.42)$$

(For a direct comparison between (2.42) and what has been reported in the literature, see [24, $K_{\text{weak}}(z_1, z_2)$ in Theorem 3] and [50, Sect. 1].) This is referred to as the weakly non-Hermitian limit. Scaling of the correlation kernel at the edge in this limit has been considered in [85, 280].

2.4 Induced GinUE

The fact that a complex Gaussian matrix G is bi-invariant with respect to multiplication by unitary matrices allows for the distribution of the singular values to be related to the eigenvalue distribution [228, 364].

Proposition 2.7 *Let G be bi-unitary invariant, and let U be a Haar distributed unitary matrix. We have that G and $(G^\dagger G)^{1/2}U$ have the same joint element distribution, and so in particular have the same distribution of eigenvalues.*

Proof By the singular value decomposition, $G = U_1 \Sigma U_2$ for some unitary matrices U_1, U_2 and where Σ is the diagonal matrix of the singular values. We then have $(G^\dagger G)^{1/2}U = U_2^\dagger \Sigma U_2 U$. By the assumed bi-unitary invariance of G, this matrix and G have the same distribution. □

The matrix

$$A := (G^\dagger G)^{1/2}U \qquad (2.43)$$

is well defined for G rectangular, and moreover the property of G being bi-unitary invariant can be generalised to this setting. Of interest is the relation between the joint element distributions of G and A [228].

Proposition 2.8 *Let G be a bi-unitary invariant rectangular $n \times N$ random matrix with joint element distribution of the functional form $g(G^\dagger G)$. The joint element distribution of the matrix A (2.43) with U a Haar distributed unitary matrix is proportional to*

$$(\det A^\dagger A)^{(n-N)} g(A^\dagger A). \qquad (2.44)$$

As a consequence, for G a rectangular complex Ginibre matrix, the eigenvalue PDF of A is

$$\frac{1}{C_{n,N}} \prod_{l=1}^{N} |z_l|^{2(n-N)} e^{-|z_l|^2} \prod_{1 \le j < k \le N} |z_k - z_j|^2, \qquad (2.45)$$

with normalisation $C_{n,N} = N! \pi^N \prod_{j=1}^N (n - N + j - 1)!$.

Proof Write $B = G^\dagger G$. With the joint element distribution of G of the functional form $g(G^\dagger G)$, it is a standard result (see e.g. [237, Eq. (3.23)]) that the joint element distribution of B is proportional to $\det B^{n-N} g(B)$. But B and $A^\dagger A$ have the same joint element distribution from the bi-unitary invariance of G, and the result just quoted with $n = N$ tells us that the Jacobian for the joint element distribution of A and B is a constant, which implies (2.44). In the particular case that G is a rectangular complex Ginibre matrix, the function g in (2.44) is the exponential $g(X) = e^{-\text{Tr} X}$. Furthermore, with the eigenvalues of A denoted by $\{z_l\}$, we have $(\det A^\dagger A)^{(n-N)} = \prod_{l=1}^N |z_l|^{2(n-N)}$. Taking these points into consideration, we see (2.45) results by following the proof of Proposition 2.1. Moreover, with $\omega(z) = \omega(|z|^2) = |z|^{2(n-N)} e^{-|z|^2}$ the weight function, that the analogue of (2.13) in that proof have the properties (2.14) requires that the normalisation equal $N! \prod_{j=0}^{N-1} 2\pi \int_0^\infty r^{2j+1} \omega(r) \, dr$. This implies the stated value of $C_{n,N}$. $\qquad\square$

The correlations for (2.45) are of the determinantal form (2.9). Denoting the corresponding correlation by $K^{iG}(w, z)$, the derivation of (2.10) shows [10, 228]

$$K_N^{iG}(w, z) = \frac{1}{\pi} e^{-(|w|^2 + |z|^2)/2} \sum_{j=1}^N \frac{(w\bar{z})^{n-N+j-1}}{(n - N + j - 1)!}$$

$$= \frac{1}{\pi} e^{-(|w|^2 + |z|^2)/2} e^{w\bar{z}} \left(\frac{\Gamma(n; w\bar{z})}{\Gamma(n)} - \frac{\Gamma(n - N; w\bar{z})}{\Gamma(n - N)} \right). \qquad (2.46)$$

We know that the eigenvalue density, $\rho_{(1),N}^{iG}(z)$ say, results by setting $w = z$ in $K_N^{iG}(w, z)$. Knowing from (2.20) that for N large $\Gamma(N; xN)/\Gamma(N)$ exhibits a transition from the value 1 for $0 < x < 1$ to the value 0 for $x > 1$, we see from (2.46) that

$$\lim_{N \to \infty} \rho_{(1),N}^{iG}(\sqrt{N}z) \Big|_{n/N = \alpha + 1} = \frac{1}{\pi} \left(\chi_{|z| < \sqrt{\alpha+1}} - \chi_{|z| < \sqrt{\alpha}} \right). \qquad (2.47)$$

Here $(\chi_{|z| < \sqrt{\alpha+1}} - \chi_{|z| < \sqrt{\alpha}})$ corresponds to an annulus of inner radius $\sqrt{\alpha}$ and outer radius $\sqrt{\alpha + 1}$. See Fig. 2.3 for the graphs illustrating the convergence (2.47). Generally, for random matrices of the form $A = UTV$, where U, V are Haar distributed and the singular values of T converge to a compactly supported probability measure, it is known that the eigenvalue PDF in the complex plane is either a disk or an annulus. This is referred to as the single ring theorem [222, 305, 306]. Scaling of (2.46) in the bulk of the annulus, or at the boundary, is straightforward and leads to the functional forms exhibited in Proposition 2.3 for the GinUE. Furthermore, in the double scaling regime where the spectrum tends to form a thin annulus of width $O(1/N)$, the scaling of (2.46) in the bulk gives rise to the weakly non-Hermitian limit (2.42) [128]. We also stress that the induced Ginibre ensemble was introduced more generally in [10], where the eigenvalue PDF is a mixture of (2.35) and (2.45).

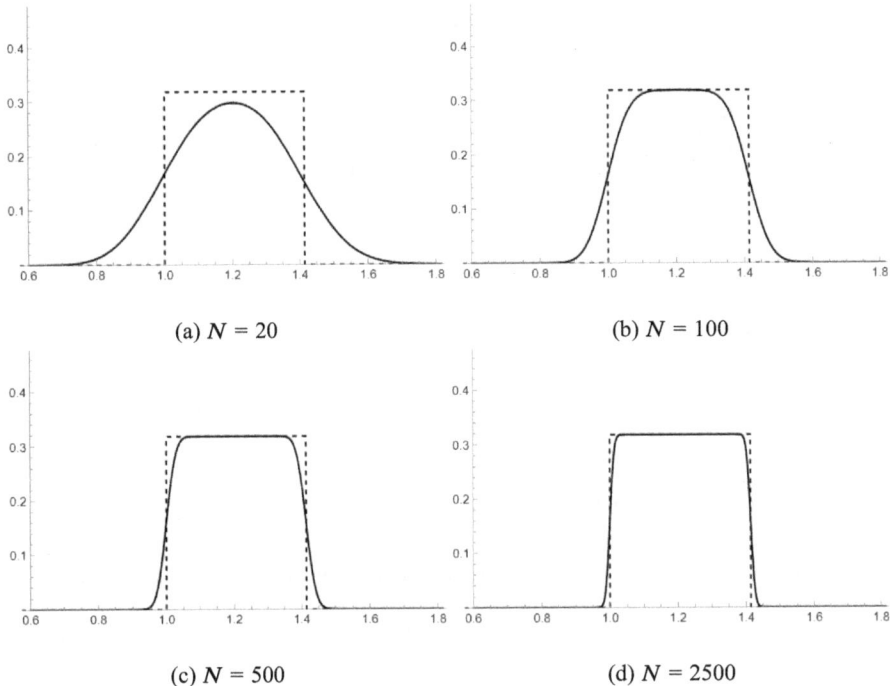

(a) $N = 20$ (b) $N = 100$

(c) $N = 500$ (d) $N = 2500$

Fig. 2.3 Graphs of $x \mapsto \rho_{(1),N}^{\mathrm{iG}}(\sqrt{N}x)$ (full line) and their comparison with $\frac{1}{\pi}(\chi_{x<\sqrt{\alpha+1}} - \chi_{\sqrt{\alpha}})$ (dashed line), where $\alpha = 1$

(The induced GinUE model is then obtained by setting the masses to zero $m_f = 0$ and $\tau = 0$.) In [10], the finite-N expression of the correlation functions as well as their scaling limits both at strong and weak non-Hermiticity were obtained.

2.5 Complex Spherical Ensemble

For G_1, G_2 matrices from the GinUE, matrices of the form $G = G_1^{-1}G_2$ are said to form the complex spherical ensemble [373]. For with $\alpha, \beta \in \mathbb{C}$ with $|\alpha|^2 + |\beta|^2 = 1$, introduce the transformed pair of matrices (C, D),

$$C := -\bar{\beta}G_2 + \bar{\alpha}G_1, \qquad D = \alpha G_2 + \beta G_1.$$

Since G_1, G_2 have independent standard complex elements, it is easy to check that the pair (G_1, G_2) has the same distribution as (C, D). As a consequence, the eigenvalues $\{z_j\}$ of G are unchanged by the fractional linear transformation

$$z \mapsto \frac{z\alpha - \bar{\beta}}{z\beta - \bar{\alpha}}.$$

This implies that upon a stereographic projection from the complex plane to the Riemann sphere, the eigenvalue distribution of $\{G\}$ is invariant under rotation of the sphere, giving rise to the name of the ensemble and telling us that on this surface the eigenvalue density is constant.

The explicit eigenvalue PDF for the complex spherical ensemble was calculated by Krishnapur [373].

Proposition 2.9 *For the complex spherical ensemble, the eigenvalue PDF is proportional to*

$$\prod_{l=1}^{N} \frac{1}{(1 + |z_l|^2)^{N+1}} \prod_{1 \leq j < k \leq N} |z_k - z_j|^2, \qquad z_l \in \mathbb{C}. \tag{2.48}$$

Proof The joint element distribution of (G_1, G_2) is proportional to $\exp(-\text{Tr}\, G_1^\dagger G_1 - \text{Tr}\, G_2^\dagger G_2)$. Substituting $G_2 = GG_1$, (this gives a Jacobian factor $|\det G_1|^N$) then integrating over G_1 gives that the element distribution of G is proportional to

$$\det(\mathbb{I} + G^\dagger G)^{-2N}. \tag{2.49}$$

Introducing now the Schur decomposition (2.2), and changing variables as in the proof of Proposition 2.1, we have from this that the element distribution of the upper triangular matrix Z therein is proportional to

$$\det(\mathbb{I} + Z^\dagger Z)^{-2N} \prod_{1 \leq j < k \leq N} |z_k - z_j|^2. \tag{2.50}$$

Here $\{z_j\}$ are the eigenvalues of G, and also the diagonal entries of Z.

It remains to integrate over the strictly upper triangular entries of Z. For this, denote the leading $n \times n$ sub-block of Z by Z_n and let the product of differentials for the strictly upper entries of Z_n be denoted by $(d\tilde{Z}_n)$. For $p \geq n$, define

$$I_{n,p}(z_1, \ldots, z_n) := \int \frac{1}{\det(\mathbb{I}_n + Z_n Z_n^\dagger)^p} (d\tilde{T}_n).$$

Writing

$$Z_n = \begin{bmatrix} Z_{n-1} & \mathbf{u} \\ \mathbf{0}_{n-1}^T & z_n \end{bmatrix}, \qquad \mathbf{u} = [Z_{jn}]_{j=1}^{n-1}, \tag{2.51}$$

it is possible to integrate out over the elements of $\mathbf{u}$ to obtain the recurrence [260, 326]

$$I_{n,p}(z_1, \ldots, z_n) = \frac{C_{n-1,p}}{(1 + |z_n|^2)^{p-n+1}} I_{n-1,p-1}(z_1, \ldots, z_{n-1}),$$

where

$$C_{n-1,p} = \int \frac{(d\mathbf{v})}{(1 + \mathbf{v}^\dagger \mathbf{v})^p}.$$

Iterating this with $n = 2N$, $p = N$ shows

$$\int \det(\mathbb{I} + Z^\dagger Z)^{-2N}\, (d\tilde{Z}_N) \propto \prod_{l=1}^{N} \frac{1}{(1 + |z_l|^2)^{N+1}},$$

which when used in (2.50) implies (2.48).

The stereographic projection from the south pole of a sphere with radius $1/2$, spherical coordinates (θ, ϕ), to a plane tangent to the north pole is specified by the equation

$$z = e^{i\phi} \tan\frac{\theta}{2}, \qquad z = x + iy.$$

Making this change of variables in (2.50) gives the PDF on the sphere proportional to

$$\prod_{1 \leq j < k \leq N} |u_k v_j - u_j v_k|^2, \qquad u = \cos(\theta/2)e^{i\phi/2}, \ v = -i\sin(\theta/2)e^{-i\phi/2}. \quad (2.52)$$

Here u, v are the Cayley–Klein parameters, and moreover $|u_k v_j - u_j v_k| = |\mathbf{r}_k - \mathbf{r}_j|$ where $\mathbf{r}_j$, $\mathbf{r}_k$ are the vector coordinates on the sphere. It is furthermore the case that minus the logarithm of this distance solves the Poisson equation on the sphere, and so (2.52) has the interpretation of the Boltzmann factor at coupling $\beta = 2$ of the corresponding Coulomb gas; see [237, Sect. 15.6] for more details.

After first removing u_k, u_j from the product of differences in (2.52), the Vandermonde determinant identity (2.12) can be used to compute the correlation functions following a strategy analogous to that used in the proof of Proposition 2.2. This calculation was first done in the context of the corresponding two-dimensional one-component plasma, as obtained by the rewrite of (2.52) analogous to (1.12) [131].

Proposition 2.10 *The n-point correlations for (2.52) are given by*

$$\rho_{(n,N)}\left((\theta_1, \phi_1), \ldots, (\theta_n, \phi_n)\right) = \left(\frac{N}{\pi}\right)^n \det\left[\left(u_j \bar{u}_k + v_j \bar{v}_k\right)^{N-1}\right]_{j,k=1,\ldots,n}. \quad (2.53)$$

Remark 2.4 1. Changing variables $\theta_j = 2r_j\sqrt{\pi/N}$ in (2.53) and taking $N \to \infty$, the bulk-scaled correlation kernel (2.18) results. Here there is no edge regime.
2. Denote by $a(X)$ an $n \times N$ ($N \times M$), $n \geq N$ ($M \geq N$) standard complex Gaussian matrix, and set $A = a^\dagger a$, $Y = A^{-1/2}X$. In terms of Y define $Z = U(YY^\dagger)^{1/2}$, which corresponds to the induced ensemble construction of Sect. 2.4. It was shown in [229] that the element PDF of Z is proportional to

$$(\det Z^\dagger Z)^{M-N} \frac{1}{\det(\mathbb{I} + Z^\dagger Z)^{n+M}}.$$

The proof of Proposition 2.9 now gives that the corresponding eigenvalue PDF is proportional to

$$\prod_{l=1}^{N} \frac{|z_l|^{2(M-N)}}{(1+|z_l|^2)^{n+M-N+1}} \prod_{1 \le j < k \le N} |z_k - z_j|^2. \tag{2.54}$$

With M, n scaled with N, $M/N \to \alpha_1 \ge 1$, $n/N \to \alpha_2 \ge 1$ the leading-order eigenvalue support now occurs in an annulus with inner and outer radii

$$r_1 = \sqrt{(\alpha_1 - 1)/\alpha_2}, \qquad r_2 = \sqrt{\alpha_1/(\alpha_2 - 1)}. \tag{2.55}$$

The associated correlation kernel K_N can be expressed in terms of the incomplete beta function $I_x(a, b) \propto \int_0^x t^{a-1}(1-t)^{b-1} dt$, normalised to equal unity for $x = 1$ as

$$K_N(w, z) = \frac{1}{\pi} \frac{|zw|^{M-N}}{((1+|z|^2)(1+|w|^2))^{(n+M-N+1)/2}} \tag{2.56}$$
$$\times (M + n - N)(1 + w\bar{z})^{n+M-N-1}\Big(I_\zeta(M - N, n) - I_\zeta(M, n - N)\Big),$$

where $\zeta = (w\bar{z})/(1 + w\bar{z})$, see e.g. [120]. Scaling of the correlation functions at either of these boundaries gives the edge kernel (2.19). Setting $w = z$ and with α_1, α_2 as in (2.55) allows for the computation of the global density [229]

$$\lim_{N \to \infty} \frac{1}{N} \rho_{(1),N}(z) = \frac{\alpha_1 + \alpha_2 - 1}{\pi(1 + r^2)^2} \chi_{r_1 \le r \le r_2}. \tag{2.57}$$

Note that for $\alpha_1, \alpha_2 \to 1^+$, this is supported on the whole complex plane, and upon a stereographic projection gives rise to a constant density on the sphere, in keeping the prediction at the end of the first paragraph of this section (Fig. 2.4).

2.6 Sub-block of a Unitary Matrix

Let U be an $(n + N) \times (n + N)$ unitary matrix, and let A be the top $N \times N$ subblock. The non-zero eigenvalues of A are then the nonzero eigenvalues of DUD, where D is the diagonal matrix with the first N diagonal entries 1, and the last n diagonal entries 0. The eigenvalues of this matrix are the same as that for UD, since in general for square matrices the eigenvalues of AB and BA coincide, and furthermore $D^2 = D$. For $\mathbf{u}$ a normalised eigenvector of UD with eigenvalue λ, computing the length squared of $UD\mathbf{u}$ gives $\mathbf{u}^\dagger D\mathbf{u} = |\lambda|^2$, where use has been made of the fact

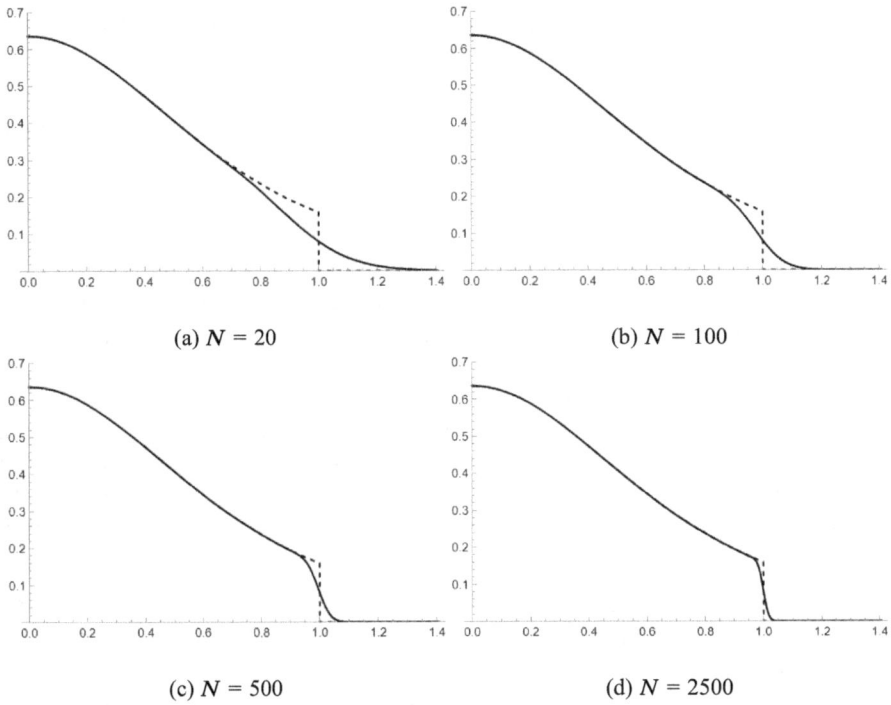

(a) $N = 20$

(b) $N = 100$

(c) $N = 500$

(d) $N = 2500$

Fig. 2.4 Illustrations of the convergence (2.57), where $\alpha_1 = 1$ and $\alpha_2 = 2$

$U^\dagger U = \mathbb{I}$. But with the entries of $\mathbf{u}$ all nonzero, the action of D reduces the length, implying $|\lambda| < 1$. With U chosen with Haar measure, it was shown by Zyczkowski and Sommers [542] that the exact eigenvalue PDF of A can be calculated.

Proposition 2.11 *For A the top $N \times N$ sub-block of an $(n + N) \times (n + N)$ unitary matrix chosen with Haar measure, the eigenvalue PDF is proportional to*

$$\prod_{l=1}^{N}(1 - |z_l|^2)^{n-1} \prod_{1 \leq j < k \leq N} |z_k - z_j|^2, \qquad |z_l| < 1. \qquad (2.58)$$

Proof Let C be the block of U in the first N columns directly below A. Then by the unitarity of U, $A^\dagger A + C^\dagger C = \mathbb{I}_N$. This implies a joint distribution in the space of general $N \times N$ and $n \times N$ complex matrices A, C proportional to the matrix delta function

$$\delta(A^\dagger A + C^\dagger C - \mathbb{I}_N) \propto \int e^{\mathrm{Tr}((iH - \mu \mathbb{I}_N)(A^\dagger A + C^\dagger C - \mathbb{I}_N))} \, (dH), \quad \mu > 0. \qquad (2.59)$$

Here in the integral form of the matrix delta function H is Hermitian; see e.g. [237, Eq. (3.27)]. Beginning with this integral form the integration of the complex matrix

C can be carried out according to [237, displayed equation below (3.27)] to give that
the element PDF of $A =: A_N$ is proportional to

$$F_n(A_N) := \int (\det(i H_N - \mathbb{I}_N))^{-n} e^{\text{Tr}((i H_N - \mathbb{I}_N)(A_N^\dagger A_N - \mathbb{I}_N))} (d H_N). \qquad (2.60)$$

This matrix integral is the starting point for the derivation of (2.59) given in [17, Appendix B], which we follow below.

In (2.60) the integral is unchanged by conjugating H with a unitary matrix V say. Choosing V to be the unitary matrix in the Schur decomposition (2.2) allows us to effectively replace A_N by Z_N throughout (2.60). Doing this, and integrating too over $\tilde{Z}_N$ (i.e. the strictly upper triangular entries of Z_N), we denote the matrix integral by $\tilde{F}_n(z_1, \ldots, z_N)$.

Substituting the decomposition (2.51) for Z_N we see that the vector $\mathbf{u}$ occurs as a quadratic form, and can be integrated over. To progress further, H_N too is decomposed by a bordering procedure of the leading $(N-1) \times (N-1)$ sub-block H_{N-1}, with the $N-1$ column (row) vector $\mathbf{w}$ ($\mathbf{w}^\dagger$) and entry h_N in the bottom right corner. Key now is a determinant identity for a block matrix

$$\det \begin{bmatrix} A & B \\ C & D \end{bmatrix} = \det(D) \det(A - B D^{-1} C),$$

familiar from the theory of the Schur complement (see e.g. [457]), applied to the first term in the integrand of (2.60) with $D = h_N$ (a scalar). A shift of the integration domain according to the additive rank 1 perturbation $H_{N-1} \mapsto H_{N-1} - i \mathbf{w} \mathbf{w}^\dagger$ gives a quadratic form in $\mathbf{w}$, which after integration reduces (2.60) to

$$\tilde{F}_n(z_1, \ldots, z_N) \propto \int (\det(i H_{N-1} - \mathbb{I}_{N-1}))^{-n} (i h_N - 1)^{-n} e^{(|z_N|^2 - 1)(i h_N - 1)}$$
$$\times e^{\text{Tr}(Z_{N-1}^\dagger Z_{N-1}(i H_{N-1} - \mathbb{I}_{N-1}))} (d H_{N-1})(d Z_{N-1}) d h_N. \qquad (2.61)$$

Here, using the residue theorem, the integral over h_N can be performed, yielding the recurrence

$$\tilde{F}_n(z_1, \ldots, z_N) \propto (1 - |z_N|^2)^{n-1} \chi_{|z_N|^2 < 1} \tilde{F}_n(z_1, \ldots, z_{N-1}).$$

Iterating gives the first factor in (2.58), while the product of differences is due to the Jacobian (2.5) for the change of variables to the Schur decomposition as computed in the proof of Proposition 2.1. $\qquad \square$

Remark 2.5 The matrix integral (2.60) can be evaluated to deduce that the element distribution of A is proportional to [167, 277] $\det(\mathbb{I} - A^\dagger A)^{n-N}$. Only for $n \geq N$ is the normalisable. This is in keeping with $A^\dagger A$ having $N - n$ eigenvalues equal to 1 for $n < N$. Starting with this expression, and thus the corresponding restriction on n and N, a derivation of (2.48) can be given which is analogous to the proof of Proposition 2.9 [260].

We have seen that the eigenvalue PDF (2.48) can, after a stereographic projection, be identified as the Boltzmann factor for a Coulomb gas model on the sphere. The concept of stereographic projection can also be applied to a pseudosphere, which refers to the two-dimensional hyperbolic space with constant negative Gaussian curvature. Doing so gives the metric specifying the Poincaré disk. Consideration of the solution of the Poisson equation on the latter allows (2.58) to be interpreted as the Boltzmann factor for a Coulomb gas model on the Poincaré disk [260].

The eigenvalue correlations for (2.58) are determinantal with correlation kernel, $K_{N,n}$ say, expressible in terms of the incomplete beta function $I_x(a, b)$. Thus [227, Sect. 3.2.3]

$$K_{N,n}(w, z) = \frac{n}{\pi} \frac{(1 - |w|^2)^{(n-1)/2}(1 - |z|^2)^{(n-1)/2}}{(1 - w\bar{z})^{n+1}} \left(1 - I_{w\bar{z}}(N, n + 1)\right). \quad (2.62)$$

In the limit $n, N \to \infty$ with $(N + n)/N = \alpha > 1$ it follows from this that

$$\lim_{N \to \infty} \frac{1}{N} \rho_{(1),N}(z) = \frac{\alpha - 1}{\pi} \frac{1}{(1 - |z|^2)^2} \chi_{|z|^2 < 1/\alpha}. \quad (2.63)$$

See Fig. 2.5 for the graphs of $\rho_{(1),N}$.

Hence there is a bulk regime, and an edge regime. The neighbourhood of the origin is typical of the bulk regime, and it follows from (2.62) that

$$\lim_{\substack{n,N \to \infty \\ n/N = \alpha}} \frac{1}{n} K_{N,n}(w/\sqrt{n}, w/\sqrt{n}) = K_\infty^b(w, z), \quad (2.64)$$

where K_∞^b is specified by (2.18). And after appropriate scaling about $|z| = \sqrt{\alpha}$, the universal edge density as given by the first term in (2.41) is obtained [227],

$$\lim_{\substack{n,N \to \infty \\ (n+N)/n = \alpha}} \alpha(1 - \alpha) \frac{1}{N} \rho_{(1),N} \left(\sqrt{\alpha} + \frac{\xi}{\sqrt{N}} \sqrt{\alpha(1 - \alpha)}\right) = \frac{1}{2\pi} \left(1 - \mathrm{erf}(\sqrt{2}\xi)\right). \quad (2.65)$$

A distinct scaling regime, referred to as close to unitary [277], or more specifically a multiplicative rank n contraction of a random unitary matrix, is obtained by taking $N \to \infty$ with n fixed in (2.62). Scaling the eigenvalues $z_k = (1 - y_k/N)e^{i\phi_k/N}$ gives [362] [227, Eq. (3.2.140)]

$$\lim_{N \to \infty} \frac{1}{N^2} K_{N,n}(z_1, z_2) = \frac{1}{\pi} \frac{(2\sqrt{y_1 y_2})^{n-1}}{(n - 1)!} \int_0^1 s^n e^{-(y_1 + y_2 + i(\phi_1 - \phi_2))s} \, ds. \quad (2.66)$$

This functional form was first obtained for the scaled correlation kernel in the setting of a rank n additive anti-Hermitian perturbation of a GUE matrix [273, Eq. (15)

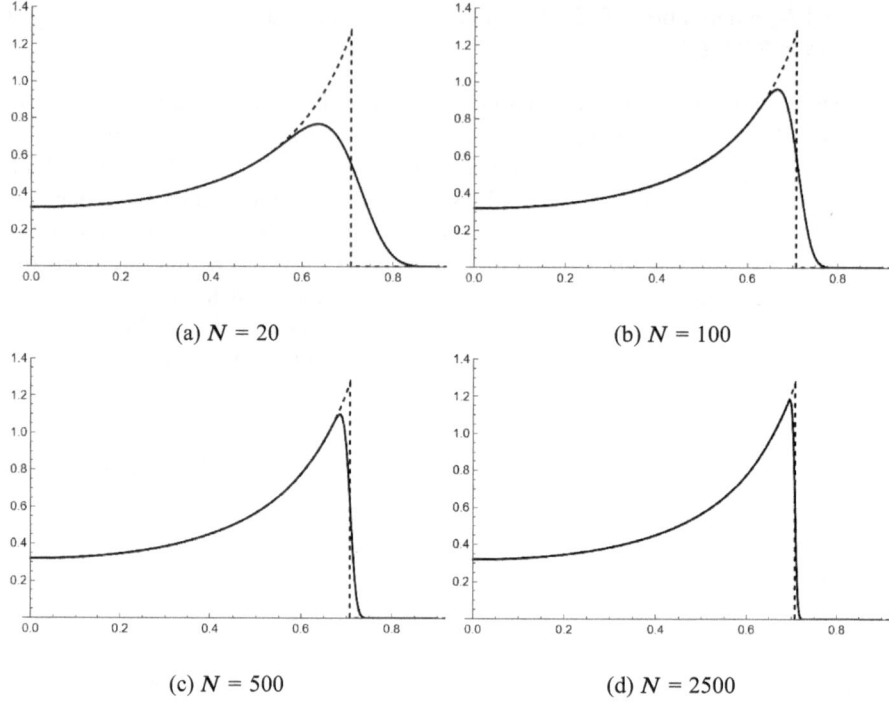

(a) $N = 20$

(b) $N = 100$

(c) $N = 500$

(d) $N = 2500$

Fig. 2.5 Illustrations of the convergence (2.63), where $\alpha = 2$

after setting $\nu(x) = 1/\pi$, identifying $\tilde{z}_k = (\phi_k + iy_k)/2$ and taking the limit $g_m \to 0$ ($m = 1, \ldots, M$) with M identified as n].

A generalisation of the setting of Proposition 2.11 is to consider the top $(N + L) \times N$ sub-block of an $(n + N + L) \times (n + N + L)$ unitary matrix chosen with Haar measure. Denote such a rectangular matrix by $A_{N,L}$, and define from this $\tilde{A}_{N,L} = V(A^{\dagger}_{N,L} A_{N,L})^{1/2}$, where V is an $N \times N$ random Haar distributed unitary matrix. It was shown in [227] that the eigenvalue PDF of $\tilde{A}_{N,L}$ is proportional to (2.48) with an additional factor of $\prod_{l=1}^{N} |z_l|^{2L}$. Denote the corresponding correlation kernel by $K_{N,n,L}$. Scaling L by requiring that $L/N = \alpha > 0$ as $N \to \infty$, scaling the eigenvalues z_k as in the above paragraph, and keeping n fixed, it was shown in [362], [227, Eq. (3.2.122)] that

$$\lim_{N \to \infty} \frac{1}{N^2} K_{N,n,L}(z_1, z_2)\Big|_{L/N=\alpha} = \frac{1}{\pi} \frac{(2\sqrt{y_1 y_2})^{n-1}}{(n-1)!} \int_{\alpha}^{\alpha+1} s^n e^{-(y_1 + y_2 + i(\phi_1 - \phi_2))s} \, ds.$$

(2.67)

Note the consistency with (2.66) in the limit $\alpha \to 0^+$.

Another class of generalisation is, for U an $(n + N) \times (n + N)$ Haar distributed unitary matrix, and A a fixed diagonal matrix with first n diagonal entries $a_1, \ldots, a_n$ and the remaining entries unity, to consider the product UA [271, 277]. With each

$a_i = 0$, this corresponds to the truncated Haar unitary model of this section. The simplest case is when $n = 1$, and furthermore we take $a_1 = a$ with $|a| < 1$. Notice that this corresponds to a multiplicative rank 1 perturbation of U — see the recent review [249] for context from this viewpoint. The eigenvalue PDF is then proportional to

$$(1 - |a|^2)^N \delta\left(|a|^2 - \prod_{l=1}^{N+1} |z_l|^2\right) \prod_{1 \leq j < k \leq N+1} |z_j - z_k|^2, \quad |z_l| < 1, \qquad (2.68)$$

(cf. (2.58)) and an explicit formula for the general k-point correlation function is known [271, 277]. The latter involves determinants but technically the state is not (unless $a = 0$) a determinantal point process as no correlation kernel can be identified. Specifically, the functional form of the density is

$$\rho_{(1),N+1}^{UA}(z) = \frac{1}{\pi} \frac{1}{(1 - |a|^2)^N} \frac{d}{dx}\left(\frac{x^N - 1}{x - 1} + \left(1 - \frac{|a|^2}{x}\right)^N\right)\Bigg|_{x=|z|^2}. \qquad (2.69)$$

For recent works relating to the UA model see [202, 255]. Let us also mention that a fixed trace constraint of the form (2.68) was also considered in [23, 24] for GinUE and its generalisations.

2.7 Products of Complex Ginibre Matrices

Consideration of the eigenvalue PDF for the product $G_1 G_2$, where G_1 (G_2) are independent $N \times p$ ($p \times N$) rectangular complex Ginibre matrices was first undertaken by Osborn [456], in the context of a study relating to quantum chromodynamics. (We also mention that the GinUE has been directly used in the QCD bulk spectrum [421].) Later [333] it was realised that this eigenvalue PDF must be the same as that for the product $\tilde{G}_1 \tilde{G}_2$ (or $\tilde{G}_2 \tilde{G}_1$), where each $\tilde{G}_i$ is an $N \times N$ complex random matrix with element distribution proportional to

$$|\det \tilde{G}_i \tilde{G}_i^\dagger|^{\nu_i} e^{-\text{Tr}\, \tilde{G}_i \tilde{G}_i^\dagger}, \quad \nu_1 = 0, \ \nu_2 = p - N; \qquad (2.70)$$

note the construction (2.43) for random matrices with this element PDF. Hence it suffices to consider the square case. Key for this is the generalised Schur decomposition (equivalent to the so-called QZ decomposition in numerical linear algebra)

$$\tilde{G}_1 = U Z_1 V, \qquad \tilde{G}_2 = V^\dagger Z_2 U^\dagger, \qquad (2.71)$$

where Z_1, Z_2 are upper triangular matrices with diagonal entries $\{z_j^{(1)}\}$, $\{z_j^{(2)}\}$ such that

$$z_j^{(1)} z_j^{(2)} = z_j \qquad (2.72)$$

with $\{z_j\}$ the eigenvalues of $\tilde{G}_1 \tilde{G}_2$.

For the change of variables (2.71) the wedge product strategy of the proof of Proposition 2.1 can again be implemented [456], [237, proof of Proposition 15.11.2] to give for the Jacobian (2.5). Substituting (2.71) in (2.70) and recalling (2.72), it follows that the eigenvalue PDF is proportional to

$$\prod_{l=1}^{N} w^{(2)}(z_l) \prod_{j<k} |z_j - z_k|^2,　\tag{2.73}$$

where

$$w^{(2)}(z) := \int_{\mathbb{C}} d^2 z_1 \, |z_1|^{2\nu_1} \int_{\mathbb{C}} d^2 z_2 \, |z_2|^{2\nu_2} \delta(z - z_1 z_2) e^{-|z_1|^2 - |z_2|^2}.$$

Moreover, it was shown in [456] that $w^{(2)}(z)$ can be expressed in terms of the modified Bessel function $K_{\nu_2-\nu_1}(2|z|)$, assuming $\nu_2 \geq \nu_1$.

Akemann and Burda [15] (in the case of all square matrices), and soon after Adhikari et al. [5] (the general rectangular case), generalised (2.73) to hold for the product of M complex Gaussian matrices. As already indicated in the case $M = 2$, following the work of Ipsen and Kieburg [333], it is now known that the rectangular case can be reduced to the square case, where the square matrices have distribution as in (2.70), with the ν_i equal to the difference (assumed non-negative) between the number of rows in G_i and the number of rows in $G_1 (= N)$. The role of the modified Bessel function, as the special function evaluating the weight $w^{(2)}(z)$ in (2.73), is now played by the Meijer G-function

$$G_{p,q}^{m,n}\left(z \left| \begin{matrix} a_1, \ldots, a_p \\ b_1, \ldots, b_q \end{matrix} \right.\right) = \frac{1}{2\pi i} \int_{\mathcal{C}} \frac{\prod_{j=1}^{m} \Gamma(b_j - s) \prod_{j=1}^{n} \Gamma(1 - a_j + s)}{\prod_{j=m+1}^{q} \Gamma(1 - b_j + s) \prod_{j=n+1}^{p} \Gamma(a_j - s)} z^s \, ds,　\tag{2.74}$$

where $\mathcal{C}$ is an appropriate contour as occurs in the inversion formula for the corresponding Mellin transform; see e.g. [414] for an extended account.

Proposition 2.12 *Consider the product $\tilde{G}_1 \cdots \tilde{G}_M$ (in any order) of complex Gaussian matrices with element PDF as given in (2.70), with each $\nu_i \geq 0$ but otherwise unrestricted. The eigenvalue PDF is proportional to*

$$\prod_{l=1}^{N} w^{(M)}(z_l) \prod_{j<k} |z_j - z_k|^2, \quad w^{(M)}(z) = G_{0,M}^{M,0}\left(|z|^2 \left| \begin{matrix} - \\ \nu_1, \ldots, \nu_M \end{matrix} \right.\right).　\tag{2.75}$$

The functional form (2.75) remains valid for the product of rectangular complex Ginibre matrices $G_1 \cdots G_M$, with ν_i specified as in the second sentence above (2.74).

Proof (Sketch) The generalised Schur decomposition (2.71) can be extended to a general number M of square matrices to give what is referred to as the periodic Schur form [99], [415, Corollary 3.2]

$$\tilde{G}_i = U_i Z_i V_i, \qquad V_i = U_{i+1}^\dagger \; (i = 1, \ldots, M-1), \; V_M = U_1^\dagger; \qquad (2.76)$$

in [15] this was deduced independently. The key features with respect to computing the eigenvalue PDF of the product $\tilde{G}_1 \cdots \tilde{G}_M$, as already discussed in the case $M = 2$, again hold true. In particular the Jacobian factor for the change of variables is given by (2.5), and the k-th diagonal entry of each Z_i multiply together to give the eigenvalue z_k of the product. It follows that the eigenvalue PDF is given by (2.75) with

$$w^{(M)}(z) \propto \int_{\mathbb{C}} d^2 z_1 \, |z_1|^{2\nu_1} \cdots \int_{\mathbb{C}} d^2 z_M \, |z_M|^{2\nu_M} \delta(z - z_1 \cdots z_M) e^{-\sum_{j=1}^M |z_j|^2}. \quad (2.77)$$

As noted in [5, 15], taking the Mellin transform of this expression leads to the Meijer G-function form in (2.75). □

An easy consequence of (2.75) is that the eigenvalue correlations are determinantal with kernel, $K_{N,M}$ say, given by [5, 15],

$$K_{N,M}(w, z) = \left(w^{(M)}(w) w^{(M)}(z) \right)^{1/2} \sum_{j=1}^N \frac{(w\bar{z})^{j-1}}{\prod_{m=1}^M \Gamma(j + \nu_m)}. \qquad (2.78)$$

Rigorous asymptotic analysis of (2.78) can be carried out [15, 409]. For example, with $w = z$ and $z \mapsto N^{M/2} z$, one obtains

$$\lim_{N \to \infty} N^{M-1} \rho_{(1),N}(N^{M/2} z) = \frac{|z|^{-2+2/M}}{\pi M} \chi_{|z|<1}, \qquad (2.79)$$

where it is assumed each ν_i is fixed. To interpret this result, form the M-th power of a single GinUE matrix G. The eigenvalues are $\{\tilde{z}_j = z_j^M\}$ with $\{z_j\}$ the eigenvalues of G. Changing variables in the circular law (2.17) $\tilde{z} = z^M$ gives the density (2.79) in the variable $\tilde{z}$. Hence, as anticipated in [112], the global eigenvalue density for the product of M independent GinUE matrices is identical to that of the M-th power of a single GinUE matrix. This same global scaling limit is also well defined if some or all of the ν_i are proportional to N. An explicit limit formula has been obtained by Liu and Wang [409]. It is found that the density has the support of an annulus if each $\mu_i := \lim_{N \to \infty} \nu_i/N$ is positive, but otherwise is again supported in a disk, albeit with a density function distinct to that in (2.79).

One observes from (2.79) a singularity at $z = 0$, not present in the circular law global density (2.17) for GinUE. A consequence is that the correlation kernel about this point, simply obtained by setting the upper terminal of the sum in (2.78) equal to infinity, is distinct from the corresponding kernel (2.18) for the bulk-scaled GinUE. On the other hand, it is shown in [409] that the bulk-scaled GinUE kernel is reclaimed by choosing the origin a distance $\alpha \sqrt{N}$ from $z = 0$, for any $0 < \alpha < 1$. Also, from

[15, 409] we know that the scaling of (2.78) about $|z| = 1$ reclaims the kernel (2.19) for the edge-scaled GinUE.

Remark 2.6 1. Required in the derivation of (2.79) from (2.78) is knowledge of the $z \to \infty$ asymptotic expansion [414]

$$G_{0,M}^{M,0}\left(z\Big|_{\nu_1, \ldots, \nu_M}^{\quad\quad}\right) \sim \frac{1}{\sqrt{M}}\left(\frac{2\pi}{z}\right)^{(M-1)/2} e^{-Mz^{1/M}} z^{(\nu_1 + \cdots + \nu_M)/M}\left(1 + \mathrm{O}(z^{-1/M})\right).$$
(2.80)

Of particular interest is the exponential term herein, which with z replaced by $|z|^2$ as required in (2.78) reads $e^{-M|z|^{2/M}}$. This suggests a modification of U in (1.12) to read

$$U_M := \frac{M}{2}\sum_{j=1}^{N}|z_j|^{2/M} - \sum_{1 \leq j < k \leq N}\log|z_k - z_j|.$$
(2.81)

Indeed, starting from U_M and repeating the working of Remark 2.2 reclaims (2.79). The correlation kernel appearing in such a calculation is called the Mittag–Leffler kernel. This terminology applies too in the more general case that a term $-c\sum_{j=1}^{N}\log|z_j|$ is included in (2.81) [60, 64]. This very case, for $c = (k - M)/M$, $(k = 1, \ldots, M)$, appears in the calculation of the joint distribution of the eigenvalues of the M-th power ($M \leq N$) of a GinUE matrix [198, Th. 1.5]; for generalisations in this direction see [201]. We also refer to [357, 358] for the appearance of particular Mittag–Leffler point process as a degenerate limit of so-called elliptic determinantal point processes, characterised by the product of differences in (1.7) being replaced by a product over Jacobi theta functions.

2. The eigenvalue PDF for a product of GinUE matrices and inverses of GinUE matrices has been shown to be of the form (2.75), but with the Meijer G-function therein replace by a different Meijer G-function [5]. The corresponding limiting eigenvalue density, in the case of equal numbers of matrices and inverses, is [301, 540]

$$\lim_{N \to \infty}\rho_{(1),N}(z) = \frac{1}{\pi M}\frac{|z|^{-2+2/M}}{(1 + |z|^{2/M})^2}.$$
(2.82)

This functional form relates to the $M = 1$ case in the same way as (2.79) relates to (2.17), being the M-th power of the so-called spherical law.

3. The case $M = 2$ of Proposition 2.12 permits a generalisation. Thus for the matrices in the product $\tilde{G}_1\tilde{G}_2$ choose

$$\tilde{G}_1 = \sqrt{1 + \tau}X_1 + \sqrt{1 - \tau}X_2, \quad \tilde{G}_2 = \sqrt{1 + \tau}X_1^\dagger + \sqrt{1 - \tau}X_2^\dagger, \quad 0 < \tau < 1,$$

where X_1, X_2 are $N \times (N + p)$ rectangular complex Gaussian matrices. The joint element distribution is then proportional to

$$\exp\left(-\frac{1}{1 - \tau^2}\mathrm{Tr}\left(\tilde{G}_1^\dagger\tilde{G}_1 + \tilde{G}_2^\dagger\tilde{G}_2 - 2\tau\,\mathrm{Re}\,\tilde{G}_1\tilde{G}_2\right)\right);$$

cf. (2.36). Results in [456] give that the eigenvalue PDF of $\tilde{G}_1\tilde{G}_2$ is of the form in (2.73) but with $w^{(2)}(z)$ now dependent on τ. Due to the shape of the resulting droplet, this gives rise to the so-called shifted elliptic law [21, Th. 1]. See also [127] for the eigenvalue correlations.

4. The product of M random matrices in the limit $M \to \infty$ is of interest from a dynamical systems viewpoint, as the scaled logarithm of the singular values gives the Lyapunov spectrum. Closely related for product matrices themselves are the stability exponents, defined as $\frac{1}{M} \log |z_k|$, $(k = 1, \ldots, N)$. In fact for bi-unitary invariant random matrices at the least, the Lyapunov and stability exponents are the same [468]. Works relating to the computation of these exponents and their related statistical properties for GinUE include [9, 16, 239, 243, 298, 331, 356, 410, 411].

2.8 Products of Truncated Unitary Matrices

The eigenvalue PDF for a truncation of a Haar distributed unitary matrix has been given in Proposition 2.11. It turns out that knowledge of the evaluation of the matrix integral (2.61), as implied in the proof of Proposition 2.11, used in conjunction with the periodic Schur form (2.76), is sufficient to allow for the determination of the eigenvalue PDF of the product of M truncated Haar distributed unitary matrices [17].

Proposition 2.13 *Consider M independent Haar distributed $(n_j + N) \times (n_j + N)$ unitary matrices, with the top $N \times N$ sub-block of each denoted by A_j $(j = 1, \ldots, M)$. The eigenvalue PDF of the product $A_1 \cdots A_M$ (in any order) is given by (2.75) with*

$$w^{(M)}(z) = G_{M,M}^{M,0}\left(|z|^2 \Big| \begin{matrix} n_1, \ldots, n_M \\ 0, \ldots, 0 \end{matrix} \right) \chi_{|z|<1}. \tag{2.83}$$

Proof (Sketch) The method of proof of Proposition 2.12, which begins with the periodic Schur decomposition (2.76), and then integrates out over the upper triangular entries of the matrices Z_j (here this latter step requires knowledge of the evaluation of the matrix integral (2.61)) gives (2.75) with

$$w^{(M)}(z) \propto \int_{\mathbb{C}} d^2 z_1 \cdots \int_{\mathbb{C}} d^2 z_M \, \delta(z - z_1 \cdots z_M) \prod_{j=1}^{M}(1 - |z_j|^2)^{n_j - 1}.$$

Taking the Mellin transform of this expression leads to the Meijer G-function form in (2.83).

Analogous to (2.78), it follows from the result of Proposition 2.13 that the eigenvalue correlations are determinantal with a kernel (to be denoted by $K_{N,M}^{\mathrm{U}}$) given by [17]

$$K_{N,M}^{U}(z_1, z_2) = \left(w^{(M)}(z_1)w^{(M)}(z_2)\right)^{1/2} \sum_{j=1}^{N} (z_1\bar{z}_2)^{j-1} \prod_{m=1}^{M}\prod_{m=1}^{M} \frac{(n_m + j - 1)!}{(j-1)!}.$$

(2.84)

Note here, that in distinction to (2.78), the $w^{(M)}$ are given by (2.83). Suppose all the n_i are equal to n, and that for large N, $n/N = \alpha > 0$. Using (2.84) in the case $z_1 = z_2$, it is derived in [17] that

$$\rho_{(1),N}(z) \sim \frac{\alpha N}{\pi M} \frac{|z|^{-2+2/M}}{(1 - |z|^{2/M})^2} \chi_{|z|<(1+\alpha)^{-M/2}};$$

(2.85)

cf. (2.82).

In the same limit, with the scaling $z \mapsto z/N^{M/2}$ (and similarly for w), the kernel (2.84) multiplied by N^{-M} has the same limit as the kernel (2.78) for the product of M GinUE matrices about the origin in the case of each matrix square (all $v_i = 0$). If instead a point z_0 with $0 < z_0 < (1 + \alpha)^{-M/2}$ is chosen before this scaling, one obtains instead the GinUE result (2.18), while at the boundary of support, generalising (2.65) it is found that [17]

$$\lim_{\substack{n,N\to\infty \\ n/N=\alpha}} \frac{M}{N} \frac{\alpha}{(1+\alpha)^{M+1}} \rho_{(1),N}\left(\frac{1}{(1+\alpha)^{M/2}} + \frac{\xi}{\sqrt{N}} \frac{\sqrt{M}\alpha^{1/2}}{(1+\alpha)^{(M+1)/2}}\right)$$

$$= \frac{1}{2\pi}\left(1 - \mathrm{erf}(\sqrt{2}\xi)\right),$$

(2.86)

thus again exhibiting the universal functional form seen at the edge scaling of the GinUE (2.29).

Also considered in [17] is a close-to-unity scaling, with $n_i = n$ all fixed ($i = 1, \ldots, M$) as $N \to \infty$. Scaling the eigenvalues $z_k = (1 - y_k/N)e^{i\phi_k/N}$, it is found that (2.66) again holds but with each occurrence of n replaced by nM.

2.9   The Distribution of the Squared Eigenvalue Moduli

All the explicit eigenvalue PDFs obtained in this chapter, excluding the elliptic Ginibre ensemble, have the form

$$\prod_{l=1}^{N} w(|z_l|^2) \prod_{1\le j<k\le N} |z_k - z_j|^2,$$

(2.87)

for some weight function w. In particular, they are invariant under rotations about the origin. An observation of Kostlan [372] for the GinUE case (1.7) of (2.87) is that the set of squared eigenvalue moduli $\{|z_j|^2\}_{j=1}^{N}$, appropriately ordered, are independently distributed (specifically as gamma random variables $\{\Gamma[j; 1]\}$); see too [145, Theorem 1.2].

Proposition 2.14 *Let $F_w(s_1, \ldots, s_N)$ denote the PDF for the distribution of $\{s_j := |z_j|^2\}_{j=1}^N$ for the PDF (2.87). We have*

$$F_w(s_1, \ldots, s_N) = \frac{1}{N!} \mathrm{Sym} \prod_{j=1}^{N} \frac{w(s_j)s_l^{l-1}}{\int_0^\infty w(s)s^{j-1}\,ds}, \tag{2.88}$$

where Sym denotes symmetrisation with respect to $\{s_j\}$.

Proof Starting with (2.11), then substituting (2.12) and its complex conjugate, we see

$$\prod_{l=1}^{N} w(|z_l|^2) \prod_{1 \le j < k \le N} |z_k - z_j|^2 = \prod_{l=1}^{N} w(|z_l|^2) \det[f_{k-1}(z_j)]_{j,k=1}^N \det[g_{k-1}(z_j)]_{j,k=1}^N,$$

where $f_l(z) = z^l$, $g_l(z) = \bar{z}^l$. According to Andréief's identity (see e.g. [246]), it follows from this that the integral over $z_l \in \mathbb{C}$ $(l = 1, \ldots, N)$, $I_{w,N}$ say, is itself given by a determinant

$$I_{w,N} = N! \det \left[\int_{\mathbb{C}} w(|z_l|^2) f_{j-1}(z) g_{k-1}(z)\,d^2z \right]_{j,k=1}^N.$$

Substituting the explicit form of f_l, g_l and changing to polar coordinates $z_l = r_l e^{i\theta_l}$ shows that only the diagonal terms are non-zero. This allows the determinant to be evaluated as

$$I_{w,N} = N! \pi^N \prod_{j=1}^{N} \int_0^\infty w(s)s^{j-1}\,ds, \tag{2.89}$$

where in each integration the change of variables $r^2 = s$ has been made. Forming now $I_{w\phi,N}/I_{w,N}$, where ϕ is a suitable test function, and assuming that (2.87) is normalisable, the result (2.88) follows. $\qquad\square$

Chapter 3
Fluctuation Formulas

Generally, a statistic of a point process refers to the stochastic variable defined by evaluating a function at the points. In this setting the class of functions of a single variable are referred to as linear statistics. One example is the counting function for eigenvalues in a specified region. The large region size form of the variance of this statistic, when proportional to the surface area, is used to specify the state as being hyperuniform. In the case of GinUE eigenvalues, a central limit theorem for the distribution can be established, and this can be strengthened for a local limit theorem of the underlying probabilities. Moreover, large deviation formulas for the latter are obtained. The counting function linear statistic is discontinuous. Smooth linear statistics with global scaling exhibit distinct large N behaviours, in particular the variance is now an $O(1)$ quantity for GinUE eigenvalues. A decomposition of this variance as a contribution from the bulk, and a contribution from the boundary, is exhibited. The chapter concludes with a discussion of GinUE eigenvalues in spatial modelling.

3.1 Counting Function in General Domains

The eigenvalues of a non-Hermitian matrix are examples of point processes in the plane. Statistical quantities characterising the point process are functions of the eigenvalues $\{z_j\}$ of the form $\sum_{j=1}^{N} f(z_j)$—referred to as linear statistics—for a given f. Such statistics are closely related to the correlation functions. Thus

$$\left\langle \sum_{j=1}^{N} f(z_j) \right\rangle = \int_{\mathbb{C}} f(z)\rho_{(1),N}(z)\, d^2z, \qquad (3.1)$$

© The Author(s) 2025
S.-S. Byun and P. J. Forrester, *Progress on the Study of the Ginibre Ensembles*, KIAS
Springer Series in Mathematics 3, https://doi.org/10.1007/978-981-97-5173-0_3

while

$$\text{Cov}\left(\sum_{l=1}^{N} f(z_l), \sum_{l=1}^{N} g(z_l)\right)$$

$$= \int_{\mathbb{C}} d^2z \int_{\mathbb{C}} d^2z'\, f(z)g(z')\left(\rho_{(2),N}^{T}(z, z') + \rho_{(1),N}(z)\delta(z - z')\right)$$

$$= -\frac{1}{2}\int_{\mathbb{C}} d^2z \int_{\mathbb{C}} d^2z'\, (f(z) - f(z'))(g(z) - g(z'))\rho_{(2),N}^{T}(z, z'), \qquad (3.2)$$

see e.g. [250, Sect. 2.1]. Here $\rho_{(2),N}^{T}(z, z') := \rho_{(2),N}(z, z') - \rho_{(1),N}(z)\rho_{(1),N}(z')$.

One of the most prominent examples of a linear statistic is the choice $f(z) = \chi_{z \in \mathcal{D}}$, where $\mathcal{D} \in \mathbb{C}$. The linear statistic is then the counting function for the number of eigenvalues in $\mathcal{D}$, $N(\mathcal{D})$ say. Let $E_N(n; \mathcal{D})$ denote the probability that there are exactly n eigenvalues in $\mathcal{D}$, so that $E_N(n; \mathcal{D}) = \Pr\left(\sum_{j=1}^{N} \chi_{z_j \in \mathcal{D}} = n\right)$. Denote the corresponding generating function (in the variable $1 - \xi$) by $\tilde{E}_N(\xi; \mathcal{D})$ so that

$$\tilde{E}_N(\xi; \mathcal{D}) = \sum_{n=0}^{N} E_N(n; \mathcal{D})(1 - \xi)^n. \qquad (3.3)$$

Note that with $1 - \xi = e^{it}$ this corresponds to the characteristic function for the probability mass functions $\{E_N(n; \mathcal{D})\}$. A straightforward calculation (see e.g. [237, Proposition 9.1.1]) shows that $\tilde{E}_N(\xi; \mathcal{D})$ can be expressed in terms of the correlation functions according to

$$\tilde{E}_N(\xi; \mathcal{D}) = 1 + \sum_{n=1}^{N} \frac{(-\xi)^n}{n!} \int_{\mathcal{D}} d^2z_1 \cdots \int_{\mathcal{D}} d^2z_n\, \rho_{(n),N}(z_1, \ldots, z_n). \qquad (3.4)$$

It can readily be checked that (3.1) and (3.2) in the special case $f(z) = g(z) = \chi_{z \in \mathcal{D}}$ are consistent with (3.4).

Specialising now to the circumstance that the PDF for the eigenvalues is of the form (2.87), a particular product formula for $\tilde{E}_N(\xi; \mathcal{D})$ can be deduced, which was known to Gaudin [286].

Proposition 3.1 *Let $\{\rho_{(n),N}\}$ in (3.4) be given by (2.9) where the correlation kernel K_N corresponds to (2.87). Let $\mathbb{K}_{N,\mathcal{D}}$ denote the integral operator supported on $z_2 \in \mathcal{D}$ with kernel $K_N(z_1, z_2)$. This integral operator has at most N non-zero eigenvalues $\{\lambda_j(\mathcal{D})\}_{j=1}^{N}$, where $0 \leq \lambda_j(\mathcal{D}) \leq 1$, and*

$$\tilde{E}_N(\xi; \mathcal{D}) = \prod_{j=1}^{N}(1 - \xi\lambda_j(\mathcal{D})). \qquad (3.5)$$

Proof Let $\{p_s(z)\}_{s=0}^{\infty}$ be a set of orthogonal polynomials with respect to the inner product $\langle f, g \rangle := \int_{\mathbb{C}} w(|z|^2) f(z) g(\bar{z}) \, d^2 z$ with corresponding normalisation denoted by h_s. Taking as a basis $\{(w(|z|^2))^{1/2} p_k(z)\}_{k=0}^{\infty}$ it is straightforward to check that

$$K_N(z_1, z_2) = (w(|z_1|^2) w(|z_2|^2))^{1/2} \sum_{s=0}^{N-1} \frac{p_s(z_1)\overline{p_s(z_2)}}{h_s}, \tag{3.6}$$

and so the eigenfunctions of $\mathbb{K}_{N,\mathcal{D}}$ are of the form $(w(|z|^2))^{1/2} \sum_{s=0}^{N-1} c_s p_s(z)$ (see e.g. [237, proof of Proposition 5.2.2]). Hence there are at most N nonzero eigenvalues, which moreover can be related to a Hermitian matrix and so must be real. In terms of these eigenvalues, the determinantal form (2.9) substituted in (3.4) implies (3.5)—this is a result from the theory of Fredholm integral operators (see e.g. [532]). Since by definition, each n-point correlation is non-negative, we see from the RHS of (3.4) that $\tilde{E}_N(\xi; \mathrm{D}) > 0$ for $\xi < 0$, and so $\lambda_j(\mathcal{D}) \geq 0$ ($j = 1, \ldots, N$). Also, the definition (3.3) tells us that $\tilde{E}_N(\xi; \mathcal{D}) > 0$ for all $\xi < 1$. This would contradict (3.5) if it was to be that any $\lambda_j(\mathcal{D}) > 1$, since in this circumstance there would be a ξ in this range such that $\tilde{E}_N(\xi; \mathcal{D})$ vanishes. $\qquad \square$

Consider $\sum_{j=1}^{N} x_j$ where $x_j \in \{0, 1\}$ is a Bernoulli random variable with $\Pr(x_j = 1) = \lambda_j(\mathcal{D})$. The characteristic function is $\prod_{j=1}^{N}(1 - \lambda_j(\mathcal{D}) + e^{it}\lambda_j(\mathcal{D}))$. With $e^{it} = 1 - \xi$ this gives the RHS of (3.5). But it has already been noted that $\tilde{E}_N(\xi; \mathcal{D})$ with ξ related to e^{it} in this way is the characteristic function for the counting statistic $\mathcal{N}(\mathcal{D})$, and hence the equality in distribution $\mathcal{N}(\mathcal{D}) \overset{\mathrm{d}}{=} \sum_{j=1}^{N} \mathrm{Bernoulli}(\lambda_j(\mathcal{D}))$ [317, 326]. From this, it follows using the standard arguments (see [223, Sect. XVI.5, Theorem 2]) that a central limit theorem holds for $\mathcal{N}(\mathcal{D}_N)$ (here the subscript N on $\mathcal{D}_N$ is to indicate that the region $\mathcal{D}$ depends on N),

$$\lim_{N \to \infty} \frac{\mathcal{N}(\mathcal{D}_N) - \langle \mathcal{N}(\mathcal{D}_N) \rangle}{(\mathrm{Var}\, \mathcal{N}(\mathcal{D}_N))^{1/2}} \overset{\mathrm{d}}{=} \mathrm{N}[0, 1], \tag{3.7}$$

valid provided $\mathrm{Var}\, \mathcal{N}(\mathcal{D}_N) \to \infty$ as $N \to \infty$; see also [172, 498, 510].

A stronger result, extending the central limit theorem (3.7), follows from the fact that (3.5) in the variable $z = 1 - \xi$ has all its zeros on the negative real axis [262].

Proposition 3.2 *In the setting of the applicability of Proposition 3.1, and with $\sigma_{N_D} := (\mathrm{Var}\, \mathcal{N}(\mathcal{D}_N))^{1/2}$, we then have that $\{E_N(k; \mathcal{D}_N)\}$ satisfy the local central limit theorem*

$$\lim_{N \to \infty} \sup_{x \in (-\infty, \infty)} \left| \sigma_{N_D} E_N([\sigma_{N_D} x + \langle \mathcal{N}(\mathcal{D}_N) \rangle]; \mathcal{D}_N) - \frac{1}{\sqrt{2\pi}} e^{-x^2/2} \right| = 0. \tag{3.8}$$

Proof The fact that the zeros of (3.5) with the variable $z = 1 - \xi$ are on the negative real axis implies, by Newton's theorem on log-concavity of the sequence of elementary symmetric functions [448], that $\{E_N(k; \mathcal{D})\}$ is log concave. It is known that log-concavity is a sufficient condition for extending a central limit theorem to a local limit theorem [84].   □

For large N, inside the disk of radius $\sqrt{N}$, the eigenvalue density for GinUE is constant and the full distribution is rotationally invariant. In such circumstances, for a two-dimensional point process in general, it is known [76] that $\operatorname{Var} \mathcal{N}(\mathcal{D}_N)$ cannot grow slower than of order $|\partial \mathcal{D}_N|$, i.e. the length of the boundary of $\mathcal{D}_N$. Thus both (3.4) and (3.5) are valid for any region $\mathcal{D}_N$ constrained strictly inside the disk of radius $\sqrt{N}$ and with a boundary of length tending to infinity with N. In fact for GinUE more precise asymptotic information is available [148, Eq. (11)], [224, Eq. (2.7)], which gives that for any $D_0 \subseteq \{z : |z| \le 1\}$,

$$\operatorname{Var} \mathcal{N}(\sqrt{N} D_0) = \sqrt{N} \frac{|\partial D_0|}{2\pi} \int_{-\infty}^{\infty} \left(\operatorname{Var} \chi_{U \le (1 + \operatorname{erf}(t/\sqrt{2}))/2} \right) dt + \mathrm{O}\left(\frac{1}{N^{1/2}}\right). \quad (3.9)$$

Here U is a random variable uniform in $[0, 1]$. A direct calculation gives that the integral evaluates to $\sqrt{\frac{1}{\pi}}$—the advantage of the form (3.9) is that it remains true if the appearance of the variance throughout is replaced by any even cumulant [148, 224]. In particular this shows that the growth of the variance with respect to the region is the smallest order possible—the corresponding point process is then referred to as being hyperuniform [289, 523, 524]. A corollary of the property of being hyperuniform, together with the fast decay of the correlations, is that the bulk-scaled GinUE exhibits number rigidity [290, 292]. This means that conditioning on the positions of the (infinite number of) points outside a region $\mathcal{D}$ fully determines the number of (but not positions of) the points inside $\mathcal{D}$, and their centre of mass.

3.1.1   Counting Function in a Disk and Scaled Asymptotics

In the special case that $\mathcal{D}_N$ is a disk of radius R centred at the origin (we write this as D_R), the polynomials in (3.6) are simply the monomials $p_s(z) = z^s$. The eigenfunctions of $\mathbb{K}_{N,\mathcal{D}}$ are also given in terms of the monomials as $\{(w(|z|^2))^{1/2} z^{j-1}\}_{j=1,\dots,N}$, and hence for the corresponding eigenvalues we have

$$\lambda_j(D_R) = \int_0^{R^2} r^{j-1} w(r)\, dr \Big/ \int_0^{\infty} r^{j-1} w(r)\, dr, \qquad j = 1, \dots, N.$$

Substituting in (3.5) and choosing $w(|z|^2) = \exp(-|z|^2)$ then shows that for the GinUE [232]

$$\tilde{E}_N(\xi; D_R) = \prod_{j=1}^{N} \left(1 - \xi \frac{\gamma(j; R^2)}{\Gamma(j)}\right), \quad (3.10)$$

where $\gamma(j; x)$ denotes the (lower) incomplete gamma function. Note that this remains valid for $N \to \infty$ in keeping with the discussion of the previous paragraph. Setting $\xi = 1$, asymptotic expansions for the incomplete gamma function can be used to deduce the $N \to \infty$ asymptotic expansion of $E_N(0; D_{\alpha\sqrt{N}})$,

$$E_N(0; D_{\alpha\sqrt{N}}) = \exp\left(C_1 N^2 + C_2 N \log N + C_3 N + C_4\sqrt{N} + \frac{1}{3}\log N + O(1)\right),$$
(3.11)

where $0 < \alpha < 1$. Here the constants $C_1, \ldots, C_4$ depend on α and are known explicitly (e.g. $C_1 = -\alpha^4/4$), being first given in [232]. The first two of these can be deduced from the large R expansion of the quantity $F_\infty(0; D_R)$, defined in Remark 3.1.1 below, given in the still earlier work [304]. The $\log N$ term was determined recently in [149], as too was the explicit form of the next order term, a constant with respect to N.

Let $\bar{D}_{\alpha\sqrt{N}}$ denote the region $\{z : |z| > \alpha\sqrt{N}\}$, i.e. the region outside the disk of radius $\alpha\sqrt{N}$, where it is assumed $0 < \alpha < 1$. Note that then $E_N(0; \bar{D}_{\alpha\sqrt{N}}) = E_N(N; D_{\alpha\sqrt{N}})$. The analogue of (3.11) has been calculated in [149], where in particular it is found that

$$C_1 = \alpha^4/4 - \alpha^2 + (1/2)\log\alpha^2 + 3/4,$$
(3.12)

the coefficient C_4 is unchanged, while the coefficient $\frac{1}{3}$ for $\log N$ seen in (3.11) is to be replaced by $-\frac{1}{4}$. (We also refer to [175] for an earlier work for which this expansion was obtained up to C_3.) One sees from (3.12) that $C_1 = 0$ for $\alpha = 1$, and the result of [149] gives that C_2, C_3 similarly vanish, giving that $E_N(N; D_{\sqrt{N}}) \sim e^{C_4\sqrt{N}}$. Extending α larger that 1 according to the precise N dependent value

$$\alpha = 1 + \frac{1}{2\sqrt{N}}\left(\sqrt{\gamma_N} + \frac{x}{\sqrt{\gamma_N}}\right), \qquad \gamma_N = \log\frac{N}{2\pi} - 2\log\log N,$$
(3.13)

gives the extreme value result $\lim_{N\to\infty} E_N(N; D_{\alpha_N\sqrt{N}}) = \exp(-\exp(-x))$ [469] (see too [145, Th. 1.3 with $\alpha = 2$] for a generalisation to the case of (2.81), considered also in [149] for $\alpha < 1$), which is the Gumbel law. Other references on fluctuations of the spectral radius under various boundary conditions include [63, 105, 128, 142, 283, 489]. Furthermore, an intermediate fluctuation regime which interpolates between the Gumbel law with the large deviation regime (3.12) was investigated in [382]. Another case considered in [149] is when D_N is specified as the outside of an annulus contained inside of the disk of radius $\sqrt{N}$. Two features of the corresponding asymptotic expansion (3.11) are: (1) the absence of a term proportional to $\log N$, and (2) the presence of oscillations of order 1 that are described in terms of the Jacobi theta function. We also refer to [54–56, 151] for further recent studies in this direction in the presence of hard edges.

For general $\mathcal{D}_N$ with $|\mathcal{D}_N| \to \infty$ the coefficient C_1 in (3.11) relates to an energy minimisation (electrostatics) problem, and similarly for the $|\mathcal{D}| \to \infty$ expansion of $E_\infty(0; \mathcal{D})$ [3, 4, 175, 207, 342]. Thus for $\mathcal{D} = D_{\alpha\sqrt{N}}$ the electrostatics problem is to compute the potential due to a uniform charge density $1/\pi$ inside a disk of radius α, with a neutralising uniform surface charge density $-\alpha/2\pi$ on the boundary. The applicability of electrostatics remains true for the asymptotic expansion of $E_N(k; D_{\alpha\sqrt{N}})$ (and $E_\infty(k; \mathcal{D})$) in the so-called large deviation regime, when $k \ll N\alpha^2$ or $k \gg N\alpha^2$. For a disk the electrostatics problem can be solved explicitly to give [45]

$$E_N(k; D_{\alpha\sqrt{N}}) \sim e^{-N^2\psi_0(\alpha;k/N)},$$
$$\psi_0(\alpha; x) = \frac{1}{4}\left|(\alpha^2 - x)(\alpha^2 - 3x) - 2x^2 \log(x/\alpha^2)\right|. \tag{3.14}$$

Note in particular that $\psi_0(\alpha; 0) = \alpha^4/4$, which is the value of $-C_1$ in (3.11), while setting $x = 1$ gives the value of $-C_1$ noted in the above paragraph. There is also a scaling regime, where $|k - N\alpha^2| = O(N^{1/2})$, for the asymptotic value of $E_N(k; D_{\alpha\sqrt{N}})$ which interpolates between (3.14) and the local central limit theorem result (3.8) [224, 383]. In the case of $E_\infty(k; D_R)$, it makes sense to consider k proportional to not only αR^2 but also to αR^γ with $\gamma > 2$. Then [224, 342]

$$E_\infty(\alpha R^\gamma; D_R) \sim e^{-\frac{1}{2}(\gamma-2)\alpha^2 R^{2\gamma} \log R(1+o(1))}.$$

Remark 3.1 1. Closely related to the probability $E_N(N; \mathcal{D})$ is the conditioned quantity $F_N(n; \mathcal{D}) := \Pr(\sum_{j=1}^{N} \chi_{z_j \in \mathcal{D}} = n | z_j = 0)$. Denote the corresponding generating function by $\tilde{F}_N(\xi; \mathcal{D})$. Proceeding as in the derivation of (3.10) shows $\tilde{F}_N(\xi; D_R) = \tilde{E}_N(\xi; D_R)/(1 - \xi(1 - e^{-R^2}))$. Thus in particular $F_N(0; D_R) = e^{R^2} E_N(0; D_R)$ [304]. Note that $-\frac{d}{dR} F(0; D_R)$ gives the spacing distribution between an eigenvalue conditioned to be at the origin, and its nearest neighbour at a distance R. The work [495] gives results relating to the PDF for the minimum of all the nearest neighbour spacings with global scaling, obtaining a scale of $N^{-3/4}$ and a PDF proportional to $x^3 e^{-x^4}$.

2. For any $p \geq 2$, the p-th cumulant $\kappa_p(D_R)$ of the number of eigenvalues in D_R can be written as

$$\kappa_p(R) = (-1)^{p+1} \sum_{j=0}^{N-1} \text{Li}_{1-p}\left(1 - \frac{1}{\lambda_j(D_R)}\right), \tag{3.15}$$

where $\text{Li}_s(x) = \sum_{k=1}^{\infty} k^{-s} x^k$ is the polylogarithm function. The formula (3.15) as well as its large N behaviour both in the bulk and at the edge were obtained in [384].

3.1.2 Counting Function in the Infinite System

We turn our attention now to the circumstance that a (possibly scaled) large N limit has already been taken, and ask about the fluctuations of the number of particles in a region $\mathcal{N}(\mathcal{D})$ for large values of $|\mathcal{D}|$, i.e. the volume of $\mathcal{D}$. The first point to note is that if the coefficient of ξ^n in (3.4) tends to zero as $N \to \infty$, then the expansion remains valid in this limit [402]. The decay is easy to establish in the determinantal case, since then (see e.g. [237, Eq. (9.13)]) $\rho_{(n),N}(z_1, \ldots, z_n) \leq \prod_{l=1}^n \rho_{(1),N}(z_l)$. Hence it is sufficient that $\int_\Omega \rho_{(1),N}(z) \, d^2z$ be bounded for $N \to \infty$. With the limiting form of (3.4) valid, the theory of Fredholm integral operators [532] tells us that the limit of (3.5) is well defined with the RHS identified as the Fredholm determinant $\det(\mathbb{I} - \xi \mathbb{K}_{\infty,\mathcal{D}})$. In (3.10) the limit corresponds to simply replacing the upper terminal of the product by ∞. We stipulate the further structure that the correlation kernel be Hermitian, as holds for the appropriately scaled form of (3.6). Then the argument of the proof of Proposition 3.1 tells us that the eigenvalues of $\mathcal{K}_{\infty,\mathcal{D}}$ are all between 0 and 1, which in turn allows the reasoning leading to (3.7) to be repeated. The conclusion is, assuming $\operatorname{Var} \mathcal{N}(\mathcal{D}) \to \infty$ as $|\mathcal{D}| \to \infty$ which as already remarked is guaranteed by the results of [76], that (3.7) remains valid with $\mathcal{D}_N$ replaced by $\mathcal{D}$, and the limit $N \to \infty$ replaced by the limit $|\mathcal{D}| \to \infty$ [172, 326, 510].

It is moreover the case that in the above setting and with these modifications the local central limit theorem of Proposition 3.2 remains valid [262]. Another point of interest is that the expansion (3.11) is uniformly valid in the variable $R = \alpha\sqrt{N}$, provided this quantity grows with N, and hence also provides the large R expansion of $E_\infty(0; D_R)$. Finally, we consider results of [395] as they apply to number fluctuations in the infinite GinUE. The plane is to be divided into squares Γ_j of area L^2 with centres at $L\mathbb{Z}^2$. Define $\Upsilon_j = \mathcal{N}(\Gamma_j)/\sqrt{\operatorname{Var} \mathcal{N}(\Gamma_j)}$. For large L, in keeping with (3.9) we have $\operatorname{Var}(\Gamma_j) \sim 2L/\pi^{3/2}$. The question of interest is the joint distribution of $\{\Upsilon_j\}$. It is established in [395] that for $L \to \infty$ this distribution is Gaussian, with covariance $\frac{1}{4}[-\Delta]_{j,k}$, where Δ is the discrete Laplacian on $\mathbb{Z}^2$. Consequently, fluctuations of $\mathcal{N}(\Gamma_j)$ induce opposite fluctuations in the regions neighbouring Γ_j.

For the infinite GinUE the exact result in terms of modified Bessel functions

$$\operatorname{Var} \mathcal{N}(D_R) = R^2 e^{-2R^2}\left(I_0(2R^2) + I_1(2R^2)\right) = \sum_{j=1}^\infty \frac{\gamma(j; R^2)}{\Gamma(j)}\left(1 - \frac{\gamma(j; R^2)}{\Gamma(j)}\right)$$

(3.16)

is known [496, Th. 1.3], [224, Appendix B]. The second expression in (3.16) also appears in [384] as a large N limit of the number variance of the finite Ginibre ensemble in the deep bulk regime. This exhibits the leading large R form $R/\sqrt{\pi}$ which is consistent with identifying $\sqrt{N}|\partial D_0|$ as $2\pi R$ on the RHS of (3.9); see also [20].

Remark 3.2 Consider the weakly non-Hermitian limit of the elliptic Ginibre ensemble specified by the correlation kernel (2.42), parametrised by $\alpha > 0$. Let $E_\infty^\alpha(0; \mathcal{D}(\chi_{|x|<s}))$ denote the probability that there are no eigenvalues in the strip

of the complex plane $\mathrm{Re}\, z < s$. It is a celebrated result of the Kyoto school that for $\alpha \to 0$ this gap probability can be expressed as a Painlevé V transcendent in sigma form (σPV function) [347]. Recently, Bothner and Little [106] extended this result to all $\alpha > 0$, with the role of the σPV function now played by a certain integro-differential Painlevé function. In [106] the same authors have obtained an analogous characterisation of $E_\infty^{\mathrm{edge},\alpha}(0, \mathcal{D}(\chi_{x>s}))$, i.e. the probability of no eigenvalues with $\mathrm{Re}\,(z) > 0$ for the edge-scaled weakly non-Hermitian limit of the elliptic Ginibre ensemble.

3.2 Smooth Linear Statistics

The theory of fluctuation formulas for GinUE in the case that $f(z)$ in (3.1) is smooth has some different features to the discontinuous case $f(z) = \chi_{z \in \mathcal{D}}$. This can be seen by considering the bulk-scaled limit, and in particular the truncated two-point correlation (2.24). From this we can compute the structure factor

$$S_\infty^{\mathrm{GinUE}}(\mathbf{k}) := \int_{\mathbb{R}^2} \left(\rho_{(2),\infty}^{b,T}(\mathbf{0}, \mathbf{r}) + \frac{1}{\pi}\delta(\mathbf{r}) \right) e^{i\mathbf{k}\cdot\mathbf{r}} \, d\mathbf{r} = \frac{1}{\pi}\left(1 - e^{-|\mathbf{k}|^2/4}\right). \quad (3.17)$$

The knowledge of the structure factor allows the limiting covariance (3.2) to be computed using the Fourier transform

$$\mathrm{Cov}^{\mathrm{GinUE}_\infty}\left(\sum f(\mathbf{r}_l), \sum g(\mathbf{r}_l) \right) = \frac{1}{(2\pi)^2}\frac{1}{\pi}\int_{\mathbb{R}^2} \hat{f}(\mathbf{k})\hat{g}(-\mathbf{k})\left(1 - e^{-|\mathbf{k}|^2/4}\right) d\mathbf{k}, \quad (3.18)$$

valid provided the integral converges. Here, with $z = x + iy$, $\mathbf{r} = (x, y)$ and the Fourier transform $\hat{f}(\mathbf{k})$ is defined by integrating $f(\mathbf{r})$ times $e^{i\mathbf{k}\cdot\mathbf{r}}$ over $\mathbb{R}^2$—thus according to (3.17) $S_\infty^{\mathrm{GinUE}}(\mathbf{k})$ is a particular Fourier transform. Now introduce a scale R so that $f(\mathbf{r}) \mapsto f(\mathbf{r}/R)$, $g(\mathbf{r}) \mapsto g(\mathbf{r}/R)$. It follows from (3.18) that

$$\lim_{R\to\infty} \mathrm{Cov}^{\mathrm{GinUE}_\infty}\left(\sum f(\mathbf{r}_l/R), \sum g(\mathbf{r}_l/R) \right) = \frac{1}{(2\pi)^2}\frac{1}{4\pi}\int_{\mathbb{R}^2} \hat{f}(\mathbf{k})\hat{g}(-\mathbf{k})|\mathbf{k}|^2 \, d\mathbf{k}, \quad (3.19)$$

again provided the integral converges. Most noteworthy is that this limiting quantity is $O(1)$. In contrast, with $f(\mathbf{r}) = g(\mathbf{r}) = \chi_{|\mathbf{r}|<1}$, and then introducing R as prescribed above, we know that (3.18) has the evaluation (3.16). As previously commented, the large R form of the latter is proportional to R, which in turn is proportional to the circumference of the disk-shaped region implied by the linear statistic $\chi_{|\mathbf{r}|<R}$.

Remark 3.3 Consider the linear statistic $A(\mathbf{x}) = -\sum_{j=1}^{N}(\log|\mathbf{x} - \mathbf{r}_j| - \log|\mathbf{r}_j|)$. In the Coulomb gas picture relating to (1.12), this corresponds to the difference in the potential at $\mathbf{x}$ and the origin. For bulk-scaled GinUE, one has from (3.18) the exact result [42]

$$\text{Var}^{\text{GinUE}_\infty} A(\mathbf{x}) = \frac{1}{2}\left(2\log|\mathbf{x}| + (|\mathbf{x}|^2 + 1)\int_{|\mathbf{x}|^2}^{\infty} \frac{e^{-t}}{t}\,dt - e^{-|\mathbf{x}|^2} + C + 1\right),$$

(3.20)

where C denotes Euler's constant. In particular, for large $|\mathbf{x}|$, $\text{Var}^{\text{GinUE}_\infty} A(\mathbf{x}) \sim \log|\mathbf{x}|$. This last point shows that the introduction of a scale R as in (3.19) would give rise to a divergence proportional to $\log R$. Such a log-correlated structure underlies a relationship between the logarithm of the absolute value of the characteristic polynomial for GinUE and Gaussian multiplicative chaos [387].

The covariance with test functions $f(\mathbf{r}) \mapsto f(\mathbf{r}/\sqrt{N})$, $g(\mathbf{r}) \mapsto g(\mathbf{r}/\sqrt{N})$ assumed to take on real or complex values is also an order-one quantity for GinUE in the $N \to \infty$ limit, upon the additional assumption that f, g are differentiable and don't grow too fast at infinity [58, 59, 235, 472].

Proposition 3.3 *Require that f, g have the properties as stated above. Let*

$$f(\mathbf{r})|_{\mathbf{r}=(\cos\theta,\sin\theta)} = \sum_{n=-\infty}^{\infty} f_n e^{in\theta}$$

and similarly for the Fourier expansion of $g(\mathbf{r})$ for $\mathbf{r} = (\cos\theta, \sin\theta)$. We have

$$\lim_{N\to\infty} \text{Cov}^{\text{GinUE}}\left(\sum_{j=1}^{N} f(\mathbf{r}_j/\sqrt{N}), \sum_{j=1}^{N} \bar{g}(\mathbf{r}_j/\sqrt{N})\right)$$

$$= \frac{1}{4\pi}\int_{|\mathbf{r}|<1} \nabla f \cdot \nabla \bar{g}\,dxdy + \frac{1}{2}\sum_{n=-\infty}^{\infty} |n| f_n \bar{g}_{-n}.$$

(3.21)

Proof (Sketch) In the method of [472], a direct calculation using (3.2), (2.18) and (2.10) allows (3.21) to be established for f, g polynomials jointly in $z = x + iy$ and $\bar{z} = x - iy$. The required integrations can be computed exactly using polar coordinates. To go beyond the polynomial case, the so-called dbar (Cauchy–Pompeiu) representation is used. This gives that for any once continuously differentiable f in the unit disk D_1, and z contained in the interior of the disk,

$$f(z) = -\frac{1}{\pi}\int_{D_1} \frac{\partial_{\bar{w}} f(w)}{w-z}\,d^2w + \frac{1}{2\pi i}\int_{\partial D_1} \frac{f(w)}{w-z}\,dw, \qquad (3.22)$$

where with $w = \alpha + i\gamma$, $\partial_{\bar{w}} = \frac{1}{2}(\frac{\partial}{\partial\alpha} + i\frac{\partial}{\partial\gamma})$. The covariance problem is thus reduced to the particular class of linear statistics of the functional form $h(z) = 1/(w-z)$. The required analysis in this case is facilitated by the use of the corresponding Laurent expansion, with only a finite number of terms contributing after integration. $\square$

Remark 3.4 1. As predicted in [235], upon multiplying the RHS by $2/\beta$, (3.21) remains valid for the Coulomb gas model (1.12) [75, 394, 491]. In the case of the elliptic GinUE, a simple modification of (3.21) holds true. Thus the domain $|\mathbf{r}| < 1$ in the first term is to be replaced by the appropriate ellipse, and the Fourier components of the second term are now in the variable η, where $(A \cos \eta, B \sin \eta)$, $0 \le \eta \le 2\pi$ parametrises the boundary of the ellipse. The results of [59, 235, 394] also cover this case.

2. In the case of an ellipse, major and minor axes A, B say, there is particular interest in the linear statistic $P_x := \sum_{j=1}^{N} x_j$ [155]. Linear response theory gives for the xx component of the susceptibility tensor χ—relating the polarisation density to the applied electric field—the formula $\chi_{xx} = (\beta/(\pi A B)) \lim_{N \to \infty} \text{Var } P_x$ (and similarly for the xy and yy components). This same quantity can be computed by consideration of macroscopic electrostatics, which gives $\chi_{xx} = (A + B)/(\pi B)$. Using (3.21) modified as in the above paragraph, the consistency of these formulas can be verified.

3. Considering further the case of elliptic GinUE, dividing by N and taking the limit $\tau \to 1$ gives the GUE with eigenvalues supported on $(-2, 2)$, and similarly for the β generalisation limiting to (1.12) restricted to this interval. For this model it is known (see e.g. [250, Eq. (3.2) with the identification $x = 2 \cos \theta$]) that

$$\lim_{N \to \infty} \text{Cov}\left(\sum_{j=1}^{N} f(x_j), \sum_{j=1}^{N} g(x_j) \right) = \frac{2}{\beta} \sum_{n=1}^{\infty} n f_n^c g_n^c,$$

where $f(x)|_{x=2\cos\theta} = f_0^c + 2 \sum_{n=1}^{\infty} f_n^c \cos n\theta$ and similarly for $g(x)$. We observe that this is identical to the final term in (3.21), modified according to the specifications of point 1. above.

4. There has been a recent application of Proposition 3.3 in relation to the computation of the analogue of the Page curve for a density matrix constructed out of GinUE matrices [164].

We turn our attention now to the limiting distribution of a smooth linear statistic. By way of introduction, consider the particular linear statistic $\frac{1}{N} \sum_{j=1}^{N} |\mathbf{r}_j|^2$ for GinUE. An elementary calculation gives that the corresponding characteristic function, $\hat{P}_N(k)$ say, has the exact functional form

$$\hat{P}_N(k) = (1 - ik/N)^{-N(N+1)/2}. \tag{3.23}$$

It follows from this that after centring by the mean, the limiting distribution is a Gaussian with variance given by (3.21) (which is this specific case evaluates to one). A limiting Gaussian form holds in the general case of the applicability of (3.21), as first proved by Rider and Virág [472].

Proposition 3.4 *Let f be subject to the same conditions as in Proposition 3.3, and denote the case $f = g$ of (3.21) by σ_f^2. For the GinUE, if f takes on complex values then as $N \to \infty$*

$$\sum_{j=1}^{N} f(\mathbf{r}_j/\sqrt{N}) - \left\langle \sum_{j=1}^{N} f(\mathbf{r}_j/\sqrt{N}) \right\rangle \xrightarrow{d} \text{N}[0, \sigma_f] + i\text{N}[0, \sigma_f],$$

*while if f is real-valued the RHS of this expression is to be replaced by $N[0, \sigma_f]$.
Moreover, this same limit formula holds for the elliptic GinUE [59] and its β generalisation [394] (both subject to further technical restrictions on f), with the variance
modified according to Remark 3.4.1.*

Proof (Comments only) The proof of [472] proceeds by establishing that the higher-
order cumulants beyond the variance tend to zero as $N \to \infty$. Essential use is made
of the rotation invariance of GinUE. The method of [59] uses a loop equation strategy,
while [394] involves energy minimisers and transport maps. For GinUE with f a
function of $|\mathbf{r}|$, a simple derivation based on the proof of Proposition 2.14 together
with a Laplace approximation of the integrals [235] (see also [120, Appendix B]). $\square$

Remark 3.5 Other settings in which Proposition 3.4 has proved to be valid include
products of GinUE matrices [171, 371] (with the additional assumption that the test
function has support strictly inside the unit circle), and for the complex spherical
ensemble of Sect. 2.5 after stereographic projection onto the sphere [88, 471].

3.3 Spatial Modelling and the Thinned GinUE

The GinUE viewed as a point process in the plane has been used to model geograph-
ical regions by way of the corresponding Voronoi tessellation [391], the positions of
objects, for example trees in a plantation [390] or the nests of birds of prey [12],
and the spatial distribution of base stations in modern wireless networks [188, 440],
amongst other examples. The wireless network application has made use of the
thinned GinUE, whereby each eigenvalue is independently deleted with probability
$(1 - \zeta), 0 < \zeta \le 1$. The effect of this is simple to describe in terms of the correlation
functions, according to the replacement

$$\rho_{(n),N}(z_1, \ldots, z_n) \mapsto \zeta^N \rho_{(n),N}(z_1, \ldots, z_n). \tag{3.24}$$

With the bulk density of GinUE uniform and is equal to $1/\pi$, we can also rescale the
position $z_j \mapsto z_j/\zeta$ so that this remains true in the thinned ensemble. For this (3.24)
is to be updated to read

$$\rho_{(n),N}(z_1, \ldots, z_n) \mapsto \rho_{(n),N}(z_1/\sqrt{\zeta}, \ldots, z_n/\sqrt{\zeta}). \tag{3.25}$$

Recalling now (2.9) and (2.18), for the bulk-scaled limit of the thinned GinUE
we have in particular

$$\rho_{(1),\infty}^{\text{tGinUE}}(z) = \frac{1}{\pi}, \qquad \rho_{(2),\infty}^{\text{tGinUE},T}(w, z) = -\frac{1}{\pi^2} e^{-|w-z|^2/\zeta}.$$

From these functional forms we see

$$\int_{\mathbb{C}} \rho_{(2),\infty}^{\text{tGinUE},T}(w,z)\,d^2z = -\frac{\zeta}{\pi} \neq -\rho_{(1),\infty}^{\text{tGinUE}}(w) \quad \text{unless } \zeta = 1,$$

where the superscript "tGinU" denotes the thinned GinUE. Equivalently, in terms of the structure factor (3.17),

$$S_{\infty}^{\text{tGinUE}}(\mathbf{0}) = \frac{1-\zeta}{\pi} \neq 0 \quad \text{unless } \zeta = 1.$$

Due to this last fact, the O(1) scaled covariance for smooth linear statistics (3.18) is no longer true, and now reads instead

$$\text{Cov}^{\text{tGinUE}}\left(\sum f(\mathbf{r}_l/R), \sum g(\mathbf{r}_l/R)\right) \underset{R\to\infty}{\sim} \frac{R^2}{(2\pi)^2}\frac{(1-\zeta)}{\pi}\int_{\mathbb{R}^2} \hat{f}(\mathbf{k})\hat{g}(-\mathbf{k})\,d\mathbf{k}. \tag{3.26}$$

This leading dependence on R^2 holds too for the counting function $f(z) = \chi_{|z|<1}$, since in distinction to (3.18) the integral now converges. Hence, in the terminology of the text introduced below (3.9), the statistical state is no longer hyperuniform. There is an analogous change to the O(1) scaled covariance (3.21), which is now proportional to N and reads

$$\text{Cov}^{\text{tGinUE}}\left(\sum_{j=1}^{N} f(\mathbf{r}_j/\sqrt{N}), \sum_{j=1}^{N} \bar{g}(\mathbf{r}_j/\sqrt{N})\right) \underset{N\to\infty}{\sim} N\frac{(1-\zeta)}{\pi}\int_{|\mathbf{r}|<1} f\bar{g}\,dxdy. \tag{3.27}$$

Notwithstanding this difference, the corresponding limiting distribution function is still Gaussian [386]. A more subtle limit, also considered in [386], is when $N \to \infty$ and $\zeta \to 1^-$ simultaneously, with $N(1-\zeta)$ fixed. The quantity (3.21) returns to being O(1), but consists of a contribution of the form (3.21), and a term characteristic of a Poisson process; see also [455].

We turn our attention now to the probabilities $\{E_N^{\text{tGUE}}(k; D_{\alpha\sqrt{N}})\}$. Upon consideration of the thinning prescription (3.25), the proof of Proposition 3.1, and (3.10) shows that the corresponding generating function is given by

$$\tilde{E}_N^{\text{tGUE}}(\xi; D_{\alpha\sqrt{\zeta N}}) = \prod_{j=1}^{N}\left(1 - \xi\zeta\frac{\gamma(j;\alpha^2 N)}{\Gamma(j)}\right). \tag{3.28}$$

Setting $\xi = 1$ in this gives the probability $E_N^{\text{tGUE}}(0; D_{\alpha\sqrt{\zeta N}})$. Note that the implied formula shows $E_N^{\text{tGUE}}(0; D_{\alpha\sqrt{\zeta N}}) = \tilde{E}_N^{\text{GUE}}(\zeta; D_{\alpha\sqrt{N}})$. The large N asymptotics of $\tilde{E}_N^{\text{GUE}}(\zeta; D_{\alpha\sqrt{N}})$, and various generalisations, are available in the literature [116, 150]. Here we present a self-contained derivation of the first two terms (cf. (3.11)).

Proposition 3.5 *For large N and with $0 < \alpha, \zeta < 1$ we have*

$$\tilde{E}_N^{\text{tGUE}}(0; D_{\alpha\sqrt{\zeta N}}) = \exp\left(\alpha^2 N \log(1 - \zeta) + \sqrt{\alpha^2 N}\, h(\zeta) + O(1)\right), \qquad (3.29)$$

where

$$h(\zeta) = \int_0^\infty \log\left(\frac{1 - (\zeta/2)(1 + \text{erf}(t/\sqrt{2}))}{1 - \zeta}\right) dt$$
$$+ \int_0^\infty \log\left(1 - (\zeta/2)(1 - \text{erf}(t/\sqrt{2}))\right) dt. \qquad (3.30)$$

Proof Our main tool is the uniform asymptotic expansion [529]

$$\frac{\gamma(M - j + 1; M)}{\Gamma(M - j + 1)} \underset{M \to \infty}{\sim} \frac{1}{2}\left(1 + \text{erf}\left(\frac{j}{\sqrt{2M}}\right)\right); \qquad (3.31)$$

cf. the leading term in (2.20). Here it is known that the error term has the structure $(1/\sqrt{M})g(j/\sqrt{2M})$, where $g(t)$ is integrable on $\mathbb{R}$ and decays rapidly at infinity. This expansion suggests we rewrite the product in (3.28) with $\xi = 1$ in the form

$$(1 - \zeta)^{[M^*]}\left(\prod_{j=1}^{M^*} \frac{1 - \zeta\gamma(j; M^*)/\Gamma(j)}{1 - \zeta}\right)\left(\prod_{j=M^*+1}^{N} (1 - \zeta\gamma(j; M^*)/\Gamma(j))\right),$$

where $M^* = [\alpha^2 N]$.

We see that the first term in this expression gives the leading-order term in (3.29). In the first product we change labels $j \mapsto M^* - j + 1$ $(j = 1, \ldots, M^*)$. In the second we change labels $j \mapsto M^* + j + 1$ $(j = 0, \ldots, N - M^* - 1)$. Now writing both these products as exponentials of sums and applying (3.31) gives, upon recognising the sums as Riemann integrals, the $O(\sqrt{\alpha^2 N})$ term in (3.29). □

The leading term in (3.29) is consistent with the general form expected for thinned log-gas systems, being of the form of the area of the rescaled excluded region, times the density, times $\log(1 - \zeta)$ [242, Conj. 10].

Remark 3.6 1. The topic of spatial modelling using Ginibre eigenvalues naturally leads to questions on the efficient simulation of the point process confined to a compact subset of $\mathbb{C}$. Practical algorithms for this task have been given in [181, 182].
2. Beyond Voronoi cells as a geometrical measure, a topological measure known as persistence diagrams has been introduced in the context of Ginibre point processes in [291].
3. A question of long-standing interest is the spectrum of an adjacency matrix for a random graph or network; see for example [196]. In the particular case of directed random networks, for which the adjacency matrix is asymmetric, relationships to bulk GinUE spectral statistics have been found [460, 537].

Chapter 4
Coulomb Gas Model, Sum Rules and Asymptotic Behaviours

Throughout this chapter we consider the Coulomb gas model (1.12) with general $\beta > 0$, for which we use the notation OCP, which stands for one-component plasma. The special case $\beta = 2$ coincides with GinUE. The study of thermodynamic properties of the OCP requires the asymptotic expansion of the logarithm of the renormalised configuration integral for large N. This holds true for the Coulomb gas model confined to general domains. A conjectured universal logarithmic term, which is proportional to the Euler characteristic of the domain, is highlighted and a conjecture for the $\sqrt{N}$ surface tension term in the case of disk geometry is also reported, as are some exact results for the energy per particle. In relation to the edge-scaled charge density, sum rules for the total charge and the dipole moment are given, and an asymptotic expansion of the density outside of the droplet is formulated, which as the multiplicative constant term involves the dimensionless free energy per particle. The final section of the chapter addresses sum rules and asymptotics associated with the truncated two-point correlation function, which is conveniently interpreted as the screening cloud about a fixed charge. Its Fourier transform gives the structure function, for which several terms in the small wavenumber expansion can be predicted. Also emphasised is distinct, slowly decaying, asymptotic behaviour of the truncated two-point correlation function at the boundary.

4.1 Asymptotics Associated with the Configuration Integral

The configuration integral for the Boltzmann factor (1.12) is specified by

$$Q_N^{\mathrm{OCP}}(\beta) = \int_{\mathbb{C}} d^2 z_1 \cdots \int_{\mathbb{C}} d^2 z_N \, e^{-(\beta/2)\sum_{j=1}^{N} |z_j|^2} \prod_{1 \leq j < k \leq N} |z_k - z_j|^{\beta}. \qquad (4.1)$$

© The Author(s) 2025
S.-S. Byun and P. J. Forrester, *Progress on the Study of the Ginibre Ensembles*, KIAS Springer Series in Mathematics 3, https://doi.org/10.1007/978-981-97-5173-0_4

From the OCP viewpoint, it is more natural to consider the renormalised quantity

$$Z_N^{D_R, \text{OCP}}(\beta) = \frac{1}{N!} A_{N,\beta} Q_N^{\text{OCP}}(\beta), \qquad A_{N,\beta} = e^{-\beta N^2 (\frac{1}{4} \log N - \frac{3}{8})}; \qquad (4.2)$$

see [237, Eq. (1.72)]. The symbol D_R on the LHS indicates an underlying disk of radius R (with $R = \sqrt{N}$ although we suppress this detail). Specifically, $A_{N,\beta}$ on the RHS of (4.2) is derived as the Boltzmann factor of the electrostatic self-energy of a smeared out uniform background, charge density $-\frac{1}{\pi}$, confined to the disk D_R, and the constant terms of its electrostatic energy when coupled to a particle inside of this disk. The quantity $Z_N^{D_R, \text{OCP}}(\beta)$ is then the partition function of the charge neutral OCP. Generally, in statistical mechanics for a stable system the dimensionless free energy, $\beta F_N(\beta) = -\log Z_N(\beta)$, is an extensive quantity, meaning that for large N it is proportional to N. For the closely related model when the particles are strictly restricted to the disk, the validity of this statement was established in [405, 486], and has been reconsidered recently using more far-reaching techniques in [393] (which for example form a platform for the study of fluctuation formulas in [394]). Making use of Proposition 2.1 the large N form of $\beta F_N^{D_R, \text{OCP}}(\beta)$ can be computed for $\beta = 2$ [520, Eq. (3.14)].

Proposition 4.1 *We have*

$$\beta F_N^{D_R, \text{OCP}}(\beta)\Big|_{\beta=2} = N\beta f(\beta)|_{\beta=2} + \frac{1}{12} \log N - \zeta'(-1) - \frac{1}{720N^2} + \mathrm{O}\left(\frac{1}{N^4}\right), \tag{4.3}$$

where

$$\beta f(\beta)|_{\beta=2} = \frac{1}{2} \log\left(\frac{1}{2\pi^3}\right). \tag{4.4}$$

Proof This relies on (2.3), identifying $\prod_{j=1}^{N-1} j! = G(N+1)$, where $G(x)$ denotes the Barnes G-function, and the knowledge of the known asymptotic expansion of $G(N+1)$ (see e.g. [226, Th. 1]). □

For the OCP on a sphere of radius R, S_R^2 say, and with $R = \frac{1}{2}\sqrt{N}$ so that the particle density is $1/\pi$, the partition function of the charge neutral system is (see e.g. [520, Eq. (2.1)])

$$Z_N^{S_R^2, \text{OCP}}(\beta) = \frac{1}{N!} N^{-N\beta/2} e^{\beta N^2/4}$$

$$\times \int_{S_R^2} d\theta_1 d\phi_1 \cdots \int_{S_R^2} d\theta_N d\phi_N \prod_{1 \le j < k \le N} |u_k v_j - u_j v_k|^\beta.$$

Here the variables $\{u_j, v_k\}$ are the Cayley–Klein parameters as in (2.52). From the analogue of Proposition 2.1 this can be evaluated exactly at $\beta = 2$ [131], implying that for large N [344], [520, Eq. (3.6)]

$$\beta F_N^{S_R^2,\text{OCP}}(\beta)\Big|_{\beta=2} = N\beta f(\beta)|_{\beta=2} + \frac{1}{6}\log N + \frac{1}{12} - 2\zeta'(-1) + \frac{1}{180N^2} + O\Big(\frac{1}{N^4}\Big),$$

$$(4.5)$$

where $\beta f(\beta)|_{\beta=2}$ is as in (4.3).

Both expansions (4.3) and (4.5) illustrate a conjectured universal property of the large N expansion of βF_N^{OCP} in the case that the droplet forms a shape with Euler index χ [344]

$$\beta F_N^{\text{OCP}}(\beta) = N\beta f(\beta) + a_\beta\sqrt{N} + \frac{\chi}{12}\log N + \cdots,$$

$$(4.6)$$

valid for general $\beta > 0$. To compare against the exact results for $\beta = 2$ it should be recalled that $\chi = 1$ for a disk and $\chi = 2$ for a sphere. An analogous calculation in annulus geometry at $\beta = 2$ gives an expansion with a term proportional to $\log N$ absent, in keeping with $\chi = 0$ [229], [125]. A further point of interest is that the large N expansions for $E_N(0; \mathcal{D}_N)$ from [149] as reviewed in Sect. 3.1.1, also exhibit simple fractions for the coefficient of $\log N$, and moreover this term is not present in the case where the eigenvalues are constrained to a single annulus.

We note that the term proportional to $\sqrt{N}$ in (4.6) has the interpretation as a surface tension, and so is expected not to be present in the case of a sphere for any β. Also, for the disk geometry, it has been conjectured (see [135, Eq. (3.2)]) that

$$a_\beta = \frac{4\log(\beta/2)}{3\pi^{1/2}},$$

$$(4.7)$$

which in particular vanishes in the special case $\beta = 2$, as is consistent with (4.5).

Remark 4.1 1. Multiplication of the configuration integral $Q_N^{\text{OCP}}(\beta)$ by $A_{N,\beta}$ as in (4.2) effectively shifts the microscopic energy U in (1.12) by a function of N. We denote this shifted, charge neutral, energy by U', and similarly in relation to the sphere. For $\beta = 2$ direct calculation of the mean is possible. Thus in the case of the OCP on the plane, one finds [494, equivalent to Eqs. (25) and (29)]

$$\langle U'\rangle^{D_R}\Big|_{\beta=2} = -\frac{1}{2}\Big(\frac{N^2}{2}(\Psi(N) - \log N) + \frac{N+1}{4} + \frac{NC}{2} - \frac{\Gamma(N+3/2)}{\Gamma(N+2)\Gamma(3/2)}$$

$$\times {}_3F_2\Big({1, N-1, N+3/2 \atop N+2, N+1}\Big|1\Big)\Big) = -\frac{CN}{4} + \frac{2\sqrt{N}}{3\sqrt{\pi}} - \frac{5}{48} + O(N^{-1/2}),$$

where C denotes Euler's constant and $\Psi(N)$ denotes the digamma function. The corresponding result for the sphere geometry at $\beta = 2$ gives the simpler formula [43, 132, 480]

$$\langle U'\rangle^{S_R^2}\Big|_{\beta=2} = \frac{N}{4}\Big(-H_N + \log N\Big) = -\frac{CN}{4} - \frac{1}{8} + O(N^{-1}),$$

where H_N denotes the harmonic numbers. The common leading-order value of the charge neutral energy per particle, $-C/4$, was known to Jancovici [337] through the formula

$$\lim_{N\to\infty} \frac{1}{N} \langle U' \rangle^{\text{OCP}} \Big|_{\beta=2} = -\frac{\pi}{2} \int_{\mathbb{R}^2} \log |\mathbf{r}| \, \rho_{(2),\infty}^{\text{b,GinUE}}(\mathbf{0}, \mathbf{r}) \, d\mathbf{r} = -\frac{C}{4},$$

where the integral is evaluated from the explicit formula for $\rho_{(2),\infty}^{\text{b,GinUE}}$ given by (2.24). Note as an expansion about $\beta = 2$, $\beta F_N(\beta) = \beta F_N(\beta)|_{\beta=2} + (\beta - 2)\langle U' \rangle + O((\beta/2 - 1)^2)$, so the above results allow (4.3) and (4.5) to be extended to first order in $(\beta - 2)$. In particular consistency is obtained with (4.6).

2. In the low temperature limit $\beta \to \infty$ the OCP particles are expected to form a triangular lattice. Recent works relating to this include [48, 65, 66, 93, 138, 394, 474, 484]. The conjectured exact value of twice the charge neutral energy per particle in this limit, with bulk density $1/(4\pi)$ (not $1/\pi$ as is natural for GinUE in the plane, rather the geometry used was a sphere of unit radius) is [108]

$$2\log 2 + \frac{1}{2}\log \frac{2}{3} + 3\log \frac{\sqrt{\pi}}{\Gamma(1/3)} = -0.0556053\ldots;$$

see [511] for a lower bound close to this value. To obtain this, a limit from the known number theoretic expression for the potential at the origin due to a triangular lattice of particles interacting by the Riesz $1/r^s$ pairwise potential is taken—the work [493] discusses analytic properties of such lattice sums and regularisations. Relevant to the different bulk density is the generalisation of (4.2) to the case of a general background charge density $-\rho_b$ inside D_R. Let this quantity be denoted by $Z_N^{D_R,\text{OCP}}(\beta; \rho_b)$. Using [237, Eq. (1.72)] a simple change of variables in the configuration integral shows

$$Z_N^{D_R,\text{OCP}}(\beta; \rho_b) = e^{N(1-\beta/4)\log \pi \rho_b} Z_N^{D_R,\text{OCP}}(\beta; 1/\pi). \tag{4.8}$$

Consequently, upon taking the logarithmic derivative with respect to β, the limiting internal energy per particle, $u_\infty^{\text{OCP}}(\beta; \rho_b)$ say, has the functional form

$$u_\infty^{\text{OCP}}(\beta; \rho_b) = -\frac{1}{4}\log \pi \rho_b + u_\infty^{\text{OCP}}(\beta; 1/\pi), \tag{4.9}$$

exhibiting a simple dependence on ρ_b. We also refer to [18, 144, 388] and references therein for recent works on the opposite, high-temperature limit $\beta \to 0$. In particular, Debye–Huckel theory predicts

$$u_\infty^{\text{OCP}}(\beta; \rho_b) \underset{\beta\to 0}{\sim} -\frac{1}{4}\log \pi \rho_b - \frac{1}{4}\log(\beta/2) - \frac{C}{2}, \tag{4.10}$$

where C denotes Euler's constant; see e.g. [481, Sect. 2.1]. Finite size corrections in the Debye–Huckel formalism are considered in [519, 525].

4.2 Sum Rules and Asymptotics for the Edge Density

First we note that for large N the leading-order support of the density is a disk of radius $\sqrt{N}$, independent of β, as follows from the potential theoretic argument of Remark 2.2.2. Using the vector coordinate $\mathbf{r} = (x, y)$, as in the second equation in (2.19) we introduce edge-scaling coordinates by writing

$$\rho_{(1),N}^{\mathrm{OCP}}((x, \sqrt{N} - y)) = \rho_{(1),\infty}^{\mathrm{e,OCP}}(y) + \frac{1}{\sqrt{N}} \mu_{(1),\infty}^{\mathrm{e,OCP}}(y) + \mathrm{O}\Big(\frac{1}{N}\Big). \qquad (4.11)$$

Here the form of the correction terms, known to be valid at $\beta = 2$ according to (2.28), are at this stage presented as an ansatz. We seek some integral identities that must be satisfied by $\rho_{(1),\infty}^{\mathrm{e,OCP}}(y)$ and $\mu_{(1),\infty}^{\mathrm{e,OCP}}(y)$. Identities of this type are referred to as sum rules.

Proposition 4.2 *We have*

$$\int_{-\infty}^{\infty} \Big(\rho_{(1),\infty}^{\mathrm{e,OCP}}(y) - \frac{1}{\pi}\chi_{y>0}\Big)\, dy = 0 \qquad (4.12)$$

and

$$\int_{-\infty}^{\infty} y\Big(\rho_{(1),\infty}^{\mathrm{e,OCP}}(y) - \frac{1}{\pi}\chi_{y>0}\Big)\, dy = \int_{-\infty}^{\infty} \mu_{(1),\infty}^{\mathrm{e,OCP}}(y)\, dy. \qquad (4.13)$$

Proof Because of the large N form of the density, we have that $\rho_{(1),N}^{\mathrm{e,OCP}}(\mathbf{r}) - \frac{1}{\pi}\chi_{|\mathbf{r}|<\sqrt{N}}$ will be concentrated near $|\mathbf{r}| = \sqrt{N}$. Now, the normalisation condition for the density gives, with the use of polar coordinates

$$\int_0^{\infty} r\Big(\rho_{(1),N}^{\mathrm{e,OCP}}(\mathbf{r}) - \frac{1}{\pi}\chi_{|\mathbf{r}|<\sqrt{N}}\Big)\, dr$$

$$= \int_{-\infty}^{\sqrt{N}} (\sqrt{N} - y)\Big(\rho_{(1),N}^{\mathrm{e,OCP}}(\sqrt{N} - y) - \frac{1}{\pi}\chi_{y>0}\Big)\, dr = 0.$$

In the second term we now substitute (4.11) and equate terms of order $\sqrt{N}$ and of order unity to zero to obtain (4.12) and (4.13). □

From the functional forms for $\rho_{(1),\infty}^{\mathrm{e,GinUE}}(y)$ and $\mu_{(1),\infty}^{\mathrm{e,GinUE}}(y)$ as read off from (2.28), we verify both of the above sum rules in this special case. Note that the integral on the LHS of (4.13) has the interpretation of the dipole moment of the excess charge in the edge boundary layer, using the Coulomb gas picture. In fact it is possible to derive a further sum rule which evaluates this dipole moment explicitly [521, Eq. (5.13)].

Proposition 4.3 *We have*

$$\int_{-\infty}^{\infty} y\Big(\rho_{(1),\infty}^{\mathrm{e,OCP}}(y) - \frac{1}{\pi}\chi_{y>0}\Big)\, dy = -\frac{1}{2\pi\beta}\Big(1 - \frac{\beta}{4}\Big). \qquad (4.14)$$

Proof From the definition, we observe

$$\left\langle \sum_{j=1}^{N} |\mathbf{r}_j|^2 \right\rangle = -\frac{\partial}{\partial c} \log \left(\int_{\mathbb{R}^2} d\mathbf{r}_1 \cdots \int_{\mathbb{R}^2} d\mathbf{r}_N \, e^{-c \sum_{j=1}^{N} |\mathbf{r}_j|^2} \prod_{1 \le j < k \le N} |\mathbf{r}_k - \mathbf{r}_j|^\beta \right)\bigg|_{c=\beta/2}.$$
(4.15)

The c dependence of the integral can be factored by a simple change of variables, allowing the integral to be replaced by $c^{-N-\beta N(N-1)/4}$, and so giving

$$\left\langle \sum_{j=1}^{N} |\mathbf{r}_j|^2 \right\rangle = \frac{2N}{\beta} + \frac{1}{2} N(N-1).$$
(4.16)

By writing the LHS as an average over the density and the use of polar coordinates, we see that this is equivalent to the sum rule

$$2\pi \int_0^\infty r^3 \left(\rho_{(1),N}^{\mathrm{OCP}}(r) - \frac{1}{\pi} \chi_{r < \sqrt{N}} \right) dr = \frac{N}{2} \left(\frac{4}{\beta} - 1 \right).$$

Proceeding now as in the proof of Proposition 4.2, by substituting (4.11), equating terms proportional to N on both sides, and making use too of (4.13), we deduce (4.14). □

One observes that the RHS of (4.14) changes sign as β increase beyond 4. In fact in the work [134] it is predicted that for general $\beta > 2$, the edge density profile of the OCP exhibits an overshoot effect where it rises before tailing off to zero. The exact evaluation of the edge density to leading-order in $\beta - 2$ in [135] lends analytic evidence to this claim, while numerical evidence from Monte Carlo simulations is presented in [138]. This edge density overshoot effect has been observed in the random matrix ensemble of even-dimensional random matrices $Z_N W$, where $Z_N = \mathbb{I}_N \otimes \begin{bmatrix} 0 & 1 \\ -1 & 0 \end{bmatrix}$ and W is a complex anti-symmetric Gaussian random matrix [243, 319]. Moreover, in [319], upon the assumption of large eigenvalue separation, an analytic calculation of the joint eigenvalue PDF gives the functional form of the OCP with $\beta = 4$.

We now turn our attention to the large N form of the global-scaled density, $\rho_{(1),N}^{\mathrm{g,OCP}}(\mathbf{r}) := \rho_{(1),N}^{\mathrm{OCP}}(\sqrt{N}\mathbf{r})$. From the theory noted at the beginning of the section, this will limit to the circular law (2.17). We ask about the leading correction term. A hint is given by (4.16), which after dividing both sides by N^2 to correspond to global coordinates tells us

$$\left\langle \frac{1}{N^2} \sum_{j=1}^{N} |\mathbf{r}_j|^2 \right\rangle_{\mathrm{g,OCP}} = \frac{1}{2} - \frac{1}{2} \left(1 - \frac{4}{\beta} \right) \frac{1}{N}.$$

Here, the subscript "g, OCP" means that the average is taken with respect to the measure (4.1). The first term $\frac{1}{2}$ is the average of the function $g(\mathbf{r}) = |\mathbf{r}|^2$ over the disk $D_{\sqrt{N}}$ with density $\frac{1}{\pi}$. The second term is a $\frac{1}{N}$ correction, so we might expect that the leading correction to the circular law is $O(1/N)$. In fact, knowledge of (4.14) is sufficient to compute the leading correction term of the large N expansion of all the moments $\langle \frac{1}{N}\sum_{j=1}^{N}|\mathbf{r}_j|^p\rangle_{\text{g,OCP}}$ ($p = 1, 2, \ldots$), allowing us to conclude [521, Eq. (5.18)] that

$$\rho_{(1),N}^{\text{g,OCP}}(\mathbf{r}) = \frac{1}{\pi}\chi_{|\mathbf{r}|<1} + \frac{1}{N}\kappa(r) + O(N^{-3/2}), \quad \kappa(r) = \frac{1}{2\pi\beta}\left(1-\frac{\beta}{4}\right)\frac{1}{r}\delta'(r-1).$$
(4.17)

Thus the correction term is concentrated on the boundary. For GinUE, it can be deduced from (2.10) that inside the droplet the corrections are exponentially small [397, Th. 1.2],[334, Lemma 3.1]. (See also [300, 335]) This same formula can also be read off from a more general formula relating to β generalised normal matrix models (see § 5.3) obtained by Zabrodin and Wiegmann [539, Eq. (5.16)]. Associated with (4.17) is the expansion [393, Eq. (1.14)]

$$\left\langle \frac{1}{N}\sum_{j=1}^{N}g(\mathbf{r}_j)\right\rangle^{\text{g,OCP}} = \frac{1}{\pi}\int_{|\mathbf{r}|<1}g(\mathbf{r})\,dxdy$$
$$+ \frac{1}{N}\frac{1}{2\pi\beta}\left(1-\frac{\beta}{4}\right)\int_{|\mathbf{r}|<1}\nabla^2 g(\mathbf{r})\,dxdy + o(N^{-1}), \quad (4.18)$$

valid for sufficiently smooth test functions g (note that rotational invariance is not assumed). In the case $\beta = 2$ this expansion can be found in [59, Th. 2.1].

As our final topic under this heading, we consider the $y \to -\infty$ asymptotic form of $\rho_{(1),\infty}^{\text{e,OCP}}(y)$. According to (2.29), at $\beta = 2$ we have

$$\rho_{(1),\infty}^{\text{e,OCP}}(y) \underset{y\to-\infty}{\sim} \frac{e^{-2y^2}}{(2\pi)^{3/2}|y|}.$$
(4.19)

As a first step to extend this to general $\beta > 0$, a large deviation formula for $\rho_{(1),N}^{\text{OCP}}(\mathbf{r})$ can be computed, which asks for the asymptotic form of $\rho_{(1),N}^{\text{OCP}}(\sqrt{N}r)$ (here polar coordinates are being used), $r > 1$ [135, Proposition 1].

Proposition 4.4 *With $\beta f(\beta)$ the dimensionless free energy per particle as in (4.6), for $r > 1$ we have*

$$\rho_{(1),N}^{\text{OCP}}(\sqrt{N}r) = \frac{e^{\beta f(\beta)}}{N^{\beta/4}}e^{-(N\beta/2)(r^2-1)}\exp\left(N\beta\log r - \frac{\beta}{2}\log(r^2-1) + o(1)\right).$$
(4.20)

Proof (Sketch) The starting point is to manipulate the definition of $\rho_{(1),N}^{\text{OCP}}$ to obtain its form written as an average

$$\rho_{(1),N+1}^{\text{OCP}}(\sqrt{N+1}\,\mathbf{r})$$

$$= (N+1)N^{\beta N/2}e^{-(N+1)\beta r^2/2}\frac{Q_N^{\text{OCP}}(\beta)}{Q_{N+1}^{\text{OCP}}(\beta)}\Big\langle \prod_{l=1}^{N}\Big|\sqrt{\frac{N+1}{N}}\mathbf{r}-\mathbf{r}_l\Big|^{\beta}\Big\rangle_{\text{OCP}^g}. \tag{4.21}$$

Here $Q_N^{\text{OCP}}(\beta)$ is the configuration integral (4.1), and the average is with respect to the global-scaled OCP, specified by the Boltzmann factor (1.12) but with the factor of $\frac{1}{2}$ multiplying the first sum in U replace by $\frac{N}{2}$. The significance of this is that upon exponentiating the product, the average can be recognised as the characteristic function for a particular linear statistics. For this, with $|\mathbf{r}| = r > 1$ Proposition 3.4 applies, telling us the leading two terms in its large N asymptotic expansion, once the corresponding mean and variance have been computed. After calculating these, (4.20) results.   □

From (4.20) we compute the limit formula

$$\lim_{N\to\infty}\rho_{(1),N}^{\text{OCP}}(\sqrt{N}r)\big|_{r=1-y/\sqrt{N}} = e^{\beta f(\beta)}\frac{e^{-\beta y^2}}{(2|y|)^{\beta/2}}, \tag{4.22}$$

which under the assumption that the large deviation formula connects to the $y \to -\infty$ tail of $\rho_{(1),\infty}^{\text{e,OCP}}(y)$ is the sought $\beta > 0$ generalisation of the $\beta = 2$ result (4.19). The consistency between (4.22) and the latter is immediate upon substituting (4.4).

4.3   Sum Rules and Asymptotics for the Two- and Higher-Point Correlations

Setting $\mathbf{k} = \mathbf{0}$ in (3.17) gives

$$\int_{\mathbb{R}^2}\Big(\rho_{(2),\infty}^{b,T}(\mathbf{0},\mathbf{r}) + \frac{1}{\pi}\delta(\mathbf{r})\Big)\,d\mathbf{r} = 0, \tag{4.23}$$

where $\rho_{(2),\infty}^{b,T}$ is given by (2.24). This constraint on $\rho_{(2),\infty}^{b,T}$ is an example of a sum rule. In physical terms, using the Coulomb gas picture, it says that the response of the system by the introduction of a charge (corresponding to the delta function) is to create a screening cloud (corresponding to $\rho_{(2),\infty}^{T}$) of opposite total charge. Consequently, (4.23) is referred to as the perfect screening sum rule. It relates to the point process being hyperuniform or equivalently incompressible—for a state that is compressible the RHS of (4.23) is not zero but rather is given in terms of the second derivative of the pressure with respect to the fugacity; see e.g. [248, Eq. (3.7)]. For states with an underlying long-range potential (as in (1.12)) the perfect screening sum rule is expected to be a necessary condition for thermodynamic stability [422]. Of similar general validity is the sum rule which results when the integrand of (4.23) is replaced by

$$q(\mathbf{r}_1, \ldots, \mathbf{r}_k, \mathbf{r}) := \rho_{(k+1),\infty}(\mathbf{r}_1, \ldots, \mathbf{r}_k, \mathbf{r})$$

$$- \frac{1}{\pi} \rho_{(k),\infty}(\mathbf{r}_1, \ldots, \mathbf{r}_k) + \sum_{j=1}^{k} \delta(\mathbf{r} - \mathbf{r}_j) \rho_{(k),\infty}(\mathbf{r}_1, \ldots, \mathbf{r}_k). \quad (4.24)$$

Furthermore, the fast decay of the correlations upon truncation (i.e. suitable subtraction by combinations of lower-order correlation as in the definition of $\rho_{(2),\infty}^T$) implies that not only does the total charge associated with q vanish, but in fact so too all the multipole moments [422].

Proposition 4.5 *Let q be as in (4.24) and set $\mathbf{r} = (x, y)$. For bulk-scaled GinUE, specified in terms of the correlation functions by (2.9) with $N \to \infty$ and the correlation kernel (2.18), we have*

$$\int_{\mathbb{R}^2} (x - iy)^p q(\mathbf{r}_1, \ldots, \mathbf{r}_k, \mathbf{r}) \, d\mathbf{r} = 0, \qquad p = 0, 1, \ldots \quad (4.25)$$

Proof For any $k \geq 1$ integrating over the term involving the sum of delta functions gives $\left(\sum_{j=1}^{k} (x_j - iy_j)^p \right) \rho_{(k),\infty}$. To then integrate over the first two terms, we expand the determinant specifying $\rho_{(k+1),\infty}$ by the final row. The term coming from the last entry is recognised as $(1/\pi)\rho_{(k),\infty}$ and so cancels. For each of the k other terms, we multiply the term coming from the j-th entry of the final row $K_\infty^b(\mathbf{r}, \mathbf{r}_j)$ times $(x - iy)^p$ down the final column containing $[K_\infty^b(\mathbf{r}_m, \mathbf{r})]_{m=1}^k$, and integrate over $\mathbf{r}$. For the latter task, polar coordinates can be used to deduce that

$$\int_{\mathbb{R}^2} (x - iy)^p K_\infty^b(\mathbf{r}_m, \mathbf{r}) K_\infty^b(\mathbf{r}, \mathbf{r}_j) \, d\mathbf{r} = (x_j - iy_j)^p K_\infty^b(\mathbf{r}_m, \mathbf{r}_j);$$

the case $p = 0$ is (2.22) with bulk scaling. After rearranging the columns, the result of the integration in each case can be identified as $-(x_j - iy_j)^p \rho_{(k),\infty}$, and thus in total cancel out with the integration over the final term. $\qquad \square$

Replacing $1/\pi$ in (4.24) by $\rho_{(1),\infty}(\mathbf{r})$ the sum rule (4.25) with $p = 0$ remains valid in the case of edge-scaling as a consequence of the validity of (2.22). For $p \geq 1$ the slow decay of the correlations along the direction of the boundary as seen in (2.26) means that the integral in (4.25) is not well defined. Specifically the $p = 0$, $k = 1$ sum rule (4.25) at the edge reads

$$\int_{\mathbb{C}} \rho_{(2),\infty}^{e,OCP,T}(z, z') \, d^2 z' = -\rho_{(1),\infty}^{e,OCP}(z). \quad (4.26)$$

This is the edge counterpart of the bulk perfect screening sum rule (4.23).

Remark 4.2 It turns out that in the GinUE case, the determinantal structure together with (4.26) can be used to show that $\rho(z) := \rho_{(1),\infty}^{e,GinUE}(z)$ (this is (2.29)) satisfies a non-linear equation of infinite order,

$$\rho(z) = \sum_{n=0}^{\infty} \frac{|\partial_z^{(n)} \rho(z)|^2}{n!};$$
(4.27)

see [61, Sect. 3.6], where (4.26) is referred to as a mass-one equation. The non-linear equation is equivalent to the special function function identity

$$\frac{1}{4}\Big(1 + \mathrm{erf}(\sqrt{2}x)\Big)\Big(1 - \mathrm{erf}(\sqrt{2}x)\Big) = \frac{e^{-4x^2}}{\pi} \sum_{n=1}^{\infty} \frac{(H_{n-1}(\sqrt{2}x))^2}{2^n n!} \qquad (x \in \mathbb{R});$$
(4.28)

see [61, Sect. 4.5] for the proof of (4.28). Together with the loop equation, the identity (4.27) is used in [61] to study the universality of normal matrix models; see Sect. 5.4.

To deduce (3.19) from (3.18) requires that

$$\lim_{R \to \infty} R^2 S_{\infty}^{\mathrm{GinUE}}(\mathbf{k}/R) = \frac{|\mathbf{k}|^2}{4\pi}.$$
(4.29)

In the general $\beta > 0$ case of the OCP we denote the bulk-scaled structure factor by $S_{\infty}^{\mathrm{OCP}}(\mathbf{k})$. A perfect screening argument extending the viewpoint which implies (4.23) (for this see e.g. [237, Sect. 15.4.1], [234, Sect. 3.2]) predicts

$$\lim_{R \to \infty} R^2 S_{\infty}^{\mathrm{OCP}}(\mathbf{k}/R) = \frac{|\mathbf{k}|^2}{2\beta\pi}.$$
(4.30)

This result, in the slightly different guise of a mesoscopic scaling limit, is consistent with the recent work [394, Th. 1 mesoscopic case, formula for the variance]. Hence, as a generalisation of (3.19), we have

$$\lim_{R \to \infty} \mathrm{Cov}^{\mathrm{OCP}}\Big(\sum f(\mathbf{r}_l/R), \sum g(\mathbf{r}_l/R)\Big) = \frac{1}{(2\pi)^2} \frac{1}{2\pi\beta} \int_{\mathbb{R}^2} \hat{f}(\mathbf{k}) \hat{g}(-\mathbf{k}) |\mathbf{k}|^2 \, d\mathbf{k},$$
(4.31)

provided the integral converges.

As previously remarked in the case of (3.19), the choice $f(\mathbf{r}) = g(\mathbf{r}) = \chi_{|\mathbf{r}|<1}$ is an example when the RHS of (4.31). Yet this case is of some interest as it corresponds to $\mathrm{Var}^{\mathrm{OCP}} \mathcal{N}(D_R)$. To determine its large R behaviour, we first note that simple manipulation of the appropriate specialisation of the first equality in (3.2) shows

$$\mathrm{Var}^{\mathrm{OCP}} \mathcal{N}(D_R) =$$

$$\int_{\mathbb{R}^2} d\mathbf{r} \, C_{\infty}^{\mathrm{OCP}}(\mathbf{r}) \int_{\mathbb{R}^2} d\mathbf{r}' \Big(\chi_{\mathbf{r}+\mathbf{r}' \in D_R} - 1\Big) \chi_{\mathbf{r}' \in D_R} + |D_R| \int_{\mathbb{R}^2} C_{\infty}^{\mathrm{OCP}}(\mathbf{r}) \, d\mathbf{r}, \quad (4.32)$$

where $C_{\infty}^{\mathrm{OCP}}(\mathbf{r})$ is the notation for the integrand of (4.23). But (4.23) itself, which is the perfect screening sum rule, tells us that the last integral in (4.32) vanishes.

Let us write $\alpha(r/R) := \frac{1}{|D_R|} \int_{\mathbb{R}^2} d\mathbf{r}' \, (\chi_{\mathbf{r}+\mathbf{r}' \in D_R} \chi_{\mathbf{r}' \in D_R})$. We see from the definition that $\alpha(r/R) = 0$ for $r \geq R$, while for $0 < r < 2R$ it is shown in [524] that

$$\alpha(x) = \frac{2}{\pi} \left(\operatorname{Arcos} x - x(1 - x^2)^{1/2} \right) = 1 - \frac{2}{\pi} x + O(x^2).$$

Consequently, for large R [424]

$$\operatorname{Var}^{\text{OCP}} \mathcal{N}(D_R) \sim -2R \int_{\mathbb{R}^2} |\mathbf{r}| C_\infty^{\text{OCP}}(\mathbf{r}) \, d\mathbf{r}. \tag{4.33}$$

This being proportional to the surface area of D_R then represents the smallest possible growth of the variance (recall the discussion (3.9), and establishes too that the OCP is hyperuniform for general $\beta > 0$.

Remark 4.3 1. In the cases of the OCP applied to the anomalous quantum Hall effect ($\beta = 2M$ with M odd), the sum rule (4.30) combined with the Feynman–Bijl formula quantifies the collective mode excitation energy from the ground state [296]. 2. The working leading to (4.33) is conditional on the validity of the perfect screening sum rule (4.23), which excluding the case $\beta = 2$ has no independent proof. A rigorous demonstration that the OCP is hyperuniform for general $\beta > 0$, independent of the validity of (4.23) has recently been established in [392]. However, the bound obtained on the variance, while decaying faster than the area of D_R as required by the definition of hyperuniform, is not sharp.

From the definition of the structure factor, (4.30) is equivalent to the moment formula

$$\int_{\mathbb{R}^2} |\mathbf{r}|^2 \rho_{(2),\infty}^{\text{OCP},T}(\mathbf{r}, \mathbf{0}) \, d\mathbf{r} = -\frac{2}{\pi\beta}, \tag{4.34}$$

known as the Stillinger–Lovett sum rule [512]; see [423] for a derivation which makes use of (4.25) for $p = 1$ and $k = 1, 2$. Higher-order moment formulas can also be derived,

$$\int_{\mathbb{R}^2} |\mathbf{r}|^4 \rho_{(2),\infty}^{\text{OCP},T}(\mathbf{r}, \mathbf{0}) \, d\mathbf{r} = -\frac{16}{\pi\beta^2} \left(1 - \frac{\beta}{4} \right), \tag{4.35}$$

and

$$\int_{\mathbb{R}^2} |\mathbf{r}|^6 \rho_{(2),\infty}^{\text{OCP},T}(\mathbf{r}, \mathbf{0}) \, d\mathbf{r} = -\frac{18}{\pi\beta^3} \left(\beta - 6 \right) \left(\beta - \frac{8}{3} \right); \tag{4.36}$$

in distinction to (4.30) the precise statement of these results depends on the underlying bulk particle density, which in keeping with GinUE has been assumed to be $\frac{1}{\pi}$. In keeping with the relationship between (4.30) and (4.34), the moment formulas can be related to the small k expansion of $S_\infty^{\text{GinUE}}(\mathbf{k})$, which must therefore read

$$\frac{2\pi\beta}{|\mathbf{k}|^2} S_\infty^{\text{OCP}}(\mathbf{k}) = 1 + \left(\frac{\beta}{4} - 1\right)\frac{|\mathbf{k}|^2}{2\beta} + \left(\frac{\beta}{4} - \frac{3}{2}\right)\left(\frac{\beta}{4} - \frac{2}{3}\right)\left(\frac{|\mathbf{k}|^2}{2\beta}\right)^2 + O(|\mathbf{k}|^6).$$

$$(4.37)$$

Evidence is given in [350] that the polynomial structure in $\beta/4$ of the coefficients in this expansion breaks down at $O(|\mathbf{k}|^6)$. This is in contrast to the power series expansion of the bulk-scaled (density $1/\pi$ for definiteness) structure factor for the log-gas on a one-dimensional domain, $S_\infty^{(\beta)}(k)$ say. Thus expanding $\pi\beta S_\infty^{(\beta)}(k)/k$ as a power series in (k/β), the j-th coefficient is a monic polynomial of degree j in $\beta/2$ [247, 258, 535].

A linear response argument can be used to derive the coefficient of $|\mathbf{k}|^2$ in the expansion (4.37); see e.g. [237, Sect. 14.1.1]. It involves the thermal pressure for the OCP, which using the simple scaling relation (4.8) takes the value $\rho_b(1 - \beta/4)$ (for a discussion of the relation between the thermal and mechanic pressure in the OCP, see [154]) which explains the appearance of this factor. Mayer high-temperature diagrammatic expansion methods and the direct correlation function were used to first derive the coefficient of $|\mathbf{k}|^4$ in the expansion (4.37), or equivalently the sixth moment condition (4.36) [350]. Later a response argument involving variations to the spatial geometry was used to give an alternative derivation of this result [136].

The use of linear response to quantify the change of charge density upon the introduction of a charge q into the OCP, computing from this the change of charge by integrating, and requiring by screening that this must equal $-q$ can be used [340] to deduce the Carnie–Chan sum rule [139, 140]

$$-\beta \int_{\mathbb{R}^2} d\mathbf{r}\left(\int_{\mathbb{R}^2} d\mathbf{r}' \log|\mathbf{r}'|\left(\rho_{(2),\infty}^{\text{OCP},T}(\mathbf{r}, \mathbf{r}') + \frac{1}{\pi}\delta(\mathbf{r} - \mathbf{r}')\right)\right) = 1. \qquad (4.38)$$

Here the integral cannot be interchanged, as according to (4.23) integrating over $\mathbf{r}$ first would give zero. The validity of (4.38) can be checked directly for $\beta = 2$ in the bulk (i.e. for GinUE in the bulk).

Proposition 4.6 *The Carnie–Chan sum rule (4.38) is valid for GinUE in the bulk.*
Proof Denote the integral over $\mathbf{r}'$ in (4.38) by $g(\mathbf{r})$. Assuming only rotation and translation invariance of $\rho_{(2),\infty}^{\text{OCP},T}$ implies

$$g(\mathbf{r}) = \log|\mathbf{r}|\left(\int_{|\mathbf{r}'|<|\mathbf{r}|} \rho_{(2),\infty}^{\text{OCP},T}(\mathbf{r}', \mathbf{0})\, d\mathbf{r}' + \frac{1}{\pi}\right) + \int_{|\mathbf{r}'|>|\mathbf{r}|} \log|\mathbf{r}'|\rho_{(2),\infty}^{\text{OCP},T}(\mathbf{r}', \mathbf{0})\, d\mathbf{r}'.$$

Specialising now to $\beta = 2$ by substituting (2.24), the use of polar coordinates and integration by parts allows the integrals to be simplified, with the result

$$g(\mathbf{r}) = -\frac{1}{\pi}\int_{|\mathbf{r}|}^\infty \frac{e^{-(r')^2}}{r'}\, dr'.$$

The integration over $\mathbf{r} \in \mathbb{R}^2$ of this can be computed by further use of polar coordinates and integration by parts to confirm (4.38). $\qquad\square$

Formal use of the convolution theorem in (4.38) shows that it reduces to (4.30) [458]. In the case of the edge geometry, this approach can be used [343] to derive from (4.38) in the edge case the edge dipole moment sum rule

$$ -2\pi\beta \int_{-\infty}^{\infty} dy \left(\int_{-\infty}^{\infty} dy' \, (y' - y) \int_{-\infty}^{\infty} dx' \, \rho_{(2),\infty}^{e,OCP,T}((0, y), (x', y')) \right) = 1, $$
(4.39)

first derived in [97]. In keeping with the analogous property of (4.38), it is not possible to interchange the integrations over y and y' in this expression (if it was the LHS would vanish due to the sign change of $y' - y$). In the case $\beta = 2$ this is readily verified using the exact result (2.25).

Proposition 4.7 *The edge dipole moment sum rule (4.39) is valid for edge-scaled GinUE.*

Proof Substituting (2.25), integration by parts gives

$$ \int_{-\infty}^{\infty} dy' \, (y' - y) \int_{-\infty}^{\infty} dx' \, \rho_{(2),\infty}^{e,GinUE,T}((0, y), (x', y')) = -\frac{1}{\sqrt{8\pi^3}} e^{-2y^2} $$

(this formula can be found in [338, Eq. (2.41)]). Now integrating over y verifies (4.39) for $\beta = 2$. □

The exact result (2.26) exhibits a slow decay of $\rho_{(2),\infty}^{T}((x_1, y_1), (x_2, y_2))$ parallel to the edge. This was first predicted by Jancovici [339, 341], who on the basis of a linear response argument relating to the screening of an oscillatory external charge density applied at the edge, obtained for general $\beta > 0$,

$$ \rho_{(2),\infty}^{T}((x_1, y_1), (x_2, y_2)) \underset{|x_1-x_2|\to\infty}{\sim} \frac{f(y_1, y_2)}{(x_1 - x_2)^2}, $$
$$ \int_{-\infty}^{\infty} dy_1 \int_{-\infty}^{\infty} dy_2 \, f(y_1, y_2) = -\frac{1}{2\beta\pi^2}. $$
(4.40)

A derivation of this using the Carnie–Chan sum rule for the OCP confined to a strip geometry is given in [340]. In keeping with this, in [345] the amplitude $f(y, y')$ in (4.40) is related to the dipole moment of the screening cloud at the edge by deriving that

$$ \int_{-\infty}^{\infty} dy' \, (y' - y) \int_{-\infty}^{\infty} dx' \, \rho_{(2),\infty}^{e,OCP,T}((0, y), (x', y')) = \pi \int_{-\infty}^{\infty} dy' \, f(y, y'). \quad (4.41) $$

Then the result for the integral over y on the RHS follows from the dipole moment sum rule (4.39).

We consider now a global scaling, so that the droplet support is the unit disk. For large N, define the charge–charge surface correlation by

$$\langle \sigma(\theta)\sigma(\theta')\rangle_N^T := \int d\mathbf{n} \int d\mathbf{n}' \, \rho_{(2),N}^{\mathrm{g,OCP},T}(\mathbf{r}, \mathbf{r}'). \tag{4.42}$$

Here $\rho_{(2),N}^{\mathrm{g,OCP},T}$ refers to $\rho_{(2),N}^{\mathrm{OCP},T}$ computed using global scaling (as specified below (4.21)), then considering the large N form with edge coordinates (these can be taken as (r, θ) with $r \approx 1$. The integration is over the normal direction $\mathbf{n}$ (here the radial direction); the more general notation has been used as this quantity can be defined in other geometries, e.g. for the elliptic GinUE [257]. The reasoning leading to (4.40) has been extended [341] to lead to the prediction

$$\langle \sigma(\theta)\sigma(\theta')\rangle_\infty^T = -\frac{1}{8\pi^2\beta \sin^2((\theta - \theta')/2)}. \tag{4.43}$$

This has been checked at $\beta = 2$ using the exact result for GinUE in [155]. Further, again for $\beta = 2$, we note that a low-energy effective conformal field theory viewpoint of this result, obtained within the Fermi gas interpretation of Sect. 6.1, has recently been given in [452].

Remark 4.4 1. The knowledge of (4.43) can be used to predict the surface contribution of the OCP generalisation of the covariance formula (3.21) [235].
2. For the bulk-scaled OCP with β an even integer, there is a constraint on the small $r = |\mathbf{r}|$ form of $\rho_{(2),\infty}^{\mathrm{b}}(\mathbf{r}, \mathbf{0})$. Thus [483]

$$\rho_{(2),\infty}^{\mathrm{b}}(\mathbf{r}, \mathbf{0}) = r^\beta e^{-\beta r^2/4} f(r^4),$$

for $f(z)$ analytic. Note from (2.24) that for GinUE,

$$f(z) = \frac{2}{\pi^2} \frac{\sinh(\sqrt{z}/2)}{\sqrt{z}}.$$

Recently, consequences have been found in [482].

Chapter 5
Normal Matrix Models

Normal matrix models, characterised by the requirement that the matrices commute with their adjoint so that their eigenvectors form a unitary matrix, permit an explicit Coulomb gas Boltzmann factor type eigenvalue PDF with $\beta = 2$, so determinantal structures hold. They permit a simplified global scaling in which the one-body potential is taken as being strictly proportional to N. The support of the corresponding large N equilibrium measure is referred to as a droplet. Perturbing the GinUE one-body potential by the real part of a polynomial leads to a relation between the moments of the droplet and the coefficients of the polynomial. Generally, the main question of interest is the dependence on the potential in characterising quantities. This is addressed in relation to the large N expansion of the partition function, and for the bulk, edge and global-scaled correlation functions. In the case of bulk and edge scaling, universality can be exhibited, meaning that the limiting correlations are in fact independent on the details of the potential, provided it corresponds to a non-vanishing density locally.

5.1 Eigenvalue PDF

A generalisation of the joint element PDF (2.36) is the functional form proportional to

$$
\exp\left(-\frac{N}{t_0}\mathrm{Tr}(JJ^\dagger) - 2\mathrm{Re}\sum_{p=2}^{M} t_p \mathrm{Tr}\,(J^p) \right), \quad t_0 > 0;
\tag{5.1}
$$

note the scaling so that there is a factor of N in the exponent. Use of the Schur decomposition (2.2) gives a separation of the eigenvalues and the strictly upper triangular elements in the exponent analogous to (2.4). This allows the latter to be integrated over, showing that the corresponding eigenvalue PDF is proportional to

© The Author(s) 2025
S.-S. Byun and P. J. Forrester, *Progress on the Study of the Ginibre Ensembles*, KIAS
Springer Series in Mathematics 3, https://doi.org/10.1007/978-981-97-5173-0_5

$$\exp\left(-\frac{N}{t_0}\sum_{j=1}^{N}\left(|z_j|^2 - 2\mathrm{Re}\sum_{p=2}^{M}t_p z_j^p\right)\right).\tag{5.2}$$

However, for $t_M \neq 0$ this is not normalisable for $M > 2$.

A simple remedy is to impose a cutoff by restricting the eigenvalues to a disk centred about the origin of radius R, and restricting to small t_0 and values of $t_2, \ldots, t_M$ so that the support Ω (assumed simply connected) of the density is contained inside this disk [217]. The mean field argument leading to (2.31) tells us that in relation (5.2), the normalised density has the constant value $1/\pi t_0$ in Ω, where the latter is such that

$$W(z) = \frac{1}{\pi}\int_{\Omega}\log|z-w|\,d^2w,\qquad W(z) = \frac{1}{2}\left(|z|^2 - 2\mathrm{Re}\sum_{p=2}^{M}t_p z^p\right),\tag{5.3}$$

for z contained inside of Ω. From this equation the coupling constants $\{t_p\}_{p=2}^{M}$ can be related to moments associated with Ω [7, 538].

Proposition 5.1 *In the above setting, for $p \geq 2$ we have*

$$t_p = \frac{1}{2\pi i p}\int_{\partial\Omega}\bar{z}z^{-p}\,dz = -\frac{1}{\pi p}\int_{\mathbb{C}\backslash\Omega}z^{-p}\,d^2z.\tag{5.4}$$

Proof By applying ∂_z to both sides of (5.3), then introducing $W(w)$ in the integrand via the identity $2\partial_w\partial_{\bar{w}}W(w) = 1$ shows

$$\partial_z W(z) = \frac{1}{\pi}\int_{\Omega}\frac{\partial_w\partial_{\bar{w}}W(w)}{z-w}\,d^2w.$$

Now, using the Cauchy–Pompeiu formula (3.22) with C_1 replaced by Ω, we see from this that

$$\int_{\partial\Omega}\frac{\partial_w W(w)}{w-z}\,dw = 0.$$

Simple manipulation then shows

$$\frac{1}{2\pi i}\int_{\partial\Omega}\frac{\bar{w}}{w-z}\,dw = \sum_{p=2}^{M}p t_p z^{p-1}.$$

The first equality of (5.4) now follows by power series expanding the LHS with respect to z. The second equality can be deduced from the first by applying the version of (3.22) valid for z outside C_1 (chosen as $\mathbb{C}\backslash\Omega$), for which the LHS is to be replaced by 0. $\square$

The simplest case of (5.2) beyond that corresponding to (2.36) is to take $M = 3$. After scaling, the choice of t_3 can be fixed. Choosing $t_3 = \frac{1}{3}$, results from [379] give that the support of Ω is contained in a disk for all $0 < t_0 \leq 1/8$, with the critical value $t_0 = 1/8$ involving three cusp singularities. Then, the boundary $\partial\Omega$ is a 3-cusped hypercycloid. For general $M \geq 2$ and Ω contained in a disk and simply connected, it is shown in [217] that $\partial\Omega$ can be parametrised by a Laurent polynomial of the form $\alpha_1 w + \alpha_0 + \cdots + \alpha_{M-1} w^{-M+1}$ with $|w| = 1$. Dropping the assumption that Ω be simply connected requires the theory of quadrature domains; see [396] and references therein.

More general than the joint eigenvalue PDF (5.1) is the functional form

$$\exp\left(-\sum_{j,k=1}^{\infty} c_{jk}\mathrm{Tr}\left(J^j(J^\dagger)^k\right)\right) =: \exp\left(-\mathrm{Tr}\, W(J, J^\dagger)\right). \qquad (5.5)$$

However, unlike the former, substituting the Schur decomposition (2.2) does not in general lead to a separation of the eigenvalues from the strictly upper triangular variables. To overcome this, attention can be restricted to the subset of $N \times N$ complex matrices having the further structure $[J, J^\dagger] = 0$, which specifies the matrices as being normal. For normal matrices, the eigenvectors can be chosen to form an orthonormal set, and so $J = UDU^\dagger$, for U unitary and D the diagonal matrix of eigenvalues. Moreover, the change of variables from J to $\{U, D\}$ gives a decomposition of measure into separate eigenvalue and eigenvector factors, with the Jacobian given by (2.5) [152]. Hence the eigenvalue PDF corresponding to (5.5) in this setting is proportional to

$$\exp\left(-\sum_{j=1}^{N} W(z_j, \bar{z}_j)\right) \prod_{1 \leq j < k \leq N} |z_k - z_j|^2. \qquad (5.6)$$

Note that for the joint element PDF (5.1), the same eigenvalue PDF (5.2) results for J specified on the full space of $N \times N$ complex matrices, as it does on the restriction to normal matrices as implied by (5.6).

5.2 Equilibrium Measure

Analogous to (5.1), to obtain a compact eigenvalue support for large N, one considers the case that the weight e^{-W} is exponentially varying in a sense that W is of order N, say $W(z, \bar{z}) = NQ(z)$ for a fixed potential $Q : \mathbb{C} \to \mathbb{R}$. Thus the eigenvalue PDF (5.6) is written as

$$e^{-H(z_1,\ldots,z_N)}, \quad H(z_1, \ldots, z_N) := \sum_{1 \leq j < l \leq N} \log \frac{1}{|z_j - z_l|^2} + N \sum_{j=1}^{N} Q(z_j). \qquad (5.7)$$

(Cf. Note that this is an extension of (1.12) to a general potential Q.) As in Remark 2.2, the macroscopic behaviour of the system (5.7) can be described using the two-dimensional Coulomb gas interpretation with the help of logarithmic potential theory. Namely, it is well known in the literature [49, 143, 144, 322, 348] that for a general Q under suitable potential theoretic assumptions, the empirical measure $\frac{1}{N} \sum \delta_{z_j}$ weakly converges to a unique probability measure μ_Q that minimises

$$I_Q[\mu] := \int_{\mathbb{C}^2} \log \frac{1}{|z - w|} \, d\mu(z) \, d\mu(w) + \int_{\mathbb{C}} Q \, d\mu. \tag{5.8}$$

One may notice that (5.8) can be interpreted as a continuum limit of the Hamiltonian H_N in (5.7) after normalisation, generalising (2.30). The probability measure μ_Q is called the equilibrium measure and its support $S_Q := \mathrm{supp}(\mu_Q)$ is called the droplet, as previously remarked. Furthermore, if Q is C^2-smooth in a neighbourhood of S_Q, by Frostman's theorem [479], μ_Q is absolutely continuous with respect to the Lebesgue measure and takes the form

$$d\mu_Q(z) = \frac{\partial_z \partial_{\bar{z}} Q(z)}{\pi} \chi_{z \in S_Q} \, d^2z. \tag{5.9}$$

In particular, for a rotationally symmetric potential $q(r) = Q(|z| = r)$, the droplet is of the form $S_Q = \{r_1 \le |z| \le r_2\}$, where (r_1, r_2) are the unique pair of constants satisfying

$$r_1 q'(r_1) = 0, \quad r_2 q'(r_2) = 2, \tag{5.10}$$

see [479, Sect. IV.6]. Here, we have assumed that $Q(z)$ is strictly subharmonic in $\mathbb{C}$, which is equivalent to the requirement that $r \mapsto r q'(r)$ is increasing in $(0, \infty)$. Let us mention that all the explicit macroscopic densities with rotation invariance in Chapter 2 can be obtained as special cases of the formulas (5.9) and (5.10). For instance, the RHS of (2.47) can be realised as the RHS of (5.9) with $Q(z) = |z|^2 - 2\alpha \log |z|$. Beyond the case when Q is radially symmetric, the determination of the droplet, also known as the two-dimensional equilibrium problem (e.g. the ellipse in Sect. 2.3), is far from being obvious even for some explicit potentials with simple form; see [21, 71, 73, 174, 376] and references therein for recent works in this direction.

5.3   Partition Functions

Continuing the discussion in Sect. 4.1, we consider the global-scaled, non-charge neutral, partition function

$$Z_N(\beta; Q) = \frac{1}{N!} \int_{\mathbb{C}} d^2z_1 \cdots \int_{\mathbb{C}} d^2z_N \, e^{-\frac{\beta}{2} H(z_1, \dots, z_N)}, \tag{5.11}$$

where $H(z_1, \ldots, z_N)$ is the Hamiltonian given in (5.7). An explicit formula for the large N expansion of $Z_N(\beta; Q)$ was predicted in [539]. Fairly recently, it was shown by Leblé and Serfaty [393] that for general $\beta > 0$ and Q, $Z_N(\beta; Q)$ admits the large N asymptotic expansion of the form

$$\log Z_N(\beta; Q) = -\frac{\beta}{2} N^2 I_Q[\mu_Q] + \left(\frac{\beta}{4} - 1\right) N \log N - \left(C(\beta) + \left(1 - \frac{\beta}{4}\right) E_Q[\mu_Q]\right) N + o(N);$$
(5.12)

see also an earlier work [485] on (5.12) up to the $O(N \log N)$ term. The term $I_Q[\mu_Q]$ appearing in the leading-order asymptotic is the energy (5.8) evaluated at the equilibrium measure μ_Q. In the case of the OCP, up to a sign it is the quantity appearing in the exponent of $A_{N,\beta}$ in (4.2) at order N^2. To see this, we note that for a radially symmetric potential $q(r) = Q(|z| = r)$ generally, it is evaluated as

$$I_Q[\mu_Q] = q(r_1) - \log r_1 - \frac{1}{4} \int_{r_0}^{r_1} r q'(r)^2 \, dr,$$
(5.13)

where r_1 and r_2 are the constants specified in (5.10). For the OCP $Q(z) = |z|^2$, which gives $I_Q[\mu_Q] = 3/4$, this indeed being the coefficient of N^2 seen in $A_{N,\beta}$. The appearance of the term $(\beta/4 - 1)N \log N$ in (5.12), whereas there is a term $\frac{1}{4}\beta N^2 \log N$ in $A_{N,\beta}$ of (4.2), due to the simple scaling $z_j \mapsto z_j/\sqrt{N}$ required to go from the OCP with global-scaled coordinates (as assumed in (5.12)) to the OCP itself (as assumed in (4.2)). The terms appearing in the $O(N)$ term of (5.12) are the entropy

$$E_Q[\mu_Q] := \int_{\mathbb{C}} \mu_Q(z) \log \mu_Q(z) \, d^2 z$$
(5.14)

associated with μ_Q and a constant $C(\beta)$ independent of the potential Q. Their sum is the free energy per particle, $\beta f(\beta; Q)$. The expansion (5.12) with quantitative error bounds is also available in the literature [75, 492].

For the random normal matrix model when $\beta = 2$, the determinantal structure (2.9) allows an explicit expression

$$Z_N(2; Q) = \prod_{j=0}^{N-1} h_j,$$
(5.15)

where h_j is the orthogonal norm in (5.18); cf. Proposition 2.1. In particular, if Q is rotationally symmetric, since $p_j(z) = z^j$, we have

$$h_j = 2\pi \int_0^\infty r^{2j+1} e^{-Nq(r)} \, dr;$$
(5.16)

cf. (2.89). Based on this knowledge together with a Laplace approximation of the integrals, the precise asymptotic expansion of the free energy up to the $O(1)$ term was derived in a recent work [125]. (See also [53].) This in particular shows that

$$\log Z_N(2; Q) = -N^2 I_Q[\mu_Q] - \frac{1}{2} N \log N + \left(\frac{\log(2\pi^2)}{2} - \frac{1}{2} E_Q[\mu_Q] \right) N$$

$$- \frac{\chi}{12} \log N + \chi \zeta'(-1) + O(1), \qquad (5.17)$$

where χ is the Euler index; cf. (4.6). Let us recall that $\chi = 0$ for the annulus ($r_0 > 0$) and $\chi = 1$ for the disk ($r_0 = 0$) geometry. Here, ζ is the Riemann zeta function and the $\chi \zeta'(-1)$ term is expected to appear universally. Furthermore, it was proposed in [539] that the $O(1)$-term is given in terms of the spectral determinant of the Laplacian. See [125, Sect. 4] and [229, Sect. 4] for some concrete examples of (5.17) associated with the matrix models discussed in Chap. 2. We also refer to [129] for a recent work establishing the free energy expansion for a non-radially symmetric potential.

5.4   Correlation Functions and Universality

Turning to the correlation functions $\rho_{(k),N}$ of (5.7), the determinantal structure (2.9) remains valid with the correlation kernel

$$K_N(z, w) = e^{-\frac{N}{2}(Q(z) + Q(w))} \sum_{j=0}^{N-1} \frac{p_j(z) \overline{p_j(w)}}{h_j}, \qquad (5.18)$$

where p_j is the monic orthogonal polynomial of degree j with respect to the weighted Lebesgue measure $e^{-N Q(z)} \, d^2 z$ and h_j is its squared orthogonal norm; cf. (3.6).

Let us first discuss the asymptotic behaviours of $\rho_{(k),N}$ in the micro-scale. Given a base point $p \in S_Q$ for which we zoom the point process, we denote by

$$\delta := \left. \frac{\partial_z \partial_{\bar{z}} Q(z)}{\pi} \right|_{z=p} \qquad (5.19)$$

the mean eigenvalue density at p; cf.(5.9). From universality principles (see e.g. [375]), one expects that for a general Q and p such that $\delta \in (0, \infty)$ (i.e. the eigenvalue density does not vanish or diverge at p), the universal scaling limit in Proposition 2.3 arises. For the bulk case when $p \in \text{Int}(S_Q)$, such a universality was established by Ameur, Hedenmalm and Makarov in [58] where they showed that for a fairly general potential Q under some mild assumptions,

$$\frac{1}{(N\delta)^k} \rho_{(k),N} \left(p + \frac{z_1}{\sqrt{N\delta}}, \dots, p + \frac{z_k}{\sqrt{N\delta}} \right) \to \det \left[K_\infty^{\text{b}}(z_j, z_l) \right]_{j,l=1}^k, \qquad (5.20)$$

uniformly on compact subsets of $\mathbb{C}$ as $N \to \infty$, where K_∞^{b} is given by (2.18). See also [87, 89]. In a sequential work [59] the theory of loop equations (or Ward's identities) was also used to show the bulk scaling limit (5.20). It says that for a given test function ψ,

$$\mathbb{E}_N W_N^+[\psi] = 0, \tag{5.21}$$

$$W_N^+[\psi] := \frac{1}{2} \sum_{j \neq k} \frac{\psi(z_j) - \psi(z_k)}{z_j - z_k} - N \sum_{j=1}^{N} [\partial Q \cdot \psi](z_j) + \sum_{j=1}^{N} \partial \psi(z_j),$$

where we have used the notations $[\partial_z Q \cdot \psi](z_j) := \partial_z Q(z)|_{z=z_j} \psi(z_j)$ and $\partial \psi(z_j) = \partial_z \psi(z)|_{z=z_j}$. The functional W_N^+ is also called the stress energy tensor in the context of conformal field theory [352, Appendix 6]. The identity (5.21) easily follows from the integration by parts

$$\mathbb{E}_N[\partial \psi(z_j)] = \mathbb{E}_N[\partial_{z_j} H(z_1, \ldots, z_N) \cdot \psi(z_j)]. \tag{5.22}$$

The approach using Ward's identities was further developed in [61] and several related works to study various scaling limits of the random normal matrix models (see e.g. [50, 62] for the bulk scaling limit at weak non-Hermiticity (2.42); [63, 490] for the edge scaling limit with boundary confinements (2.67); [60, 64] for normal matrices with Mittag–Leffler type singularities as in Sects. 2.4 and 2.7) and we refer to [22, Remark 2.9] for an expository summary of this strategy. In particular, in [61], the rescaled version of Ward's identity was introduced and used to show the edge universality: for $p \in \partial S_Q$,

$$\frac{1}{(N\delta)^k} \rho_{(k),N}\left(p + i\mathbf{n}\frac{z_1}{\sqrt{N\delta}}, \ldots, p + i\mathbf{n}\frac{z_k}{\sqrt{N\delta}}\right) \to \det\left[K_\infty^e(z_j, z_l)\right]_{j,l=1}^{k}, \tag{5.23}$$

where $\mathbf{n}$ is the outer normal vector at p as in Proposition 2.6 and K_∞^e is given by (2.19). However, the result in [61] has an additional assumption that the limiting correlation function is translation invariant along the real axis. This assumption is intuitively natural but hard to rigorously show in general (except e.g. for the rotationally symmetric potential Q). The edge universality was later then shown by Hedenmalm and Wennman for a wide class of potentials Q in [324], where they developed an asymptotic theory for general planar orthogonal polynomials; see also [321] for an alternative approach to derive the main result in [324] and [323] for a theory developed for the orthogonal polynomials associated with non-exponentially varying weight. The asymptotic results in [324] together with (5.18) then leads to (5.23) by the Riemann sum approximation.

We now turn our attention to the asymptotic behaviours of $\rho_{(k),N}$ in the macroscale. Let us begin with the 1-point function $\rho_{(1),N}$. For the bulk case when z is the interior of the droplet, the asymptotic behaviour of $\rho_{(1),N}$ is well known in the literature, see [47, 87] and references therein. It says that under the suitable assumptions on Q, there are real-analytic functions B_j such that

$$\pi \rho_{(1),N}(z) = N \Delta Q(z) + \frac{1}{2} \Delta \log \Delta Q(z) + N^{-1} B_2(z) + \cdots + N^{-k+1} B_k(z) + O(N^{-k}), \tag{5.24}$$

where $\Delta := \partial_z \partial_{\bar{z}}$ is one quarter of the usual Laplacian. Notice here that the leading-order asymptotic of (5.24) is implied by the Laplacian growth (5.9). As a concrete example, we consider the induced spherical ensemble (2.54) with $M = \alpha_1 N$ and $n = \alpha_1 N - 1$, which can be realised as (5.7) with

$$Q(z) = (\alpha_1 + \alpha_2 - 1) \log(1 + |z|^2) - 2(\alpha_1 - 1) \log |z|.$$

On the other hand, it follows from (2.56) that with $\zeta = |z|^2/(1 + |z|^2)$,

$$\pi \rho_{(1),N}(z) = \frac{|z|^{2(M-N)}}{(1 + |z|^2)^2} (M + n - N) \Big(I_\zeta(M - N, n) - I_\zeta(M, n - N) \Big). \quad (5.25)$$

Then using the well-known asymptotic behaviours of the incomplete beta function, one can observe that for $r_1 < |z| < r_2$, where r_1, r_2 are given in (2.55), the asymptotic behaviour (5.24) holds with

$$\Delta Q(z) = \frac{\alpha_1 + \alpha_2 - 1}{(1 + |z|^2)^2}, \qquad \Delta \log \Delta Q(z) = -\frac{2}{(1 + |z|^2)^2}.$$

Next, we consider the off-diagonal asymptotic behaviour of the correlation kernel. For the Ginibre ensemble, equivalently, for the random normal matrix model (5.7) with $Q(z) = |z|^2$, the associated correlation kernel K_N in (5.18) satisfies

$$K_N(z, w) = \sqrt{\frac{2N}{\pi}} (z\bar{w})^N e^{N - \frac{N}{2}(|z|^2 + |w|^2)} S(z, w) \Big(1 + O(\frac{1}{N}) \Big), \qquad (z \neq w) \tag{5.26}$$

where $S(z, w)$ is the exterior Szegö kernel

$$S(z, w) := \frac{1}{2\pi} \frac{1}{z\bar{w} - 1}. \tag{5.27}$$

From a viewpoint of Proposition 2.2, the asymptotic behaviour (5.26) can be realised as a uniform expansion of the incomplete gamma function $z \mapsto \Gamma(N; Nz)$, which is available in the literature in some particular domains; see e.g. [529] for $|\arg(z - 1)| < 3\pi/4$. In a recent work [57], generalising the classical results, it was shown that the asymptotic behaviour (5.26) remains valid as long as $z\bar{w}$ is outside the Szegö curve $\{z \in \mathbb{C} : |z| \leq 1, |z\, e^{1-z}| = 1\}$. Note in particular that if $|z| = |w| = 1$, then (5.26) reads

$$K_N(z, w) \overset{c}{\sim} \sqrt{\frac{2N}{\pi}} S(z, w) \Big(1 + O(\frac{1}{N}) \Big), \tag{5.28}$$

where $\overset{c}{\sim}$ means that the asymptotic expansion holds up to a sequence of cocycles (in this case $(z\bar{w})^N$), which cancel out when forming a determinant (2.9). From a statistical physics point of view, the behaviour (5.28) indicates that there are strong correlations among the particles on the boundary of the droplet, which also shows the slow decay of correlations at the boundary. For GinUE this is explicit in (4.43). For elliptic GinUE, the phenomenon was studied in [257], as a test of the generalisation to more general shaped droplets, when the RHS of (4.43) is predicted to be given in terms of a certain Green's function for an electrostatics problem outside of the droplet, which acts as a macroscopic conductor [341], [234, Eq. (3.29)]. Furthermore, it was obtained by Ameur and Cronvall [57] that for a general class of potentials Q, the associated correlation kernel K_N satisfies

$$K_N(z, w) \sim \sqrt{2N} \left(\frac{\partial_z \partial_{\bar{z}} Q(z)}{\pi} \right)^{1/4} \left(\frac{\partial_w \partial_{\bar{w}} Q(w)}{\pi} \right)^{1/4} S(z, w)(1 + o(1)). \quad (5.29)$$

For this, the use of a general theory on the orthogonal polynomial due to Hedenmalm and Wennman [324] was made. We also refer to [25, 51, 52, 130] for more recent studies in this direction.

Remark 5.1 Let $\{p_j^{(N,R)}(z)\}_{j=0,1,...}$ be the orthogonal polynomials with respect to the inner product $\langle f|g \rangle := \int_{C_R} f(z)g(\bar{z})e^{-2NW(z)/t_0}$, where $W(z)$ as in (5.3) and it is assumed that the limiting eigenvalue support Ω corresponding to (5.2) is contained in D_R. Consider the probability distribution $\mathcal{P}_k^{(N,R)}$ specified by the eigenvalue PDF proportional to

$$\prod_{l=1}^{k} e^{-2NW(z_l)/t_0} \prod_{1 \le j < l \le k} |z_l - z_j|^2. \quad (5.30)$$

As a result of the underlying determinantal structure, one has the formula for $p_k^{(N,R)}(z)$ as an expectation with respect to $\mathcal{P}_k^{(N,R)}$ (see e.g. [237, proof of Proposition 5.1.3]; in fact such formulae were known to Heine [513]):

$$p_k^{(N,R)}(z) = \left\langle \prod_{l=1}^{k} (z - z_l) \right\rangle_{\mathcal{P}_k^{(N,R)}}.$$

For $k/N = x$ as $k, N \to \infty$ it has been conjectured [216] that the zeros of $p_k^{(N,R)}(z)$ accumulate on certain arcs Σ contained in Ω, with corresponding measure μ_x^*. Assuming this, it follows that for $z \in \mathbb{C} \backslash \Omega$

$$\frac{1}{\pi t_0} \int_{\Omega} \log|z - w| \, d^2w = \int_{\Sigma} \log|z - s| \, d\mu_x^*. \quad (5.31)$$

This identifies Σ as the so-called mother body or potential theoretic skeleton of Ω. A number of works give further developments along these lines, especially in relation to the strong asymptotics of the planar orthogonal polynomials based on Riemann–Hilbert analysis. In the case of (5.1) with a cubic potential $p = 3$, references include [95, 96, 379, 425]. For the induced Gaussian-type weight

$$2W(z)/t_0 = |z|^2 - 2c \log|z - a|,$$

the associated Riemann–Hilbert problem was first derived in [71], where the authors also performed a detailed analysis for the case $c = O(1)$. This Riemann–Hilbert problem was further used in [72, 92, 398] to obtain strong asymptotics of the orthogonal polynomials when $c = O(1/N)$. These results were later utilised in [180, 387, 531] to study the characteristic polynomials of the Ginibre matrix. Recently, the higher-order Riemann–Hilbert problems for a more general potential of the form

$$2W(z)/t_0 = |z|^2 - 2 \sum_{j=1}^{M} c_j \log |z - a_j|$$

were constructed in [86, 399] as well, and the case $c_j = O(1/N)$ was further analysed in [400].

Chapter 6
Further Theory and Applications

In addition to an analogy with the Boltzmann factor of the OCP with inverse temperature $\beta = 2$, the GinUE eigenvalue PDF permits an interpretation as the absolute value squared of the ground state wave function for spinless fermions in the plane subject to a perpendicular magnetic field. More generally, for β twice an odd positive integer m the OCP Boltzmann factor occurs as proportional to the absolute value squared of Laughlin trial wavefunction in the theory of the fractional quantum Hall effect with filling fraction $1/m$. In relation to the topic of chaotic behaviour in dissipative quantum systems, statistical properties of the now complex energy spectrum can be compared with those of bulk GinUE eigenvalues. This leads to consideration of the so-called dissipative spectral form factor, the second moment of the absolute value of the statistics Tr X^k, and simple dissipators associated with Lindblad operators, all of which can be calculated exactly for GinUE matrices. The remaining two sections of this chapter are on the singular values, and eigenvectors, respectively of GinUE matrices. Consideration of the singular values is natural in relation to the questions relating to the determinant of GinUE matrices, while there have been recent applications of the statistical properties of eigenvectors to the eigenstate thermalisation hypothesis for non-Hermitian Hamiltonians.

6.1 Fermi Gas Wave Function Interpretation

We have seen that the rewrite of (1.7), written in the exponential form (1.12), allows for the GinUE eigenvalue PDF to be interpreted as the Boltzmann factor for a particular Coulomb gas. If instead of an exponential form we use (2.12) to rewrite (1.7) as

$$\left| \prod_{j=1}^{N} e^{-|z_j|^2/2} \det[z_j^{k-1}]_{j,k=1,\ldots,N} \right|^2, \tag{6.1}$$

© The Author(s) 2025
S.-S. Byun and P. J. Forrester, *Progress on the Study of the Ginibre Ensembles*, KIAS Springer Series in Mathematics 3, https://doi.org/10.1007/978-981-97-5173-0_6

then we are led to an interpretation as the absolute value squared of a ground state free Fermi quantum many-body wave function. Thus inside the absolute value of (6.1) is a Slater determinant of single-body wave functions $\{\phi_l(z)\}_{l=0,\dots,N-1}$ with $\phi_l(z) = e^{-|z|^2/2}z^l$. What remains then is to identify the corresponding one-body Hamiltonian for quantum particles in the plane which have these single-body wave functions for the lowest energy states.

The appropriate setting for this task is a quantum particle confined to the xy-plane subject to a perpendicular magnetic field, $(0, 0, B)$, $B > 0$. Fundamental to this setting is the vector potential $\mathbf{A}$, related to the magnetic field by $\nabla \times \mathbf{A} = (0, 0, B)$. The so-called symmetric gauge corresponds to the particular choice $\mathbf{A} = (-By, Bx, 0) =: (A_x, A_y, 0)$, which henceforth will be assumed. Physical quantities in this setting are m (the particle mass), e (particle charge), $\hbar$ (Planck's constant), c (speed of light), which together with B are combined to give $\omega_c := eB/mc$ (cyclotron frequency) and $\ell := \sqrt{\hbar c/eB}$ (magnetic length).

Defining the generalised momenta and corresponding raising and lowering operators by

$$\Pi_u = -i\hbar\frac{\partial}{\partial u} + \frac{e}{c}A_u \ (u = x, y), \qquad a^\dagger = \frac{\ell}{\sqrt{2\hbar}}(\Pi_x + i\Pi_y), \qquad a = (a^\dagger)^\dagger,$$

allows the quantum Hamiltonian to be written in the harmonic oscillator like form $H_B = \hbar\omega_c(a^\dagger a + \frac{1}{2})$ [166]. Important too are the quantum centre of orbit operators and associated raising and lowering operators

$$U = u - \frac{\ell^2}{\hbar}\Pi_u \ (U = X, Y; \ u = x, y), \qquad b^\dagger = \frac{1}{\sqrt{2}\ell}(X - iY), \qquad b = (b^\dagger)^\dagger,$$

for which $X^2 + Y^2 = 2\ell^2(b^\dagger b + \frac{1}{2})$. The operators $\{a, a^\dagger\}$ commute with $\{b, b^\dagger\}$, implying that H and $X^2 + Y^2$ permit simultaneous eigenstates. A complete orthogonal set can be constructed using the raising operators according to

$$|n, m\rangle = \frac{(a^\dagger)^n(b^\dagger)^m}{\sqrt{n!m!}}|0, 0\rangle, \tag{6.2}$$

with eigenvalues of H equal to $(n + \frac{1}{2})\hbar\omega_c$ and eigenvalue of $X^2 + Y^2$ equal to $(2m + 1)\ell^2$. The ground state $|0, 0\rangle$ is characterised by $a|0, 0\rangle = b|0, 0\rangle = 0$, which can be checked to have the unique solution $|0, 0\rangle \propto e^{-(x^2+y^2)/4\ell^2}$. From this, application of $(b^\dagger)^m$ gives $|0, m\rangle \propto \bar{z}^m e^{-|z|^2/4\ell^2}$, $z = x + iy$. Forming a Slater determinant with respect to the first N eigenstates of this type gives (6.1) with $\ell^2 = 1/2$. Generally, states with quantum number $n = 0$ and thus belonging to the ground state are said to be in the lowest Landau level. One remarks that the largest eigenvalue of $X^2 + Y^2$ is then $N - 1/2$, which is in keeping with the squared radius of the leading-order support in the circular law.

The above theory of a quantum particle in the plane subject to a perpendicular magnetic field can be recast to apply to a rotating quantum particle in the plane [325, 384]. There, for an appropriate rotation frequency, the infinite degenerate lowest Landau level in the magnetic interpretation becomes a unique ground state wave function. It is further true that the elliptic GinUE PDF (2.35) admits an interpretation as the absolute value squared of state in the lowest Landau level, and furthermore the corresponding orthogonal polynomials (2.37) can be constructed using a Bogolyubov transformation of $\{b, b^\dagger\}$ [257]. Also, the PDF on the sphere (2.52) permits an interpretation as the absolute value squared of the ground state wave function for a free Fermi gas on the sphere subject to a perpendicular magnetic field [311]. Another point of interest relates to the N-body Fermi ground state corresponding to the quantum Pauli Hamiltonian in the plane with a perpendicular inhomogeneous magnetic field $B(x, y)$. The Hamiltonian H_B defined above then is to be multiplied by the 2×2 identity matrix, and the spin coupling term $-(g\hbar/2m)B(x, y)\text{diag}(1/2, -1/2)$ added. With $B(x, y) = -\frac{1}{2}\nabla^2 W(x, y)$ for some real-valued W, and with the assumption $\Phi := \int B(x, y)\, dx dy < \infty$, the ground state for this model (which is spin polarised with all spins up) permits an exact solution for $g = 2$ [8]. The ground state of normalisable eigenfunctions has degeneracy $[\Phi/2\pi\hbar] =: N$, with basis of eigenfunctions $\{z^j e^{W(x,y)/2\hbar}\}_{j=0}^{N-1}$. This implies the Fermi many-body ground state (6.1) with $e^{-|z_j|^2/2}$ replaced by $e^{W(x_j, y_j)/2\hbar}$ [7].

A single particle state formed as a linear superposition of $\{|0, m\rangle\}_{m=0}^N$ has the form

$$\phi(z) = \frac{1}{N\pi^{1/2}}e^{-|z|^2/2}p(\bar{z}), \quad p(z) := \sum_{n=0}^{N}\frac{\alpha_n}{\sqrt{n!}}z^n,$$

with $\mathcal{N} = (\sum_{j=0}^{N}|\alpha_n|^2)^{1/2}$. Following [252] and choosing each α_n as an independent standard complex Gaussian one has that $p(z)$ is an example of a complex Gaussian random polynomial, where the variance of each coefficient is $1/n!$. Nearly ten years before Ginibre's calculation of the GinUE eigenvalue PDF (1.7), Hammersley [315] obtained the explicit functional form of the zero PDF of $p(z)$, which was found to be proportional to

$$\frac{\prod_{1 \le j < k \le N}|z_k - z_j|^2}{(\sum_{j=0}^{N}|e_j|^2)^N}, \tag{6.3}$$

where e_j is the j-th elementary symmetric polynomial in $\{z_j\}$. One immediately notices that the product term $\prod_{1 \le j < k \le N}|z_k - z_j|^2$ is common to both (1.7) and (6.3). However, while (1.7) gives rise to a determinantal point process, it turns out that the point process implied by (6.3) is permanental [316]. Nonetheless, there are many striking similarities, including a circular law for the global eigenvalue density, a rapidly decaying simple functional form for the bulk truncated two-particle correlation

$$\rho_{(2)}^T(z_1, z_2) = \frac{1}{\pi^2}\left(f(|z_1 - z_2|^2/2) - 1\right), \quad f(x) = \frac{1}{2}\frac{d^2}{dx^2}(x^2\coth x),$$

and a slowly decaying truncated two-point correlation at the boundary; see [252, 316]. In the limit $N \to \infty$ this Gaussian random polynomial forms what is referred to as the Gaussian analytic function [326].

Remark 6.1
1. Upon stereographic projection of the sphere to the plane, it is possible to write the quantum Hamiltoniacge particle in a constant perpendicular magnetic field in a form unified with the original planar case [203]. This involves the Kähler metric and potential, and permits a viewpoint which carries over to further generalise the space to higher-dimensional complex manifolds in $\mathbb{C}^m$. A point of interest is that doing so gives, for the bulk scaling limit of the corresponding N particle lowest Landau level state, the natural higher-dimensional analogue of the kernel (2.19) [89, 90].
2. The squared wavefunction for higher Landau levels (say the r-th) has been shown to give rise to the determinantal point process with bulk-scaled kernel

$$K_\infty^r(w, z) = L_r^0(|w - z|^2)e^{w\bar{z}}e^{-(|w|^2+|z|^2)/2};$$

see e.g. [497, Proposition 2.5]. Allowing for mixing between Landau levels up to and including level r leads to squared wave functions giving rise to the same determinantal point process except for the replacement of Laguerre polynomials $L_r^0 \mapsto L_r^1$ in the kernel [309]. For finite N both kernels are built up from particular polyanalytic functions; see e.g. [2]. Extending [384], the precise mapping between the rotating fermions in the higher Landau levels and the polyanalytic Ginibre ensemble was established in [380]. Furthermore, its full counting statistics and generalisations to finite temperature were obtained in [381, 505], while [1, 148] relates the variance of the counting statistics to the entanglement entropy of the region.
3. In the theory of the fractional quantum Hall effect, constructing an anti-symmetric state with filling fraction of the lowest Landau level $\nu = 1/m$, for m an odd integer, plays a crucial role. To accomplish this, Laughlin [389] proposed the ground state wave function proportional to

$$\prod_{l=1}^{N} e^{-|z_l|^2/4\ell^2} \prod_{1 \le j < k \le N} (\bar{z}_k - \bar{z}_j)^m; \tag{6.4}$$

note that with the assumption that m is odd, this is anti-symmetric as required for fermions. Moreover, it belongs to the lowest Landau level as follows from the theory in the text below (6.2). The absolute value squared of (6.4) coincides with the Boltzmann factor (1.12) with $\beta = 2m$, and the scaling $z_l \mapsto z_l/\sqrt{2m\ell^2}$. From potential theoretic/Coulomb gas reasoning, the bulk density is therefore $1/(2m\pi\ell^2)$. The factor of m in the denominator is in precise agreement with the requirement that the filling fraction be equal to $1/m$.
4. The ground state N-body free spinless Fermi gas in the plane, without a magnetic field but confined by a radial harmonic potential, is also an example of a determinantal point process for which exact calculations are possible; see the recent review [179].

However, its statistical state is distinct from that of GinUE. Thus with a global scaling so that the support is the unit disk, the density profile is the $d = 2$ Thomas–Fermi functional form $\frac{2}{\pi}(1 - |z|^2)\chi_{|z|<1}$, in contrast to the circular law (2.17). The bulk-scaled two-point correlation function (bulk density $1/4\pi$) is given in terms of the J_1 Bessel function

$$\rho_{(2),\infty}^{\text{hF}}(z_1, z_2) = \left(\frac{1}{4\pi}\right)^2 \left(1 - \left(\frac{2J_1(|z_1 - z_2|)}{|z_1 - z_2|}\right)^2\right),$$

in contrast to (2.24). This gives a decay proportional to $1/|z_1 - z_2|^3$ of $\rho_{(2),\infty}^{\text{hF},T}$. Also, the edge-scaled correlation kernel now involves Airy functions [178], rather than the error function seen in (2.19). Universality results relating to many-body free Fermi ground states in dimension $d \geq 2$ have recently been obtained [184]. We highlight in particular the macroscopic fluctuation theorem for the linear statistic $G = \sum_j g(\mathbf{r}_j/R)$, with g assumed sufficiently smooth and absolutely integrable, in the $R \to \infty$ limit [184, Th. III.2]:

$$\frac{G - \frac{R^d \omega_d}{(2\pi)^d} \int_{\mathbb{R}^d} g(\mathbf{r}) \, d^d \mathbf{r}}{\sigma_d R^{(d-1)/2}} \xrightarrow{\text{d}} N[0, \Sigma(g)], \quad (\Sigma(g))^2 = \int_{\mathbb{R}^d} |\hat{g}(\mathbf{r})|^2 |\mathbf{r}| \, d^d \mathbf{r}. \quad (6.5)$$

Here $\omega_d = \pi^{d/2}/\Gamma(1 + d/2)$ is the volume of the Euclidean ball in $\mathbb{R}^d$, $\omega_d/(2\pi)^d$ is the bulk density, $\sigma_d^2 := \omega_{d-1}/(2\pi)^d$ and the Fourier transform has the definition $\hat{g}(\xi) = \frac{1}{(2\pi)^{d/2}} \int_{\mathbb{R}^d} e^{-i\xi \cdot \mathbf{r}} g(\mathbf{r}) \, d^d \mathbf{r}$. Note in particular that in contrast to (3.19), the variance of G now diverges with the scale R.

5. The functional form (6.3) consists of the product over differences times a many-body term, whereas the GinUE eigenvalue PDF (1.7) is the same product over differences times a single-body term. Another occurrence of the former structure, now with the product over differences raised to the power of β as for the OCP, has recently appeared as the eigenvalue PDF for a particular tridiagonal non-Hermitian random matrix [438].

6.2 Quantum Chaos Applications

The pioneering works of Wigner and Dyson relating to the Hermitian random matrix ensembles was, as noted in Chap. 1, motivated by seeking a model for the (highly excited) energy levels of a complex quantum system. Later, in the 1980s, as a fundamental contribution to the then emerging subject of quantum chaos, Bohigas et al. [98] identified the correct meaning of a complex quantum system not by the number of particles but rather as one for which the underlying classical mechanics is chaotic. To test this prediction on say the numerically generated spectrum of a quantum billiard system, the energy levels (beyond some threshold to qualify as being highly excited) were first unfolded so that their local density became unity, and then their numerically determined statistical properties were compared against random matrix predictions for the appropriate symmetry class; see e.g. [307]. Most popular among the statistical properties have been the variance for the number of eigenvalues in a large interval, and the distribution of the spacing between successive eigenvalues.

A natural extension of these advances is to inquire about the spectrum of a dissipative chaotic quantum system, which due to the loss of energy need not be real. This question was taken up by Grobe, Haake and Sommers [304] for the specific model of a damped periodic kicked top. The quantum dynamics are specified by a subunitary density operator. It is the spectrum of this operator, which after unfolding, and considering only those eigenvalues in the upper half plane away from the real axis (there is a symmetry which requires that the eigenvalues come in complex conjugate pairs—see the recent paper [32] for a discussion of this point in a random matrix context) that was compared in [304] in a statistical sense to GinUE. Following from precedents in the Hermitian case, in the statistical quantity measured was the distribution of the radial spacing between closest eigenvalues, to be denoted by $P^{s,\text{GinUE}}(r)$ with the normalisation $\int_0^\infty P^{s,\text{GinUE}}(r)\,dr = 1$. This is the quantity $F_\infty(0; D_r)$ of Remark 3.1.1. Recalling (3.10) we therefore have

$$P^{s,\text{GinUE}}(r) = -\frac{d}{dr}e^{r^2}\prod_{j=1}^{\infty}\left(1 - \frac{\gamma(j; r^2)}{\Gamma(j)}\right). \tag{6.6}$$

It follows that for small r, $P^{s,\text{GinUE}}(r) \sim 2r^3$, while it follows from (3.11) and the statement in the final sentence of the first paragraph of Sect. 3.1.2 that for large r, $\log P^{s,\text{GinUE}}(r) \sim -r^4/4$. A numerical plot can be obtained from the functional form (6.6). For the moments the formula $\langle r^p \rangle = p \int_0^\infty r^{p-1} e^{r^2} E_\infty(0; r)\,dr$ holds true. In particular, for the mean we calculate $\langle r \rangle = 1.142929\ldots$.

A variation of the closest neighbour spacing for an eigenvalue at z is the complex ratio $(z^c - z)/(z^{nc} - z)$, where z^c is the closest neighbour to z, and z^{nc} is the next closest neighbour [477]. An approximation, with fast convergence properties to the large N form, has been given recently in [204].

Also very recently a non-Hermitian Hamiltonian realisation of GinUE has been obtained in the context of a proposed non-Hermitian q-body Sachdev–Ye–Kitaev (SYK) model, with N Majorana fermions—N large and tuned mod 8—and $q > 2$ and tuned mod 4 [281]. On another front, again very recently, the emergence of GinUE behaviours in certain models of many-body quantum chaotic systems in the space direction has been demonstrated [499]. Of interest in both these lines of study is the so-called dissipative (connected) spectral form factor

$$K_N^c(t, s) = \frac{1}{N}\text{Cov}\left(\sum_{j=1}^{N} e^{i(x_j t + y_j s)}, \sum_{j=1}^{N} e^{-i(x_j t + y_j s)}\right).$$

Making use of the first formula in (3.2) and the finite N form of (2.23), this can be evaluated in terms of the hypergeometric function $_1F_1$ [404, Eq. (3)] (corrected in [282, Appendix A]; note too that both those references use global-scaled variables, whereas we do not).

Proposition 6.1 *We have*

$$K_N^c(t, s) = 1 - \frac{1}{N} \sum_{m,n=0}^{N-1} \frac{(t^2 + s^2)^{|m-n|/2}}{n!m!2^{|m-n|}}$$

$$\times \left(\frac{\max(m, n)!}{|m - n|!} {}_1F_1\left(\max(m, n) + 1, |m - n| + 1; -\frac{t^2 + s^2}{4}\right) \right)^2.$$

(6.7)

In particular,

$$\lim_{N \to \infty} K_N^c(t, s) = 1 - e^{-(t^2+s^2)/4};$$

(6.8)

cf. (3.17).

Remark 6.2 From [404] one has the large N form

$$\frac{1}{N} K_N^c\left(\frac{t}{\sqrt{N}}, \frac{s}{\sqrt{N}}\right) \sim \frac{1}{N} + 4\frac{J_1(|\tau|^2)}{|\tau|^2} - \frac{1}{N}e^{-|\tau|^2/4N}, \quad \tau := t + is,$$

extending the limit formula (6.8) and exhibiting a so-called slope-dip-ramp-plateau graphical form; see also [163].

Also of interest in the many-body quantum chaos application is the GinUE average of $|\mathrm{Tr}\, X^k|^2$ for positive integer k [499].

Proposition 6.2 *We have*

$$\left\langle |\mathrm{Tr}\, X^k|^2 \right\rangle_{\mathrm{GinUE}} = \frac{1}{(k+1)(N-1)!} \left((k + N)! - \frac{N!(N-1)!}{(N-k-1)!} \right).$$

In particular

$$\lim_{\substack{N,k \to \infty \\ k/N=x}} \frac{1}{kN^k} \left\langle |\mathrm{Tr}\, X^k|^2 \right\rangle_{\mathrm{GinUE}} = \frac{2\sinh(x^2/2)}{x^2}.$$

Proof (Sketch) In [499] the average is reduced to $\int_{\mathbb{C}} dz_1 \int_{\mathbb{C}} dz_2 \, \rho_{(2),N}(z_1, z_2) z_1^k \bar{z}_1^k$; see too the earlier work [374]. The evaluation of a more general quantity can be found in [268, Corollary 4]. $\qquad\square$

We conclude this section with a brief account of the use of the GinUE in an ensemble theory of Lindblad dynamics [133, 189, 478]. This relates to the evolution of the density matrix ρ_t for an N-level dissipative quantum system in the so-called

Markovian regime, specified by the master equation $\dot{\rho}_t = \mathcal{L}(\rho_t)$. Here the operator $\mathcal{L}$ assumes a special structure identified by Lindblad [406], and by Gorini, Kossakowski, and Sudarshan [299]. Specifically, $\mathcal{L}$ consists of the sum of two terms, the first corresponding to the familiar unitary von Neumann evolution, and the second to a dissipative part, being the sum over operators D_L (referred to as simple dissipators), represented as $N^2 \times N^2$ matrices according to

$$D_L = 2L \otimes_T L^\dagger - L^\dagger L \otimes_T \mathbb{I}_N - \mathbb{I}_N \otimes_T L^\dagger L,$$

for some $N \times N$ matrix L. Here $A \otimes_T B := A \otimes B^T$, where $\otimes$ is the usual Kronecker product. In an ensemble theory, there is interest in $F_N(t) := \frac{1}{N^2} \langle \mathrm{Tr}\, e^{tD_L} \rangle_L$ [133].

Proposition 6.3 *Let L be chosen from GinUE with global scaling. We have*

$$\lim_{N\to\infty} F_N(t) = e^{-4t} \Big(I_0(2t) + I_1(2t) \Big)^2.$$

Proof (Sketch) Following Can [133], using a diagrammatic calculus, it is first demonstrated that

$$\lim_{N\to\infty} \frac{1}{N^2} \langle D_L^k \rangle = \lim_{N\to\infty} \frac{(-1)^k}{N^2} \Big\langle \Big(L^\dagger L \otimes_T \mathbb{I}_N + \mathbb{I}_N \otimes_T L^\dagger L \Big)^k \Big\rangle_{L \in \mathrm{GinUE}}.$$

The average on the RHS, in terms of the eigenvalues $\{x_j\}$ of $L^\dagger L$, reads

$$\Big\langle \sum_{j,l=1}^N (x_j + x_l)^k \Big\rangle_{L^\dagger L}$$

and consequently

$$\lim_{N\to\infty} F_N(t) = \lim_{N\to\infty} \Big\langle \frac{1}{N^2} \sum_{j,l=1}^N e^{-t(x_j + x_l)} \Big\rangle_{L^\dagger L} = \lim_{N\to\infty} \Big(\Big\langle \frac{1}{N} \sum_{j=1}^N e^{-tx_j} \Big\rangle_{L^\dagger L} \Big)^2.$$

The latter is the mean of a linear statistic in the ensemble $\{L^\dagger L\}$ (complex Wishart matrices; see e.g. [237, Sect. 3.2]). Using the Marchenko–Pastur law for the global density of this ensemble (see e.g. [237, Sect. 3.4.1]), the stated result follows. $\square$

Remark 6.3 (*Classification of non-Hermitian matrices*) It was commented in the Introduction that, in distinction to Dyson's viewpoint based on symmetry considerations, Ginibre's study [293] was not similarly motivated. Nowadays however, it is recognised that a symmetry viewpoint is fundamental to topologically driven effects in non-Hermitian quantum physics [67]. Starting with [91, 418] and continuing in [359], a classification scheme based on symmetries with respect to the involutions of transpose, complex conjugation and Hermitian conjugation, and in which the (anti-)commutation relation involves unitary matrices satisfying certain quadratic relations in terms of these involution, has been given. For example, defining the block unitary matrix $P = \mathrm{diag}\,(\mathbb{I}_N, -\mathbb{I}_N)$, and requiring that the matrix ensemble $\{A\}$ have the (anti-)symmetry $A = -PAP$, gives that each A has the form

$$A = \begin{bmatrix} 0_{N \times N} & X \\ Y & 0_{N \times N} \end{bmatrix} \tag{6.9}$$

for some square matrices X, Y. Denoting the eigenvalues of the matrix product XY as $\{-z_j^2\}$, one sees that the eigenvalues of A are $\{\pm i z_j\}$. In the realm of non-Hermitian classifications, this is referred to as the chiral class [456]. Additionally, the chiral models corresponding to GinSE and GinOE were studied in [11] and [38], respectively.

In keeping with the viewpoint of this section, a basic question concerns signatures of the symmetry in the eigenvalue spectrum. For example, in (6.9), with X, Y GinUE matrices, are the bulk-scaled eigenvalues of A statistically distinct from individual GinUE matrices? We know from the results quoted in the paragraph above Remark 2.6 that the answer in this case is no. However, the answer to this question is yes, if instead the symmetry is that $A = A^T$, for the independent entries of A standard complex Gaussians. This was demonstrated in [313] by a numerical study of the nearest neighbour spacing distribution, and the relevance to Lindblad dynamics discussed.

6.3 Singular Values

One recalls that for a complex square matrix X the squared singular values are the eigenvalues of $X^\dagger X$. For a general ensemble of non-Hermitian matrices $\{X\}$, motivation to study the singular values comes from various viewpoints. For example, in Remark 2.6.4, singular values (specifically of product matrices) appeared in the context of Lyapunov exponents. As other example, one recalls that plus/minus of the singular values are the eigenvalues of the $2N \times 2N$ Hermitian matrix

$$H = \begin{bmatrix} 0_{N \times N} & X \\ X^\dagger & 0_{N \times N} \end{bmatrix}. \tag{6.10}$$

The importance of this in relation to the eigenvalues of X is that the resolvent of the modification of (6.10) obtained by replacing each X by $X - z\mathbb{I}_N$ is fundamental to the study of the circular law for the spectral density beyond the Gaussian case; see e.g. [100, Sect. 4.1]. Another piece of theory is that the condition number κ_N associated with X is equal to the ratio of the smallest to the largest singular value [210]. We remark too that from the identity $|\det X| = |\det X^\dagger X|^{1/2}$ the distribution of the modulus of $\det X$ is determined by the singular values.

For the GinUE, the squared singular values $\{s_j\}_{j=1}^N$ say are known to have for their joint distribution a PDF proportional to

$$\prod_{j=1}^{N} e^{-s_j} \prod_{1 \leq j < k \leq N} (s_k - s_j)^2, \quad s_j \in \mathbb{R}_+; \tag{6.11}$$

see e.g. [237, Proposition 3.2.2 with $\beta = 2$, $n = m = N$]. This is an example of the Laguerre unitary ensemble (LUE). After scaling by N, almost surely the largest squared singular value has the limiting value 4 [459]. However, after the same scaling, a simple change of variables in (6.11) integrated from (s, ∞) in each variable reveals that the smallest singular value is an exponential random variable with rate parameter N^2. Putting these facts together implies that for large N, κ_N/N is distributed according to the heavy tailed distribution with PDF $\frac{8}{x^3} e^{-4/x^2} \chi_{x>0}$ [210]. Also, for $n \times N$ ($n \geq N$) rectangular GinUE matrices it is proved in [153] that $\langle \log \kappa_N \rangle < \frac{N}{|n-N|+1} + c$, where $c = 2.24$, for any $N \geq 2$.

Let $P_N(t)$ denote the PDF for the distribution of $|\det X|^2$ for GinUE matrices. Making use of knowledge of the PDF of squared singular values (6.11) shows that the Mellin transform of $P_N(t)$ is equal to the multiple integral

$$\frac{1}{C_N} \int_0^\infty ds_1 \cdots \int_0^\infty ds_N \prod_{j=1}^{N} s_j^{s-1} e^{-s_j} \prod_{1 \leq j < k \leq N} (s_k - s_j)^2 = \prod_{j=0}^{N-1} \frac{\Gamma(s+j)}{\Gamma(1+j)}. \tag{6.12}$$

Here the normalisation C_N is such that the expression equals unity for $s = 1$, while the evaluation of the multiple integral follows as a special case of the Laguerre weight Selberg integral; see e.g. [237, Proposition 4.7.3]. As noted in [269, Eq. (2.17)] (see also [475, Proposition 2.2]), it follows immediately from this that

$$|\det X|^2 \overset{\mathrm{d}}{=} \prod_{l=1}^{N} \frac{1}{2} \chi_{2l}^2. \tag{6.13}$$

(Proposition 2.14 regarding the independent distribution of the absolute values of the eigenvalues.) In words this says that the absolute value squared of the determinant of GinUE matrices is equal in distribution to the product of N independent chi-squared distributions, with degrees of freedom $2, 4, \ldots, 2N$, each scaled by a factor of 2. Starting from (6.13), and defining the global-scaled GinUE matrices X^g by $X^g = \frac{1}{\sqrt{N}} X$, the distribution of $\log |\det X^g|^2$ can be shown to have leading-order mean $-N$, variance $\log N$, and after recentring and rescaling satisfy a central limit theorem [475, Th. 3.5]. For a general linear statistic $\sum_{j=1}^{N} f(z_j)$ of global-scaled GinUE matrices, the leading-order mean is $\frac{N}{\pi} \int_{|z|<1} f(z) \, d^2z$. For $f(z) = \log |z|^2$, this gives the stated value of $-N$. Also, we notice that substituting this choice of $f(z)$ in the variance formula implied by (3.21) gives $\frac{1}{\pi} \int_{|r|<1} \frac{1}{x^2+y^2} \, dxdy$, which is not integrable at the origin, in keeping with the variance actually diverging as $\log N$.

There is an alternative viewpoint on the result (6.13) which does not require knowledge of the joint distribution of the singular values (6.11), nor the evaluation of the multiple integral (6.12). The idea, used in both [269, 475] and which goes back to Bartlett [74] in the case of real Gaussian matrices, is to decompose X in terms of its QR (Gram–Schmidt) decomposition. The matrix of orthonormal vectors Q constructed from the columns of X will for $X \in \text{GinUE}$, be a Haar distributed unitary matrix, which we denote by U. The matrix $R = [r_{jk}]_{j,k=1}^{N}$ is upper triangular with diagonal elements real and positive. One notes

$$\det X^{\dagger} X = \prod_{j=1}^{N} r_{jj}^2, \tag{6.14}$$

and so it suffices to have knowledge on the distribution of $\{r_{jj}\}_{j=1}^{N}$ for X.

Proposition 6.4 *Let $\{r_{jj}\}_{j=1}^{N}$ denote the diagonal elements in the QR decomposition of a GinUE matrix X. We have*

$$r_{jj}^2 \overset{\mathrm{d}}{=} \frac{1}{2} x_{2j}^2. \tag{6.15}$$

Proof The QR decomposition $X = UR$ gives the corresponding decomposition of measure (see e.g. [237, Proposition 3.2.5])

$$(dX) = \prod_{j=1}^{N} r_{jj}^{2(N-j)+1} (dR)(U^{\dagger} dU),$$

where as anticipated $(U^{\dagger} dU)$ is recognised as the Haar measure on the space of complex unitary matrices. The element distribution of GinUE matrices is proportional to $e^{-\operatorname{Tr} X^{\dagger} X} = e^{-\sum_{1 \le j \le k \le N} |r_{jk}|^2}$. The various factorisations imply that integrating over U and the off-diagonal elements of R only changes the normalisation. We then read off that each r_{jj} has a distribution with PDF proportional to $r^{2(N-j)+1} e^{-r^2}$, which implies (6.15). □

Using (6.15) in (6.14) reclaims (6.13).

Remark 6.4 1. The fact that the QR decomposition of a Ginibre matrix gives rise to a Haar distributed random unitary matrix provides a practical realisation of the latter. However, the QR decomposition of in-built mathematical software may not generate the triangular matrix R with strictly positive entries, which affects the properties of Q. Further details, and how this can be fixed, are discussed in [213].
2. Since with $\{z_j\}$ the eigenvalues of X, $|\det X|^2 = \prod_{j=1}^{N} |z_j|^2$, the fact that the Mellin transform of the distribution of this quantity is given by the product of gamma functions in (6.12) implies

$$\left\langle \prod_{l=1}^{N} |z_l|^{2(s-1)} \right\rangle_{\text{GinUE}}^{g} = N^{N(s-1)} \prod_{j=0}^{N-1} \frac{\Gamma(s+j)}{\Gamma(1+j)}. \tag{6.16}$$

Here the superscript "g" indicates the use of global scaling coordinates $z_l \mapsto \sqrt{N} z_l$. We observe that the knowledge of the induced GinUE normalisation $C_{n,N}$ in Proposition 2.8 provides a direct derivation of (6.16). For large N this ratio of gamma functions can be written in terms of the Barnes G-function according to $\frac{G(N+s)}{G(N+1)G(s)}$; see [237, Eq. (4.183)]. Known asymptotics for ratios of the Barnes G-function (see e.g. [237, Eq. (4.185)]) then gives that for large N, and with $s = \gamma/2 + 1$ for convenience,

$$\left\langle \prod_{l=1}^{N} |z_l|^\gamma \right\rangle_{\text{GinUE}}^{\text{g}} \sim N^{\gamma^2/8} e^{-(\gamma/2)N} \frac{(2\pi)^{\gamma/4}}{G(1+\gamma/2)}. \tag{6.17}$$

This is the special case $z = 0$ of an asymptotic formula for $\langle \prod_{l=1}^{N} |z - z_l|^\gamma \rangle_{\text{GinUE}}^{\text{g}}$ given by Webb and Wong [531, Th. 1.1].

3. It is a standard result in random matrix theory (see e.g. [459]) that the density of singular values in (6.11), after the global scaling $s_j \mapsto s_j N$, has the particular Marchenko–Pastur form $\rho_{(1),\infty}^{\text{MP}}(x) = \frac{1}{2\pi}(\frac{4-x}{x})^{1/2} \chi_{0<x<4}$. The k-th moment of the density is given in terms of the specific mixed moment of a global-scaled Ginibre matrix $\tilde{G}$, $\langle \text{Tr}(\tilde{G}^\dagger G)^k \rangle$. The calculation of these moments for large N relates to free probability—see the recent introductory text [464] for the main ideas—and to combinatorics as is seen from the fact that $\int_0^4 x^k \rho_{(1),\infty}^{\text{MP}}(x)\, dx = C_k$, where C_k denotes the k-th Catalan number. Works on mixed moments of Ginibre matrices include [176, 191, 312, 530]. An explicit example is the average in Proposition 6.2. The scaling limit therein where both N and k become large with their ratio fixed is called the BMN large N limit; see e.g. [40, Sect. 6]. Mixed moments of the pseudo inverse of rectangular GinUE are calculated in [168].

4. The squared singular values as specified by the PDF (6.11) form a determinantal points process, being a special case of the LUE; see e.g. [237, Chaps. 3 and 5]. This is similarly true of the squared singular values of the various extensions of GinUE considered above: for example in the case of the spherical model and truncated unitary matrices, it is the classical Jacobi unitary ensemble (JUE) which arises, while the singular values of products of GinUE matrices, or of truncated unitary matrices, gives rise to a class of determinantal point processes called Pólya ensembles [270, 365, 366, 378]. A determinantal point process also results from the squared singular value of the so-called shifted GinUE—matrices $G + A$ for $G \in$ GinUE and A fixed [81, 263, 377]. A notable exception is the singular values of elliptic GinUE matrices, which form a Pfaffian point process [351]. Similarly, with $G \in$ GinUE and $U \in U(N)$ chosen with Haar measure, the squared singular values of $X; = (\mathbb{I}_N + U)G$ form a Pfaffian point process [259]. The random matrix X occurs in the construction of the Bures–Hall measure on the space of density matrices [541].

5. Remark 6.3 drew attention to a symmetry based classification of non-Hermitian matrices. The recent work [360] discusses properties of the corresponding singular values.

6.4 Eigenvectors

Associated with the set of eigenvalues $\{z_j\}$ of a Ginibre matrix G are two sets of eigenvectors—the left eigenvectors $\{\ell_j\}$ such that $\ell_j^T G = x_j \ell_j^T$, and the right eigenvectors $\{\mathbf{r}_j\}$ such that $G\mathbf{r}_j = x_j \mathbf{r}_j$. These are not independent, but rather (upon suitable normalisation), form a biorthogonal set

$$\ell_i^T \mathbf{r}_j = \delta_{i,j}. \tag{6.18}$$

This property follows from the diagonalisation formula $G = XDX^{-1}$, where the columns of X are the right eigenvectors, D the diagonal matrix of eigenvectors, and the rows of X^{-1} are the left eigenvectors. For nonzero scalars $\{c_i\}$ we see that (6.18) is unchanged by the rescalings $\mathbf{r}_j \mapsto c_j \mathbf{r}_j$ and $\ell_j \mapsto (1/c_j)\ell_j$.

For $N \times 1$ column vectors $\mathbf{u}, \mathbf{v}$, define the inner product $\langle \mathbf{u}, \mathbf{v} \rangle := \bar{\mathbf{u}}^T \mathbf{v}$. The so-called overlap matrix has its elements O_{ij} expressed in terms of this inner product according to

$$O_{ij} := \langle \ell_i, \ell_j \rangle \langle \mathbf{r}_i, \mathbf{r}_j \rangle. \tag{6.19}$$

Note that this is invariant under the mappings noted in the final sentence of the above paragraph, and for fixed i and summing over j gives 1. Also, it follows from (6.19) that the diagonal entries relate to the lengths

$$O_{jj} = ||\ell_j||^2 ||\mathbf{r}_j||^2. \tag{6.20}$$

The square root of this quantity is known as the eigenvalue condition number; see the introduction to [107] and [173, Sect. 1.1] for further context and references. Significant too is the fact that the overlaps (6.20) appear in the specification of a Dyson Brownian motion extension of GinUE [107, 220, 303].

Statistical properties of $\{O_{ij}\}$ for GinUE were first considered by Chalker and Mehlig [146, 147]. By the Schur decomposition (2.2), instead of a GinUE matrix G, we may consider an upper triangular matrix Z with the eigenvalues $\{z_j\}$ of G on the diagonal, and standard complex Gaussians, as off-diagonal entries. For the eigenvalue z_1, the triangular structure shows that $\ell_1 = (1, b_2, \ldots, b_N)^T$ and $\mathbf{r}_1 = (1, 0, \ldots, 0)^T$ where for $p > 1$ and $b_1 = 1$, $b_p = \frac{1}{z_1 - z_p} \sum_{q=1}^{p-1} b_q Z_{qp}$. From this last relation, it follows that with $\ell_1^{(n)} = (1, b_2, \ldots, b_n)^T$ for $n < N$ we have

$$||\ell_1^{(n+1)}||^2 = ||\ell_1^{(n)}||^2 \left(1 + \frac{1}{|z_1 - z_{n+1}|^2} \left| \sum_{q=1}^n \tilde{b}_q Z_{q(n+1)} \right|^2 \right), \qquad \tilde{b}_q := \frac{b_q}{\sqrt{\sum_{q=1}^n |b_q|^2}}. \tag{6.21}$$

This has immediate consequences in relation to O_{11} as shown by Bourgade and Dubach [107].

Proposition 6.5 *Let the eigenvalues $\{z_j\}$ be given. We have*

$$O_{11} \overset{d}{=} \prod_{n=2}^{N} \left(1 + \frac{|X_n|^2}{|z_1 - z_n|^2} \right), \tag{6.22}$$

where each X_n is an independent complex standard Gaussian. Furthermore, it follows from this that after averaging over $\{z_2, \ldots, z_N\}$

$$O_{11}\Big|_{z_1=0} \overset{d}{=} \frac{1}{B[2, N-1]}, \tag{6.23}$$

where $B[\alpha, \beta]$ refers to the beta distribution.

Proof A product formula for O_{11} follows from (6.20), the fact that $\|\mathbf{r}_1\| = 1$, and by iterating (6.21). This product formula is identified with the RHS of (6.22) upon noting that a vector of independent standard complex Gaussians dotted with any unit vector (here $(\tilde{b}_1, \ldots, \tilde{b}_n)$) has a distribution equal to a standard complex Gaussian.

In relation to (6.23) a minor modification of the proof of Proposition 2.14 shows that conditioned on $z_1 = 0$, the ordered squared moduli $\{|z_j|^2\}_{j=2}^N$ are independently distributed as $\{\Gamma[j; 1]\}_{j=2}^N$. Noting too that with each X_j a standard complex Gaussian, $|X_j|^2 \overset{d}{=} \Gamma[1; 1]$ it follows that

$$O_{11}\Big|_{z_1=0} \overset{d}{=} \prod_{n=2}^{N} \left(1 + \frac{\tilde{X}_n}{Y_n} \right), \qquad \tilde{X}_n \overset{d}{=} \Gamma[1; 1], \ Y_n \overset{d}{=} \Gamma[n; 1].$$

Next, we require the knowledge of the standard fact that $Y_n/(Y_n + \tilde{X}_n) \overset{d}{=} B[n, 1]$. Furthermore (see e.g. [237, Exercises 4.3 q.1]), for $x \overset{d}{=} B[\alpha + \beta, \gamma]$, $y \overset{d}{=} B[\alpha, \beta]$, we have that $xy \overset{d}{=} B[\alpha, \beta + \gamma]$, which tells us that with $b_n \overset{d}{=} B[n, 1]$ we have $\prod_{n=2}^{N} b_n \overset{d}{=} B[2, N-1]$.

Dividing both sides of (6.23) by N we see that the $N \to \infty$ is well defined since $NB[2, N-1] \to \Gamma[2, 1]$. After scaling the GinUE matrix $G \mapsto G/\sqrt{N}$ so that the leading eigenvalue support is the unit disk, an analogous limit formula has been extended from $z_1 = 0$ to any $z_1 = w$, $|w| < 1$ in [107]. Thus

$$\frac{O_{11}\Big|_{z_1=w_1}}{N(1 - |w_1|^2)} \overset{d}{\to} \frac{1}{\Gamma[2; 1]}. \tag{6.24}$$

In words, with O_{11} corresponding to the condition number, one has that the instability of the spectrum is of order N and is more stable towards the edge. Another point of interest is that the PDF for $1/\Gamma[2, 1]$ is $\chi_{t>0} e^{-1/t}/t^3$, which is heavy tailed, telling us that only the zeroth and first integer moments are well defined. We remark that limit theorems of the universal form (6.24) have been proved in the case of the complex spherical ensemble of Sect. 2.5, and for a sub-block of a Haar distributed unitary matrix [200]; see also [451].

The $1/t^3$ tail implied by (6.24) has been exhibited from another viewpoint in the work of Fyodorov [272]. There the joint PDF for the overlap non-orthogonality $O_{jj} - 1$, and the eigenvalue position z_j, was computed for finite N. The global-scaled limit of this quantity, $\mathcal{P}^g(t, w)$ say, was evaluated as [272, Eq. (2.24)]

$$\mathcal{P}^g(t, w) = \frac{(1 - |w|^2)^2}{\pi t^3} e^{-(1-|w|^2)/t}, \quad |w| < 1. \tag{6.25}$$

Note that for the first moment in t this gives

$$\int_0^\infty t \mathcal{P}^g(t, w) \, dt = \frac{1}{\pi}(1 - |w|^2), \tag{6.26}$$

in keeping with a prediction from [146, 147]. This was first proved in [530].

An explicit formula for the large N form of the average of the overlap (6.19) in the off-diagonal case (say $(i, j) = (1, 2)$), with the GinUE matrix scaled $G \mapsto G/\sqrt{N}$, and conditioned on $z_1 = w_1$, $z_2 = w_2$ with $|w_1|, |w_2| < 1$ is also known [39, 107, 146, 147, 173]:

$$\left\langle O_{12}\Big|_{z_1=w_1, z_2=w_2} \right\rangle \underset{N \to \infty}{\sim} -\frac{1}{N} \frac{1 - w_1 \bar{w}_2}{\pi^2 |w_1 - w_2|^4} \left(\frac{1 - (1 + N|w_1 - w_2|^2) e^{-N|w_1 - w_2|^2}}{1 - e^{-N|w_1 - w_2|^2}} \right). \tag{6.27}$$

This formula is uniformly valid down to the scale $N|w_1 - w_2| = O(1)$. The large N form of the average value of the diagonal overlap for products of M global-scaled GinUE matrices has been considered in [78, 113], with the result

$$\lim_{N \to \infty} \frac{1}{N} \left\langle O_{11}\Big|_{z_1=w} \right\rangle = \frac{1}{\pi} |z|^{-2+2/M}(1 - |z|^{2/M}) \chi_{|z|<1}; \tag{6.28}$$

cf. (2.79). We remark that the average values of O_{11} and O_{12}, conditioned on multiple eigenvalues are shown to have a determinantal form in [39, Th. 1], thus exhibiting an integrable structure for these eigenvector statistics.

For general complex non-Hermitian matrices with independently distributed entries of the form $\xi_{jk} + i\zeta_{jk}$, where each $\xi_{jk}, \tilde{\zeta}$ is an identically distributed zero mean real random variable of unit variance (this class is sometimes referred to as complex non-Hermitian Wigner matrices; see e.g. [28] and Sect. 6.5 below), a line of research in relation to the normalised eigenvectors is to quantify the similarity with a complex vector drawn from the sphere embedded in $\mathbb{C}^N$ with uniform distribution. Recent references on this include [413, 417]. For random vectors on the sphere, there are bounds on the size of the components which rule out gaps in the spread of the size of the components, referred to in [476] as no-gaps localisation. As noted in [417,

Sect. 1.1], the bi-unitary invariance of GinUE matrices implies individual eigenvectors are distributed uniformly at random from the complex sphere, and thus with probability close to one have that the j-th largest modulus of the entries is bounded above and below by a positive constant times $\sqrt{N-j}/N$, for j in the range from $N/2$ up to N minus a constant time $\log N$.

Remark 6.5 1. The overlap $O_{11}|_{z=w}$ as specified by (6.24) shows itself in a construction of a reduced density matrix ρ_A associated with Ginibre eigenvectors, conditioned on the event that the eigenvalue z is equal to w. Thus from [164, Eq. (18)] we have

$$\rho_A = \frac{\mathbb{I}_A}{N_A} + \sqrt{\frac{1-|w|^2}{\Gamma[2,1]}}G, \tag{6.29}$$

where $1 \ll N_A \ll N_B$, $N_A + N_B = N$, and G is an $N \times N$ global-scaled Ginibre matrix. Relevant to the study of [164] is the eigenvalue density implied by (6.29). Appropriately shifting the functional form of the circular law (2.32) and averaging over $\Gamma[2,1]$ shows that in the variable $x = |N_A^{-1} - z|/\sqrt{1-|w|^2}$ the density is given by [164, Eq. (21)]

$$\mu(x) = \frac{1}{\pi}\left(2 - e^{-1/x^2}\left(\frac{1+2x^2+2x^4}{x^4}\right)\right). \tag{6.30}$$

2. Another appearance of O_{11} (and O_{12}) is in the study of the eigenstate thermalisation hypothesis (ETH) for non-Hermitian Hamiltonians [165]. In this topic, with A a fixed $N \times N$ matrix, of interest is the statistical properties of $\langle \boldsymbol{\ell}_i, A\mathbf{r}_j \rangle$. With $\mathbb{E}_U$ denoting the expectation with respect to the matrix U in the Schur decomposition (2.2), it is found, restricting attention to the case $i = j$ for simplicity, that

$$\mathbb{E}_U \langle \boldsymbol{\ell}_i, A\mathbf{r}_i \rangle = \frac{\mathrm{Tr}\, A}{N},$$
$$\mathbb{E}_U |\langle \boldsymbol{\ell}_i, A\mathbf{r}_i \rangle|^2 = \frac{|\mathrm{Tr}\, A|^2}{N} + \frac{O_{11}}{N^2}\left(\mathrm{Tr}\, A^\dagger A - \frac{|\mathrm{Tr}\, A|^2}{N}\right).$$

With $\mathrm{Tr}\, A = O(N)$, in light of (6.24) this shows that the fluctuations are of the same order as the mean, which is interpreted as a violation of the ETH.

3. A so-called non-Hermitian Rosenweig–Porter model can be defined by weighting each of the off-diagonal entries of a GinUE matrix by $N^{-\gamma/2}$, where $\gamma > 0$. As for the Hermitian case, of interest is the inverse partition ratio $I_q(j)$, defined as

$$I_q(j) = \sum_{p=1}^{N}\left|\langle \mathbf{e}_p, \mathbf{r}_j \rangle\langle \boldsymbol{\ell}_j, \mathbf{e}_p \rangle\right|^q, \qquad q > 1/2,$$

where $\mathbf{e}_p$ denotes the p-th standard basis vector in $\mathbb{C}^N$. It is found in [177] that

$$I_q(j) \asymp N^{1-q}, \quad \gamma < 1 \qquad I_q(j) = O(1), \quad \gamma > 1,$$

independent of j. In particular, $I_q(j) = O(1)$ has the interpretation of the eigenstates being localised.

6.5 Non-Hermitian Wigner Ensembles

As remarked at the beginning of the paragraph below (6.28), the term (complex) non-Hermitian Wigner matrices refers to the class of non-Hermitian random matrices with independently distributed entries of the form $\xi_{jk} + i\zeta_{jk}$, where each ξ_{jk}, $\tilde{\zeta}$ is an identically distributed zero mean real random variable of unit variance. When the latter is the real normal N[0, $1/\sqrt{2}$], the GinUE is obtained. However, more generally, non-Hermitian Wigner matrices do not share with GinUE the property of allowing for explicit formulas in relation to the joint eigenvalue PDF or the correlation functions. Nonetheless, for large N many statistical properties of non-Hermitian Wigner matrices are in fact independent of the detail of the particular distribution of the elements, beyond it having mean zero and unit variance. This property is referred to as universality, and generally is a prominent theme in random matrix theory [219].

In the specific case of the universality of the circular law functional form for the global density, references on this have been given below (2.17). As noted in the first paragraph of Sect. 6.4, important to establishing this result is control of the resolvent associated with a shifted version of the matrix H (6.10). The work [46] establishes the near optimal convergence of the spectral radius s_N say, improved on in [162] to the statement that for any $\epsilon > 0$ there is a $C_\epsilon > 0$ such that

$$\limsup_N \Pr\left(\left|s_N - 1 - \sqrt{\frac{\gamma_N}{4N}}\right| \geq \frac{C_\epsilon}{\sqrt{N \log N}}\right) \leq \epsilon,$$

where γ_N is as in (3.13). An analysis based on the resolvent associated with (6.10) has been used to establish local universality of the edge correlation functional form associated with the correlation kernel (2.19), where it is required that the expected value of the square of the entries vanish. In such work it is not convergence to the explicit functional form that is established, but rather convergence to the case of standard complex Gaussian entries.

Consider now the global-scaled covariance of smooth linear statistics $f, \bar{g}$, which in the case of GinUE is given by (3.21), with a normal Gaussian fluctuation as specified by Proposition 3.4. In the case of complex non-Hermitian Wigner matrices, with the expected value of the square of the entries random variable $\xi + i\zeta$ equally zero, the RHS of (3.21) aquires the extra term [160] (see also the review [156])

$$\kappa_4 \left(\frac{1}{\pi} \int_{|\mathbf{r}|<1} f(\mathbf{r}) \, dxdy - f_0 \right) \left(\frac{1}{\pi} \int_{|\mathbf{r}|<1} \bar{g}(\mathbf{r}) \, dxdy - \bar{g}_0 \right).$$

Here κ_4 is the fourth cumulant of a single entry, $\kappa_4 := \langle |\xi + i\zeta|^4 \rangle - 2$ (note that this vanishes for GinUE) and f_0, $\bar{g}_0$ are as those in the statement of Proposition 3.3. Note that this extra term vanishes if f or g are real analytic. Similarly, the $\beta = 2$ case of the expansion (4.18) must be modified by the additive term [160]

$$-\frac{\kappa_4}{\pi N} \int_{|z|<1} f(\mathbf{r})(2|z|^2 - 1) \, d^2z.$$

Should the variance of the entries of a non-Hermitian Wigner matrix not be well defined due to the entries being heavy tailed, distinct statistical properties to those described above are expected. While explicit formulas are generally not available, introducing a parameter α as in the theory of Lévy α-stable laws it is known that the limiting global density decays proportionally to $|z|^{2(\alpha-1)} e^{-(\alpha/2)|z|^\alpha}$ for large $|z|$; see [100, Sect. 6].

Aspects of universality in relation to eigenvector overlaps are investigated at a numerical level in [107]. In particular, evidence is presented for the validity of the inverse gamma distribution (6.24) for a general class of complex non-Hermitian Wigner ensembles.

Part II
GinOE and GinSE

Chapter 7
Eigenvalue Statistics for GinOE and Elliptic GinOE

For GinUE matrices the (complex) Schur decomposition underpinned our ability to compute the eigenvalue PDF. This has a real analogue, although it is necessary to condition on the number of real eigenvalues. Consequently, the eigenvalue PDF is no longer absolutely continuous, but rather breaks up into sectors depending on the number of real eigenvalues. Nonetheless, as for real symmetric GOE matrices, structures giving rise to a Pfaffian point process result, the further development of which requires identifying particular skew-orthogonal polynomials. Questions relating to the distribution of real eigenvalues have a rich content, the development of which is facilitated by the identification of a simpler determinantal structure for the generating function, which permits the establishment of a local central limit theorem. Remarkably, the Pfaffian point process specifying the correlations of the real eigenvalues coincides with the those for the annihilation process on the real line $A + A \to \emptyset$, which has consequences with respect to determining the large gap asymptotics. An extension of GinOE also considered in this chapter is elliptic GinOE, which is defined as a linear combination of real symmetric and real anti-symmetric Gaussian matrices. Similar structures to those for GinOE result, but now with the possibility of studying the weakly non-symmetric limit. An application is given to equilibria counting.

7.1 Eigenvalue PDF for GinOE

Dyson's derivation of the GinUE eigenvalue PDF consisted of the following main steps (see Sect. 2.1):

(i) Decompose a GinUE matrix G using the Schur decomposition $G = UZU^{\dagger}$. Here U is a unitary matrix, and Z is an upper triangular matrix with the eigenvalues $\{z_j\}$ of G on the diagonal.

(ii) Decompose the measure for the independent elements of G, both real and imaginary parts, using the coordinates from (i). It is found that the dependence

© The Author(s) 2025
S.-S. Byun and P. J. Forrester, *Progress on the Study of the Ginibre Ensembles*, KIAS
Springer Series in Mathematics 3, https://doi.org/10.1007/978-981-97-5173-0_7

on U, $\tilde{Z}$ (the strictly upper triangular elements of Z) and $\{z_j\}$ factorises, and that the Jacobian equals $\prod_{j<k} |z_k - z_j|^2$.

(iii) Rewrite the weight for the joint element PDF using the coordinates of (i),
$e^{-\text{Tr}\, G^\dagger G} = e^{-\sum_{j=1}^N |z_j|^2 - \sum_{j<k} |\tilde{Z}_{jk}|^2}$.

(iv) From the decomposition in (ii) and (iii) observe that the dependence on the elements of $\tilde{Z}$ factorises and hence only contributes to the normalisation after integration over these variables to leave the functional form (1.7).

Following [211] (see also [237, Sect. 15.10]), in the case of $A \in$ GinOE, conditioned to have k real eigenvalues (k same parity as N), the appropriate Schur decomposition in step (i) reads $A = QRQ^T$. Here Q is a real orthogonal matrix, while R is upper block triangular. The first k diagonal elements of R are the scalars $\{\lambda_j\}_{j=1}^k$—the real eigenvalues—while the next $(N-k)/2$ diagonal elements are the 2×2 matrices

$$X_j := \begin{bmatrix} x_j & b_j \\ -c_j & x_j \end{bmatrix}, \qquad b_j, c_j > 0,$$

where $x_j \pm iy_j$ ($y_j = \sqrt{b_j c_j}$) are the complex eigenvalues.

Step (ii) seeks to decompose the measure for the elements of A. A factorisation again results, with the Jacobian equalling $2^{(N-k)/2} |\tilde{\Delta}|$, where $\tilde{\Delta}$ is as in (1.10) but with the difference between each pair $x_j \pm iy_j$ omitted, times the additional factor $\prod_{l=k+1}^{(N+k)/2} |b_l - c_l|$. For step (iii) we calculate

$$e^{-\text{Tr}\, AA^T/2} = e^{-\sum_{i<j} r_{ij}^2/2} e^{-\sum_{j=1}^k \lambda_j^2/2} e^{-\sum_{j=1}^{(N-k)/2} (x_j^2 + y_j^2 + \delta_j^2/2)},$$

where $\delta_j = b_j - c_j$ and $\{r_{ij}\}$ are the off-diagonal elements. To carry out step (iv) and thus integrate out all variables except the eigenvalues, it is convenient to change variables from $\{b_j, c_j\}$ to $\{y_j, \delta_j\}$ according to $db_j dc_j = (2y_j/\sqrt{\delta_j^2 + 4y_j^2}) dy_j d\delta_j$. Integrating over δ_j in this is responsible for the factors $e^{y_j^2} \text{erfc}(\sqrt{2}y_j)$ in (1.10), while integrating over the other variables only contributes to the normalisation.

7.2   Coulomb Gas Perspective

The factor $|\Delta|$ in (1.10) can be written in exponential form

$$\left| \Delta(\{\lambda_l\}_{l=1,\ldots,k} \cup \{x_j \pm iy_j\}_{j=1,\ldots,(N-k)/2}) \right|$$

$$= \exp\left(\sum_{1 \le j < p \le k} \log |\lambda_p - \lambda_j| + \sum_{j=1}^k \sum_{s=1}^{(N-k)/2} \log |z_s - \lambda_j||\bar{z}_s - \lambda_j| + \sum_{a,b=1}^{(N-k)/2} \log |z_a - \bar{z}_b| \right)$$

$$\times \exp\left(2 \sum_{1 \le a < b \le (N-k)/2} \log |z_b - z_a||z_b - \bar{z}_a| \right). \tag{7.1}$$

This permits the interpretation as a Boltzmann factor for a classical two-dimensional Coulomb gas [245, 284]. Relevant to this is the fact that the solution of the two-dimensional Poisson equation $\nabla_{\mathbf{r}}^2 \phi(\mathbf{r}, \mathbf{r}') = -2\pi \delta(\mathbf{r} - \mathbf{r}')$ with the Neumann boundary condition along the x-axis $\frac{\partial}{\partial y} \phi(\mathbf{r}, \mathbf{r}')|_{y \to 0^+} = 0$ is, with the use of complex coordinates, given by

$$\phi(\mathbf{r}, \mathbf{r}') = -\log\left(|z - z'| \, |z - \bar{z}'|\right). \tag{7.2}$$

How to approximate (7.2) in terms of different dielectric constants for $y > 0$ and $y < 0$ is discussed in [237, Sect. 15.9]; its effect is to give rise to an image particle of identical charge at the reflection point $\bar{z}'$ of z' about the real axis.

We see that (7.1) contains terms corresponding to the sum over pairs of the potential (7.2) for the complex coordinates $\{z_j\}_{j=1}^{(N-k)/2}$ in the upper half plane, interacting at dimensionless inverse temperature $\beta = 2$. In addition there is an interaction energy between k real coordinates and the complex coordinates. However, this is only consistent with (7.2) if the real eigenvalues are weighted by assigning their charge to equal $1/2$. Assuming this, after noting from (7.2) that before weighting the k real coordinates themselves interact via the pair potential $-\log|\lambda - \lambda'|^2$, the correct term in the Boltzmann factor is obtained.

We turn our attention now to the one-body terms involving the weight in (1.10), which when written out in full read

$$e^{-\sum_{j=1}^{k} \lambda_j^2/2} e^{-\sum_{j=1}^{(N-k)/2}(x_j^2+y_j^2)} \prod_{j=1}^{(N-k)/2} e^{2y_j^2} \operatorname{erfc}(\sqrt{2}y_j). \tag{7.3}$$

From a two-dimensional Coulomb gas viewpoint, we see that the first two exponential terms result from a coupling between the charges (particles on the real line having charge $1/2$) confined to the semi-disk $|z| < \sqrt{N}$, $y > 0$, and with a neutralising background of uniform density $\rho = 1/\pi$ filling the semi-disk. In the random matrix problem, this implies that the global-scaled eigenvalues obey the circular law (2.17). The final term can be interpreted as the coupling of the complex coordinates to a smeared out charge on the real axis; asymptotically each factor decays as $1/y_j$. This is then cancelled by the term in (7.1) corresponding to the interaction between complex coordinate and its image.

7.3 Generalised Partition Function and Probabilities

Integrating (1.10) over the specified range of the eigenvalues gives the probability that a GinOE matrix has precisely k real eigenvalues, $p_{k,N}^{\mathrm{GinOE}}$ say. However, to perform the integrations in general, more theory is required—that of skew-orthogonal polynomials—which is to be developed below. An exception is the case $k = N$ [211].

Proposition 7.1 *We have* $p_{N,N}^{\mathrm{GinOE}} = 2^{-N(N-1)/4}$.

Proof According to (1.10) with $k = N$,

$$
p_{N,N}^{\mathrm{GinOE}} = \frac{C_N^{\mathrm{g}}}{N!} \int_{-\infty}^{\infty} d\lambda_1 \cdots \int_{-\infty}^{\infty} d\lambda_N \, e^{-\sum_{j=1}^{N} \lambda_j^2/2} \prod_{1 \le j < k \le N} |\lambda_k - \lambda_j|.
$$

This multiple integral—the $\beta = 1$ case of what is referred to as Mehta's integral—has a known evaluation in terms of products of gamma functions (see [237, Proposition 4.7.1]), which implies the result.   □

Pfaffian structures associated with integrations over (1.10) are most readily revealed by considering the so-called generalised partition function

$$
Z_{k,(N-k)/2}[u, v] = \Big\langle \prod_{l=1}^{k} u(\lambda_l) \prod_{l=1}^{(N-k)/2} v(x_l, y_l) \Big\rangle. \tag{7.4}
$$

Proposition 7.2 *Let* $\{p_{l-1}(x)\}_{l=1,\ldots,N}$ *be a set of monic polynomials, with* $p_{l-1}(x)$ *of degree* $l - 1$. *With the weight* $\omega^{\mathrm{g}}(x)$ *as in (1.9) let*

$$
\alpha_{j,k}^{\mathrm{g}} = \int_{-\infty}^{\infty} dx \, u(x) \int_{-\infty}^{\infty} dy \, u(y) \, (\omega^{\mathrm{g}}(x)\omega^{\mathrm{g}}(y))^{1/2} p_{j-1}(x) p_{k-1}(y) \mathrm{sgn}\,(y - x),
$$

$$
\beta_{j,k}^{\mathrm{g}} = 2i \int_{\mathbb{R}_+^2} dx dy \, v(x, y) \omega^{\mathrm{g}}(z)
$$

$$
\times \Big(p_{j-1}(x + iy) p_{k-1}(x - iy) - p_{k-1}(x + iy) p_{j-1}(x - iy) \Big).
$$

For k, N *even and with* C_N^{g} *as in (1.9) we have*

$$
Z_{k,(N-k)/2}[u, v] = C_N^{\mathrm{g}}[\zeta^{k/2}]\mathrm{Pf}[\zeta \alpha_{j,l}^{\mathrm{g}} + \beta_{j,l}^{\mathrm{g}}]_{j,l=1,\ldots,N}, \tag{7.5}
$$

where $[\zeta^p]f(\zeta)$ *denotes the coefficient of* ζ^p *in* $f(\zeta)$. *With*

$$
Z_N[u, v] := \sum_{k=0}^{N/2} Z_{2k,(N-2k)/2}[u, v], \tag{7.6}
$$

it follows that

$$
Z_N[u, v] = C_N^{\mathrm{g}} \mathrm{Pf}[\alpha_{j,k}^{\mathrm{g}} + \beta_{j,k}^{\mathrm{g}}]_{j,k=1,\ldots,N}. \tag{7.7}
$$

Proof The strategy to deduce (7.5) is to combine two integration methods involving determinants due to de Bruijn [110]. Details can be found in [237, proof of Proposition 15.10.3].   □

Remark 7.1 The case $u = v$ of (7.7) is due to Sinclair [502], who also considered the modification required for N odd. Subsequently, the N odd case of GinOE was considered in more detail in [104, 264, 509]. Introducing $v_k := \int_{-\infty}^{\infty} e^{-x^2/2} p_{k-1}(x) \, dx$, the N odd version of (7.7) reads

$$Z_N[u, v] = C_N^g \text{Pf} \begin{bmatrix} [\alpha_{j,k} + \beta_{j,k}] & [v_j] \\ [-v_l] & 0 \end{bmatrix}_{j,k=1,\ldots,N}. \tag{7.8}$$

However, below, for efficiency of presentation, we will always take N to be even as we further develop the theory relating to GinOE.

Define $Z_N(\zeta) := \sum_{k=0}^{N/2} \zeta^k Z_{2k,(N-2k)/2}[1, 1]$, which is the generating function for the probabilities $p_{2k,N}^{\text{GinOE}} = Z_{2k,(N-2k)/2}[1, 1]$ of there being exactly $2k$ real eigenvalues. Choosing $p_j(x)$ to be even for j even and odd for j odd, the Pfaffian in (7.7) can then be written as a determinant of half the original size [31],

$$Z_N(\zeta) = C_N^g \det \left[\alpha_{2j-1,2k}|_{u=1} + \beta_{2j-1,2k}|_{v=1} \right]_{j,k=1,\ldots,N/2}. \tag{7.9}$$

Moreover, further refining the choice of the p_j to be skew-orthogonal—see the following section for this notion and their expansion as monomials—allows (7.9) to be written in the explicit form [355]

$$Z_N(\zeta) = \det \left[\delta_{j,k} + \frac{(\zeta - 1)}{\sqrt{2\pi}} \frac{\Gamma(j + k - 3/2)}{\sqrt{\Gamma(2j - 1)\Gamma(2k - 1)}} \right]_{j,k=1,\ldots,N/2}. \tag{7.10}$$

As an application, the first term of the conjectured asymptotic formula

$$\frac{1}{\sqrt{N}} \log p_{N,0}^{\text{GinOE}} = -\frac{1}{\sqrt{2\pi}} \zeta\left(\frac{3}{2}\right) + \frac{C}{\sqrt{N}} + \cdots, \tag{7.11}$$

where $\zeta(x)$ denotes the Riemann zeta function and

$$C = \log 2 - \frac{1}{4} + \frac{1}{4\pi} \sum_{n=2}^{\infty} \frac{1}{n} \left(-\pi + \sum_{p=1}^{n-1} \frac{1}{p(n-p)} \right) \approx 0.0627, \tag{7.12}$$

as implied by results of [244] for the probability of a large gap in the real spectrum of bulk-scaled GinOE, was rigorously established. Beyond the case $k = o(N)$, the leading-order asymptotic behaviour of $\log p_{2k,N}^{\text{GinOE}}$, with $k = O(N)$, was derived in [185] using a Coulomb gas approach. A known arithmetic property of $\{p_{2k,N}^{\text{GUE}}\}$, namely that each member of the sequence is of the form $p_{2k,N} = r + s\sqrt{2}$, with r and s rational numbers which in reduced form consist of powers of 2 in the denominator [211], is (after minor manipulation) also evident from (7.10).

Another application of (7.10) is to differentiate with respect to ζ and set $\zeta = 1$. This gives for the expected number of real eigenvalues, $E_N^r := \sum_{k=0}^{N/2} 2k p_{2k,N}$, the explicit formulas [212]

$$E_N^{\mathrm{r}} = \sqrt{\frac{2}{\pi}} \sum_{k=1}^{N/2} \frac{\Gamma(2k - 3/2)}{\Gamma(2k - 1)} = \frac{1}{2} + \sqrt{\frac{2}{\pi}} \frac{\Gamma(N + 1/2)}{\Gamma(N)} \, {}_2F_1\left(\begin{matrix} 1, -1/2 \\ N \end{matrix} \Big| \frac{1}{2}\right),$$

(7.13)

where the validity of the hypergeometric expression can be checked by recurrence. The latter, which holds too for N odd, has the utility of implying the large N asymptotic expansion

$$E_N^{\mathrm{r}} - \frac{1}{2} \underset{N \to \infty}{\sim} \sqrt{\frac{2N}{\pi}} \left(1 - \frac{3}{8N} - \frac{3}{128N^2} + \cdots\right).$$

(7.14)

(We refer to [115, 123] for recent works on extending this to the general spectral moments.) We will see later (working below Proposition 7.6) that the leading-order value $\sqrt{2N/\pi}$ is consistent with the bulk density of real eigenvalues equalling the value $1/\sqrt{2\pi}$, and being supported on the interval $[-\sqrt{N}, \sqrt{N}]$ to leading-order. Relating to this, we already know from the Coulomb gas viewpoint of Sect. 7.2 that the density of complex eigenvalues is $1/\pi$ supported on the disk of radius $\sqrt{N}$. We also mention that the leading-order asymptotic of (7.14) has been extended to a class of i.i.d. real random matrices [515].

The variance $(\sigma_N^{\mathrm{r}})^2$ of the distribution of the real eigenvalues can, using (7.10), be expressed in terms of a double summation over gamma functions; however, the large N asymptotic form is not easy to then deduce. Later, in Proposition 7.7, an alternative method will be used which shows $(\sigma_N^{\mathrm{r}})^2 \sim (2 - \sqrt{2})E_N^{\mathrm{r}}$, and so in particular the variance diverges as $N \to \infty$. This fact, together with a corollary of (7.10) regarding the zeros of $Z_N(\zeta)$, can be used to deduce that upon centring and scaling, the probability distribution for the number of real eigenvalues satisfies a local central limit theorem. This is in the spirit of Proposition 3.2, although to our knowledge a local central limit theorem in this context has not appeared previously in the literature. For the corresponding central limit theorem in a more general setting, see [500] and Sect. 7.7 below.

Proposition 7.3 *We have that* $\{p_{2k,N}^{\mathrm{GinOE}}\}$ *satisfies the local central limit theorem*

$$\lim_{N \to \infty} \sup_{x \in (-\infty, \infty)} \left| \sigma_N^{\mathrm{r}} \, p_{2k,N} \, |_{2k=[\sigma_N^{\mathrm{r}}x + E_N^{\mathrm{r}}]} - \frac{1}{\sqrt{2\pi}} e^{-x^2/2} \right| = 0.$$

(7.15)

Proof It is established in [355] that the matrix formed by the ratio of gamma functions in (7.10) is positive definite, and hence the zeros of $Z_N(\zeta)$ are all real. Moreover, they are negative real since $Z_N(\zeta)$ is the generating function for the probabilities $\{p_{2k,N}^{\mathrm{GinOE}}\}$. As remarked above, we know too that for $N \to \infty$ the variance of this probability distribution diverges. Combining these two facts gives, upon appealing to [84, Th. 2], the stated result.   □

Remark 7.2 It has been commented in Sect. 7.2 that the global density of the eigenvalues obeys the circular law in the limit $N \to \infty$. However, the fact that the expected value of real eigenvalues is proportional to $\sqrt{N}$, whereas the corresponding distribution function takes on non-zero values for all (even) values of the number up to N creates, for finite N, a so-called Saturn effect whereby the support of the real eigenvalues visibly overshoots the circular law; see [69, Fig. 1].

7.4 Skew-Orthogonal Polynomials

In (7.7) set $\gamma_{j,k} := \alpha_{j,k}|_{u=1} + \beta_{j,k}|_{v=1}$. We see that associated with $\gamma_{j,k}$ is the skew inner product

$$
\langle f, g \rangle_{s,0}^{\mathrm{g}} := \int_{-\infty}^{\infty} dx \int_{-\infty}^{\infty} dy \, (\omega^{\mathrm{g}}(x)\omega^{\mathrm{g}}(y))^{1/2} f(x)g(y)\mathrm{sgn}\,(y-x),
$$
$$
+ 2i \int_{\mathbb{C}_+} d^2 z \, \omega^{\mathrm{g}}(z)\Big(f(z)g(\bar{z}) - g(z)f(\bar{z}) \Big). \quad (7.16)
$$

Here the subscripts on the inner product indicate that it is skew symmetric and associated with the GinOE. A Pfaffian is well defined for anti-symmetric matrices of even size only. The analogue of a diagonal form in this setting is the direct sum of $N/2$ two-by-two anti-symmetric matrices. We would like to choose the polynomials $\{p_{l-1}(x)\}$ in the definition of $[\gamma_{j,k}]$ so that it takes on such a direct sum form. Equivalently, we seek a monic polynomial basis that skew-diagonalises the skew inner product (7.16). This requires that

$$
\gamma_{2j,2k} = \gamma_{2j-1,2k-1} = 0, \qquad \gamma_{2j-1,2k} = -\gamma_{2k,2j-1} = r_{j-1}\delta_{j,k}, \qquad (7.17)
$$

where $\{r_{j-1}\}$ (the skew norms) are the nonzero elements in the block diagonal form

$$
[\gamma_{j,k}] = \oplus_{j=1}^{N/2} \begin{bmatrix} 0 & r_{j-1} \\ -r_{j-1} & 0 \end{bmatrix}.
$$

If these relations are satisfied, the polynomials are said to be skew-orthogonal.

One observes that (7.17) does not uniquely determine the polynomials. Thus the odd polynomials $p_{2n+1}(z)$ can be replaced by $p_{2n+1}(z) + cq_{2n}(z)$ for any constant c (see e.g. [237, Sect. 6.1.1]). On the other hand, the existence of $\{q_m\}_{m \geq 0}$ follows from a Gram–Schmidt skew-orthogonalisation procedure [26, Th. 2.4]. Nevertheless, since this procedure requires us to evaluate certain Pfaffians, it is not very useful for the actual computation of the skew-orthogonal polynomials.

Crucial for determining the skew-orthogonal polynomials relating to the skew inner product (7.16), and various generalisations associated with ensembles related to the GinUE to be discussed below, are their realisations as $2n \times 2n$ GinOE matrix averages [33, Eqs. (4.6)–(4.7)]:

$$p_{2n}(z) = \langle \det(z\mathbb{I}_{2n} - G) \rangle, \quad p_{2n+1}(z) = z p_{2n}(z) + \langle \det(z\mathbb{I}_{2n} - G) \operatorname{Tr} G \rangle. \quad (7.18)$$

The matrix averages in (7.18) are simple to evaluate [254].

Proposition 7.4 *Let the random matrix* $G = [g_{jk}]$ *such that the average of distinct pairs is the same as the product of the averages of the individual entries. Suppose that the distribution of each* g_{jk} *is the same as that for* $-g_{jk}$. *Then the formulas of (7.18) simplify,*

$$p_{2n}(z) = z^{2n}, \quad p_{2n+1}(z) = z^{2n+1} - \langle \operatorname{Tr} G^2 \rangle z^{2n-1}. \quad (7.19)$$

Also, in the case of GinOE, $\langle \operatorname{Tr} G^2 \rangle = 2n$ *which allows (7.19) to be written*

$$p_{2n}^{g}(z) = z^{2n}, \quad p_{2n+1}^{g}(z) = -e^{z^2/2} \frac{d}{dz} e^{-z^2/2} p_{2n}(z). \quad (7.20)$$

For the normalisation we have

$$r_{n-1}^{g} = 2\sqrt{2\pi}\,\Gamma(2n-1). \quad (7.21)$$

Proof The expansion of $\det(z\mathbb{I}_{2n} - G)$ gives a term z^{2n} plus terms involving lower-order powers of z, the coefficients of which are linear with respect to any single matrix element. Averaging over the matrix element using the assumed invariance of the distribution by negation must give zero. In relation to $\langle \det(z\mathbb{I}_{2n} - G) \operatorname{Tr} G \rangle$, the assumed invariance of the distribution by negation implies that a nonzero value will result only for terms in the expansion of $\det(z\mathbb{I}_{2n} - G)$ which contain single powers of the matrix elements, these terms being in total $z^{2n-1} \operatorname{Tr} G$. The value of $\langle \operatorname{Tr} G^2 \rangle$ for G a member of $2n \times 2n$ GinOE is immediate from the fact that the elements all have unit variance. For the normalisation r_n, from the block diagonal form we have $\operatorname{Pf}[\gamma_{j,k}] = \prod_{j=0}^{N/2-1} r_j$. Substituting in (7.7), and noting that as a result of its interpretation as a sum over probabilities we have $Z_N[1, 1] = 1$, allows (7.21) to be verified. $\qquad \square$

Remark 7.3

1. The skew inner product

$$\alpha_{j,k} := \int_{-\infty}^{\infty} dx \int_{-\infty}^{\infty} dy \, (\omega^{g}(x)\omega^{g}(y))^{1/2} p_{j-1}(x) p_{k-1}(y) \operatorname{sgn}(y-x),$$

is well known in the theory of the Gaussian orthogonal ensemble (GOE). The corresponding skew-orthogonal polynomials are then given in terms of Hermite polynomials (see e.g. [6])

$$p_{2j}^{\text{GOE}}(x) = 2^{-2j} H_{2j}(x), \quad p_{2j+1}^{\text{GOE}}(x) = -e^{x^2/2} \frac{d}{dx}\left(e^{-x^2/2} p_{2j}^{\text{GOE}}(x)\right);$$

cf. (7.19) and (7.20).

2. Let $\phi(z) = |\frac{1}{2}(z + \sqrt{z^2 - 4})|^{-2s}$, $s > N$ and define the skew inner product $\langle f, g \rangle^{\phi}_{s,0}$ as in (7.16) but with $\omega^{g}(z)$ replaced by $\phi(z)$ throughout. This arose in a study of the so-called Mahler measure of random polynomials [503], and the corresponding skew-orthogonal polynomials were required. Without a random matrix underpinning, there is no meaning to (7.18). Nonetheless, in [503, Th. 2.1], the skew-orthogonal polynomials have been explicitly determined in terms of a single Chebyshev polynomial (for $p_{2n}(z)$) and a sum of two Chebyshev polynomials (for $p_{2n+1}(z)$).

7.5   Correlation Functions for the Real Eigenvalues

From a statistical mechanics viewpoint, the eigenvalues of GinOE matrices form a two-component system consisting of the real eigenvalues, and of the complex eigenvalues. Both form Pfaffian point processes. Details relating to the real–real correlations are considered here, with the complex–complex and complex–real correlations treated in the next section.

As a starting point, use will be made of the definition (7.4) of the generalised partition function $Z_N[u, v]$, containing arbitrary functions $u(x)$ and $v(x, y)$. Generally, for integrations involving arbitrary functions, the remaining factors of the integrand can be extracted by functional differentiation,

$$\frac{\delta}{\delta a(x)} \int_{-\infty}^{\infty} a(y) f(y) \, dy = f(x). \tag{7.22}$$

All correlations can be obtained from $Z_N[u, v]$ in (7.6) using the operation of functional differentiation. As an explicit example, for the m-point correlation function of the real eigenvalues, $\rho^{r}_{(m),N}$ say, we have

$$\rho^{r}_{(m),N}(x_1, \ldots, x_m) = \frac{\delta^m}{\delta u(x_1) \cdots \delta u(x_m)} Z_N[u, v] \bigg|_{u=v=1}. \tag{7.23}$$

This can be checked from (7.22) and the definition of $\rho^{r}_{(m),N}$ as the sum over (1.10) for $k = m, \ldots, N$, with $\lambda_l = x_l$ ($l = 1, \ldots, m$), each term weighted by the combinatorial factor $k!/(k - m)!$, and the variables $\{\lambda_l\}_{l=m+1,\ldots,k}$ each integrated over $\mathbb{R}$. Starting with (7.7), and choosing the polynomials therein to have the skew-orthogonality property (7.17), a Pfaffian formula for $\rho^{r}_{(m),N}$ can be deduced [104, 266, 506].

Proposition 7.5 *Let* $\{p_j(x)\}$ *be the skew-orthogonal polynomials for GinOE as specified in (7.20), and for* x *real set* $\omega(x) = \omega^{g}(x)$ *as implied by (1.9) (specifically then* $\omega(x) = e^{-x^2}$). *Use these polynomials and this weight to define* $\Phi_k(x) = \int_{-\infty}^{\infty} \mathrm{sgn}(x - y) p_k(y)(\omega(y))^{1/2} \, dy$, *which in turn is used to define*

$$S_N^{\mathrm{r}}(x, y) = \sum_{k=0}^{N/2-1} \frac{(\omega(y))^{1/2}}{r_k} \left(\Phi_{2k}(x) p_{2k+1}(y) - \Phi_{2k+1}(x) p_{2k}(y) \right),$$

$$D_N^{\mathrm{r}}(x, y) = \frac{1}{2} \frac{\partial}{\partial x} S^{\mathrm{r}}(x, y), \qquad \tilde{I}_N^{\mathrm{r}}(x, y) = \mathrm{sgn}(y - x) - 2 \int_x^y S^{\mathrm{r}}(x, z) \, dz. \quad (7.24)$$

(An equivalent form of $\tilde{I}^{\mathrm{r}}(x, y)$ is to replace the integral therein by $\int_{-\infty}^{\infty} \mathrm{sgn}(y - z) S^{\mathrm{r}}(x, z) \, dy$.) We have

$$\rho_{(m),N}^{\mathrm{r}}(x_1, \ldots, x_m) = \mathrm{Pf}\,[\mathcal{K}_N^{\mathrm{r}}(x_j, x_k)]_{j,k=1,\ldots,m},$$

$$\mathcal{K}_N^{\mathrm{r}}(x, y) := \begin{bmatrix} D_N^{\mathrm{r}}(x, y) & S_N^{\mathrm{r}}(x, y) \\ -S_N^{\mathrm{r}}(y, x) & \tilde{I}_N^{\mathrm{r}}(x, y) \end{bmatrix}. \quad (7.25)$$

Before presenting the proof, it is instructive to outline a derivation of the determinantal expression for the m-point eigenvalue correlation function of GinUE (cf. (2.9) and Proposition 2.2), the main steps of which can be generalised to GinOE [432]. These steps are:

(i) Show that for the monic polynomials $p_j(z) = z^j$,

$$\left\langle \prod_{l=1}^{N} u(z_l) \right\rangle_{\mathrm{GinUE}}$$

$$= \det\left[\delta_{j,k} + \frac{1}{\langle p_{j-1}, p_{j-1} \rangle} \int_{\mathbb{C}^2} e^{-|z|^2} (u(z) - 1) z^{j-1} \bar{z}^{k-1} \, d^2z \right]_{j,k=1}^{N},$$

where $\langle p_{j-1}, p_{k-1} \rangle := \int_{\mathbb{C}^2} e^{-|z|^2} z^{j-1} \bar{z}^{k-1} \, d^2z$.

(ii) With $\mathbb{I}$ denoting the identity operator, use the operator theoretic identity

$$\det(\mathbb{I} + AB) = \det(\mathbb{I} + BA), \quad (7.26)$$

valid whenever the determinant is well defined (see [183]) to rewrite the formula in (i) as

$$\left\langle \prod_{l=1}^{N} u(z_l) \right\rangle_{\mathrm{GinUE}} = \det(\mathbb{I} + K_u),$$

where K_u is the integral operator on $\mathbb{C}$ with kernel $(u(z_2) - 1) K_N(z_1, z_2)$, with $K_N(z_1, z_2)$ given by (2.9).

(iii) Use functional differentiation to now extract the m-point eigenvalue correlation function from the Fredholm expansion [532]

$$\det(\mathbb{I} + K_u) = 1 + \sum_{k=1}^{N} \frac{1}{k!} \int_{\mathbb{C}} dz_1 \, u(z_1) \cdots \int_{\mathbb{C}} dz_N \, u(z_N) \det[K_N(z_j, z_l)]_{j,l=1,\ldots,k}.$$

Proof of Proposition 7.5 In relation to steps (i) and (ii), we follow [237, Proof of Proposition 6.3.6]. Starting with (7.7), we set $v = 1, u = 1 + \hat{u}$ and choose $\{p_{l-1}(x)\}$ to have the skew-orthogonality property (7.17). Also, we introduce the notations $\psi_j(x) = e^{-x^2/2} p_{j-1}(x)$ and $\varepsilon[f](x) = \int_{-\infty}^{\infty} \operatorname{sgn}(x - y) f(y)\, dy$. Then, with

$$\gamma_{j,k} := \alpha_{j,k}|_{u=1} + \beta_{j,k}|_{v=1},$$

we have

$$\alpha_{jk} + \beta_{jk} = \gamma_{jk}$$
$$- \int_{-\infty}^{\infty} \left(\hat{u}(x)\psi_j(x)\varepsilon[\psi_k](x) - \hat{u}(x)\psi_k(x)\varepsilon[\psi_j](x) - \hat{u}(x)\psi_k(x)\varepsilon[\hat{u}\psi_j](x) \right) dx.$$

Next, we introduce the further notation $G_{2j-1}(x) = \psi_{2j}(x)$, $G_{2j}(x) = -\psi_{2j-1}(x)$, multiply the even rows by -1 and use the facts that $[\gamma_{jk}]$ has the block diagonal form as specified below (7.17) and that the square of the Pfaffian is the corresponding determinant, to deduce

$$(Z_N[1 + \hat{u}, 1])^2 = \prod_{j=1}^{N} r_{j-1}^2 \det \left[\delta_{j,k} + \frac{1}{r_{[(j-1)/2]}} \right.$$
$$\times \left. \int_{-\infty}^{\infty} \left(\hat{u}(x)G_j(x)\varepsilon[\psi_k](x) - \hat{u}(x)\psi_k(x)\varepsilon[G_j](x) - \hat{u}(x)\psi_k(x)\varepsilon[\hat{u}G_j](x) \right) dx \right].$$
$$(7.27)$$

This accomplishes step (i).

For step (ii), the key observation, due to [526], is that the determinant in (7.27) can be written as $\det(\mathbb{I}_N + AB)$ for A and B appropriate matrix operators. Specifically, A is the $N \times 2$ matrix-valued integral operator, with row j of the kernel corresponding to the pair

$$\frac{1}{r_{[(j-1)/2]}} \left(-\hat{u}(y)\varepsilon[G_j](y) - \hat{u}(y)\varepsilon[\hat{u}G_j](y), \, \hat{u}(y)G_j(y) \right).$$

The operator B is a certain matrix multiplication. Specifically, the matrix has size $2 \times N$ matrix with first row $\psi_k(y)$, and second row $\varepsilon[\psi_k](y)$.

We are now ready to apply (7.26). With A and B as above we see that the operator $\mathbb{I} + BA$ is the 2×2 matrix integral operator

$$\begin{bmatrix} 1 - \sum_{j=1}^{N} \left(\tilde{\psi}_j \otimes f\varepsilon G_j + \tilde{\psi}_j \otimes f\varepsilon(fG_j) \right) & \sum_{j=1}^{N} \tilde{\psi}_j \otimes fG_j \\ -\sum_{j=1}^{N} \left(\varepsilon\tilde{\psi}_j \otimes f\varepsilon G_j + \varepsilon\tilde{\psi}_j \otimes f\varepsilon(fG_j) \right) 1 + \sum_{j=1}^{N} \varepsilon\tilde{\psi}_j \otimes fG_j \end{bmatrix}$$
$$= \begin{bmatrix} 1 - \sum_{j=1}^{N} \tilde{\psi}_j \otimes f\varepsilon G_j & \sum_{j=1}^{N} \tilde{\psi}_j \otimes fG_j \\ -\sum_{j=1}^{N} \varepsilon\tilde{\psi}_j \otimes f\varepsilon G_j - \varepsilon f & 1 + \sum_{j=1}^{N} \varepsilon\tilde{\psi}_j \otimes fG_j \end{bmatrix} \begin{bmatrix} 1 & 0 \\ \varepsilon f & 1 \end{bmatrix}. \quad (7.28)$$

Here $\tilde{\psi}_j := \psi_j / r_{[(j-1)/2]}$ and the notation $a \otimes b$ denotes the integral operator with kernel $a(x)b(y)$. The determinant of the second matrix in the second expression is equal to 1, and so it can be replaced by the elementary anti-symmetric matrix obtained by setting $z = 0$, $w = 1$ in (1.1), without changing its value. Doing this we can identify the kernel of the matrix integral operator as being given by an anti-symmetric matrix so the square root of the determinant is a Pfaffian. Hence

$$Z_N[1 + \hat{u}, 1] = \mathrm{Pf}\left(\begin{bmatrix} 0 & \mathbb{I} \\ -\mathbb{I} & 0 \end{bmatrix} + \begin{bmatrix} D^r(x, \cdot)[\hat{u}] & S^r(x, \cdot)[\hat{u}] \\ -S^r(\cdot, x)[\hat{u}] & \tilde{I}^r(x, \cdot)[\hat{u}] \end{bmatrix} \right), \tag{7.29}$$

where the Pfaffian is defined in terms of the product of the eigenvalues. In the second matrix the operators act by integration over the non-specified variable, weighted by $\hat{u}$, and otherwise have kernels given by the corresponding entries in (7.25).

We are now up to stage (iii) of the outlined strategy. Analogous to the theory of Fredholm determinants, this quantity—a Fredholm Pfaffian—can be expanded in a series of Pfaffians of scalar matrices to read [466]

$$Z_N[1 + \hat{u}, 1] = 1 + \sum_{m=1}^{N} \frac{1}{m!} \int_{-\infty}^{\infty} dx_1 \, \hat{u}(x_1) \cdots \int_{-\infty}^{\infty} dx_m \, \hat{u}(x_m)$$
$$\times \mathrm{Pf} \, [\mathcal{K}_N^r(x_j, x_k)]_{j,k=1,\ldots,m}. \tag{7.30}$$

Here the 2×2 anti-symmetric matrix kernel $\mathcal{K}_N^r$ is determined by the kernel of the second matrix integral operator in (7.29). We note that the formula (7.23) remains valid with each $u(x)$ in the functional derivative replaced by $\hat{u}$, with the latter now set equal to zero after the derivatives have been computed. Applying this modified formula to (7.30), we read off the sought Pfaffian formula (7.25) for the correlations. $\square$

With the polynomials $\{p_{j-1}(x)\}$ given according to (7.20), the matrix elements in (7.24) can be made explicit, with knowledge of $S^r(x, y)$ sufficing.

Proposition 7.6 *We have*

$$S_N^r(x, y) = \frac{e^{-(x^2+y^2)/2}}{\sqrt{2\pi}} \sum_{k=0}^{N-2} \frac{(xy)^k}{k!} + \frac{e^{-y^2/2}}{2\sqrt{2\pi}} \frac{y^{N-1} \Phi_{N-2}(x)}{(N-2)!}$$
$$= \frac{e^{-(x^2+y^2)/2}}{\sqrt{2\pi}} \left(e^{xy} \frac{\Gamma(N-1; xy)}{\Gamma(N-1)} + \frac{1}{2} (\sqrt{2}y)^{N-1} e^{x^2/2} \mathrm{sgn}(x) \frac{\gamma(N/2 - 1/2; x^2/2)}{\Gamma(N-1)} \right). \tag{7.31}$$

Proof The key to obtaining the first equality is to first note from the derivative form of $p_{2n+1}(z)$ given by (7.20) that

$$\Phi_{2k+1}(x) = -2e^{-x^2/2} x^{2k},$$

which implies the summation identity

$$-\sum_{k=0}^{N/2-1} \frac{1}{r_k} \Phi_{2k+1}(x) p_{2k}(y) = \frac{1}{\sqrt{2\pi}} e^{-x^2/2} \sum_{k=0}^{N/2-1} \frac{(xy)^{2k}}{(2k)!}. \tag{7.32}$$

Also needed is the further derivative formula

$$\frac{2(k+1)}{r_{k+1}} p_{2k+2}(x) - \frac{1}{r_k} p_{2k}(x) = -\frac{1}{2\sqrt{2\pi}} \frac{1}{(2k+1)!} e^{x^2/2} \frac{d}{dx} e^{-x^2/2} x^{2k+1}, \tag{7.33}$$

which implies the further summation identity

$$\sum_{k=0}^{N/2-1} \frac{1}{r_k} \Phi_{2k}(x) p_{2k+1}(y) = \frac{1}{\sqrt{2\pi}} e^{-x^2/2} \sum_{k=0}^{N/2-2} \frac{(xy)^{2k+1}}{(2k+1)!} + \frac{1}{r_{N/2-1}} \Phi_{N-2}(x) y^{N-1}. \tag{7.34}$$

Adding (7.32) and (7.34) establishes the first equality.

In relation to the second equality, by identifying terms with the first equality, the task becomes to show

$$\Phi_{N-2}(x) = 2^{(N-1)/2} \text{sgn}(x)\gamma(N/2-1/2; x^2/2).$$

Since from the definition in Proposition 7.5,

$$\Phi_{N-2}(x) = 2 \int_0^x e^{-y^2/2} y^{N-2} \, dy,$$

this is readily verified.   □

Setting $x = y$ in (7.31) gives the density of real eigenvalues

$$\rho^r_{(1),N}(x) = \frac{1}{\sqrt{2\pi}} \left(\frac{\Gamma(N-1; x^2)}{\Gamma(N-1)} + \frac{1}{2}(\sqrt{2}x)^{N-1} e^{-x^2/2} \text{sgn}(x) \frac{\gamma(N/2-1/2; x^2/2)}{\Gamma(N-1)} \right). \tag{7.35}$$

Now setting $x = \sqrt{N}\tilde{x}$ corresponds to a global scaling of the density of real eigenvalues. One can verify that the large N limit is determined entirely by the first term, with the simple result

$$\lim_{N\to\infty} \rho^r_{(1),N}(\sqrt{N}\tilde{x}) = \frac{1}{\sqrt{2\pi}} \chi_{|\tilde{x}|<1}. \tag{7.36}$$

Knowledge of (7.36) indicates that local limit theorems associated with (7.31) will depend on the choice of origin. In keeping with this, we find that with the latter chosen on the real axis at $v\sqrt{N}$ with $|v| < 1$ the final term does not contribute, while the ratio of the incomplete gamma function in the first term equals unity, giving for the bulk limit

$$S_\infty^{r,b}(x, y) := \lim_{N \to \infty} S_N^r(v\sqrt{N} + x, \sqrt{N} + y) = \frac{1}{\sqrt{2\pi}} e^{-(x-y)^2/2}. \qquad (7.37)$$

Setting $x = y$ in this gives for bulk density the value $\rho_{(1),\infty}^{r,b}(x) = 1/\sqrt{2\pi}$, as already anticipated below (7.14). The result (7.37) gives for the bulk limiting kernel in (7.25) the functional form

$$K_\infty^{r,b}(x, y) = \begin{bmatrix} \frac{1}{2\sqrt{2\pi}}(y - x)e^{-(x-y)^2/2} & \frac{1}{\sqrt{2\pi}}e^{-(x-y)^2/2} \\ -\frac{1}{\sqrt{2\pi}}e^{-(x-y)^2/2} & \mathrm{sgn}(x - y)\mathrm{erfc}(|x - y|/\sqrt{2}) \end{bmatrix}. \qquad (7.38)$$

The result (7.37) breaks down for $v = \pm 1$, as $\pm\sqrt{N}$ are the (leading-order) boundaries of the support of the real eigenvalues. Then the uniform asymptotic expansion [529]

$$\frac{\gamma(M - j + 1; M)}{\Gamma(M - j + 1)} \underset{M \to \infty}{\sim} \frac{1}{2}\left(1 + \mathrm{erf}\left(\frac{j}{\sqrt{2M}}\right)\right) \qquad (7.39)$$

gives for the edge scaling limit

$$\lim_{N \to \infty} S_N^r(\sqrt{N} + x, \sqrt{N} + y) = \frac{1}{\sqrt{2\pi}}\left(\frac{1}{2}e^{-(x-y)^2/2}\left(1 - \mathrm{erf}\frac{x+y}{\sqrt{2}}\right) + \frac{e^{-y^2}}{2\sqrt{2}}(1 + \mathrm{erf}\,x)\right). \qquad (7.40)$$

In particular, setting $x = y$ gives for the edge-scaled density

$$\rho_{(1),\infty}^{r,e}(x) = \frac{1}{2\sqrt{2\pi}}\left(1 - \mathrm{erf}(\sqrt{2}x) + \frac{e^{-x^2}}{\sqrt{2}}(1 + \mathrm{erf}\,x)\right). \qquad (7.41)$$

For $x, y \to -\infty$ (7.40) reclaims the bulk limiting form (7.37).

For finite N, the density of real eigenvalues $\rho_{(1),N}^r(x) = S_N^r(x, x)$ was first calculated in [212]. The starting point was the formula

$$\rho_{(1),N}^r(x) = \frac{e^{-x^2/2}}{2^{N/2}\Gamma(N/2)}\langle|\det(G - x\mathbb{I}_{N-1})|\rangle, \qquad (7.42)$$

where the average is over $(N - 1) \times (N - 1)$ GinOE matrices. This can be deduced from (1.10), although in [212] a method called eigenvalue deflation was used, whereby a Householder transformation is applied to reduce all entries in the final column, except the last entry, to zero. The formula $E_N^r = \int_{-\infty}^{\infty} \rho_{(1),N}^r(x)\,dx$ can then be used to deduce (7.13).

Remark 7.4

1. Up to proportionality the term involving the summation in the first equality of (7.31) is recognised as the correlation kernel $K_N(w, z)$ for GinUE (2.10), with w, z real and $N \mapsto N - 1$. Using this notation, the results of Proposition 7.4 for the skew-orthogonal polynomials substituted into the definition of S_N^r in (7.24) show that it is furthermore true that [254, Sect. 4.2]

$$S_N^{r,g}(x, y) = \frac{1}{2\sqrt{2\pi}} \int_{-\infty}^{\infty} (y - s)\mathrm{sgn}(x - s) K_{N-1}(y, s)\, ds. \tag{7.43}$$

2. The entry $D_N^r(x, y)$ of K_N^r in (7.25) can be multiplied by any scalar $c \neq 0$, provided that the entry $\tilde{I}_N^r(x, y)$ is multiplied by the scalar $1/c$ without changing the value of $\rho_{(m),N}^r$. This follows from the fact for B a $2m \times 2m$ symmetric matrix, and A a $2m \times 2m$ anti-symmetric matrix, $\mathrm{Pf}(BAB^T) = \det(B)\mathrm{Pf}(A)$ (for a derivation see e.g. [237, Exercises 6.1 q.1]).

7.6 Correlation Functions for Complex Eigenvalues

For finite N the complex–complex correlation function is formally the same as (7.25), but now with the entries of the 2×2 correlation kernel $\mathcal{K}_N^c$, but now with the need to only introduce a single scalar kernel $S_N^c(w, z)$. For this purpose, set $q_j(z) = (\omega^g(z))^{1/2} p_j(z)$, where $\{p_j(z)\}$ are the skew-orthogonal polynomials of Proposition 7.4. The definition of the scalar kernel is then

$$S_N^c(w, z) = 2i \sum_{j=0}^{N/2-1} \frac{1}{r_j}\left(q_{2j}(w)q_{2j+1}(\bar{z}) - q_{2j+1}(w)q_{2j}(\bar{z}) \right), \tag{7.44}$$

with the matrix kernel then being given by [104, 266], [432, Sect. 4.5]

$$\mathcal{K}_N^c(w, z) = \begin{bmatrix} -i\,S_N^c(w, \bar{z}) & S_N^c(w, z) \\ -S_N^c(z, w) & i\,S_N^c(\bar{w}, z) \end{bmatrix}. \tag{7.45}$$

The results of Proposition 7.4 substituted in (7.44) shows

$$S_N^c(w, z) = \frac{i(\bar{z} - w)}{\sqrt{2\pi}}\left(\mathrm{erfc}(\sqrt{2}u)e^{2u^2}\mathrm{erfc}(\sqrt{2}y)e^{2y^2} \right)^{1/2} K_{N-1}(w, z), \tag{7.46}$$

where K_{N-1} corresponds to the correlation kernel for GinUE (2.10) with $N \mapsto N - 1$ and $y = \mathrm{Im}\, z$, $u = \mathrm{Im}\, w$. In particular, this implies

$$\rho_{(1),N}^c(z) = S_N^c(z, z) = \sqrt{\frac{2}{\pi}}\, y\, \mathrm{erfc}(\sqrt{2}y)e^{2y^2} \frac{\Gamma(N - 1; x^2 + y^2)}{\Gamma(N - 1)}, \tag{7.47}$$

which was first derived in [211] by (effectively) obtaining a complex analogue of (7.42),

$$\rho_{(1),N}^c(z) = 2y\omega^g(z) \frac{C_{N-1}^g}{C_N^g}\left\langle |\det(G - z\mathbb{I}_{N-1})|^2 \right\rangle, \tag{7.48}$$

with the average over $(N-1) \times (N-1)$ GinUE matrices. (This approach can also be utilised for the elliptic Ginibre ensembles [37].) The global scaling limit of (7.47) is readily computed to give [211, Th. 6.3]

$$\lim_{N \to \infty} \rho^c_{(1),N}(\sqrt{N}z) = \begin{cases} \frac{1}{\pi}, & |z| < 1, \\ 0, & |z| > 1, \end{cases} \tag{7.49}$$

in keeping with the circular law (2.17). Moreover, in the neighbourhood of the boundary, but away from the real axis, the edge scaling reclaims the edge-scaled GinUE result (2.19) with $(x_1, y_1) = (x_2, y_2)$); see [104].

Away from the boundary but still in the vicinity of the real axis, the bulk-scaled limit of the correlation kernel is [104, 266]

$$\mathcal{K}^c_\infty(w, z) = \frac{1}{\sqrt{2\pi}} (\text{erfc}(\sqrt{2}v)\text{erfc}(\sqrt{2}y))^{1/2} \tag{7.50}$$

$$\times \begin{bmatrix} (z-w)e^{-(w-z)^2/2} & i(\bar{z}-w)e^{-(w-\bar{z})^2/2} \\ i(z-\bar{w})e^{-(\bar{w}-z)^2/2} & -(\bar{z}-\bar{w})e^{-(\bar{w}-\bar{z})^2/2} \end{bmatrix}, \tag{7.51}$$

where $w = u + iv$, $z = x + iy$ with $v, y > 0$. In particular, we read off from the latter of these the formula for the bulk density of complex eigenvalues

$$\rho^c_{(1),\infty}(z) = \sqrt{\frac{2}{\pi}} \, \text{erfc}(\sqrt{2}y) y e^{2y^2}, \tag{7.52}$$

which tends to the constant $1/\pi$ as $y \to \infty$, as required by the circular law (7.49).

The matrix kernels determining the real–real and complex–complex correlations both appear in the formula for the (k_1, k_2)-point correlation function for k_1 real eigenvalues $x_1, \ldots, x_{k_1}$ and k_2 complex eigenvalues $z_1, \ldots, z_{k_2}$. Thus from [104, 266, 509] we have that this has the Pfaffian form

$$\rho_{(k_1,k_2),N}(x_1, \ldots, x_{k_1}; z_1, \ldots, z_{k_2})$$

$$= \text{Pf} \begin{bmatrix} [\mathcal{K}^r_N(x_j, x_l)]^{k_1}_{j,l=1} & [\mathcal{K}^{r,c}_N(x_j, z_l)]_{\substack{j=1,\ldots,k_1 \\ l=1,\ldots,k_2}} \\ \left(-[\mathcal{K}^{r,c}_N(x_j, z_l)]_{\substack{j=1,\ldots,k_1 \\ l=1,\ldots,k_2}}\right)^T & [\mathcal{K}^c_N(z_j, z_l)]^{k_2}_{j,l=1} \end{bmatrix}$$

generalising (7.25), where each of $\mathcal{K}^r_N, \mathcal{K}^{r,c}_N, \mathcal{K}^c_N$ is a 2×2 correlation kernel. The bulk scaling limits of $\mathcal{K}^r_N, \mathcal{K}^c_N$ are known from (7.38) and (7.50) respectively. In relation to $K^{r,c}_\infty$ we have [104, Cor. 9]

$$\mathcal{K}^{r,c}_\infty(x, w) = \frac{1}{\sqrt{2\pi}} (\text{erfc}(\sqrt{2}v))^{1/2} \begin{bmatrix} (w-x)e^{-(x-w)^2/2} & i(\bar{w}-x)e^{-(x-\bar{w})^2/2} \\ -e^{-(x-w)^2/2} & -ie^{-(x-\bar{w})^2/2} \end{bmatrix}.$$

Remark 7.5 The formula (7.48) has the generalisation [37, 509]

$$S_N^c(w, z) = i(\bar{z} - w)\Big(\omega^g(w)\omega^g(z)\Big)^{1/2} \frac{C_{N-1}^g}{C_N^g}\Big\langle \det\Big((G - w\mathbb{I}_{N-1})(\bar{G} - \bar{z}\mathbb{I}_{N-1})\Big)\Big\rangle.$$
(7.53)

With the matrix in the determinant of this expression generalised by the inclusion of a term $|\mu|^2\mathbb{I}_{N-1}$, and G now drawn from a general non-Hermitian ensemble, the corresponding average has been used in [111] to probe the eigenvalue density of G.

7.7 Fluctuation Formulas for Real Eigenvalues

The variance $(\sigma_N^r)^2$ of the distribution of the real eigenvalues can be expressed in terms of $\rho_{(2),N}^r(x, y)$, and its explicit expression as implied by (7.25) used to deduce its large N form [266].

Proposition 7.7 *In terms of the leading large N form of E_N^r, $E_N^r \sim \sqrt{2N/\pi}$ as given by (7.14), we have*

$$(\sigma_N^r)^2 \sim (2 - \sqrt{2})E_N^r.$$
(7.54)

Proof With $\rho_{(2),\infty}^T(z_1, z_2) := \rho_{(2),\infty}(z_1, z_2) - \rho_{(1),\infty}(z_1)\rho_{(1),\infty}(z_2)$, we have

$$(\sigma_N^r)^2 = \int_{-\infty}^{\infty} dx \int_{-\infty}^{\infty} dy \, (\rho_{(2),N}^{r,T}(x, y) + \rho_{(1),N}(x)\delta(x - y));$$
(7.55)

cf. first formula of (3.2). From the qualitative knowledge that $\rho_{(2),N}^{r,T}(x, y)$ is to leading-order translationally invariant for $x, y \in (-\sqrt{N}, \sqrt{N})$ with corresponding density $1/\sqrt{2\pi} =: \rho^r$ it follows that

$$(\sigma_N^r)^2 \sim E_N^r\Big(1 + \frac{1}{\rho^r}\int_{-\infty}^{\infty} \rho_{(2),\infty}^{r,T}(0, y)\, dy\Big).$$
(7.56)

We read off from (7.25) that

$$\rho_{(2),\infty}^{r,T}(x, y) = -S_\infty^r(x, y)S_\infty^r(y, x) + D_\infty^r(x, y)\tilde{I}_\infty(x, y),$$
(7.57)

where the explicit form of the functions herein are the matrix elements of (7.38). This allows the integral to be computed to give (7.54). □

The asymptotic formula (7.54) can be generalised to the setting of scaled linear statistics $F := \sum_{j=1}^N f(x_j/\sqrt{N})$ [231, 500].

Proposition 7.8 *Subject only to a mild growth condition on f, we have*

$$\text{Var } F \sim \Big(\frac{1}{2}\int_{-1}^1 f^2(x)\, dx\Big)(2 - \sqrt{2})E_N^r.$$
(7.58)

Proof (Sketch) The variance is given by the RHS of (7.55) with factors

$$f(x/\sqrt{N})f(y/\sqrt{N})$$

also in the integrand. A simple change of variables, and knowledge that correlations decay as Gaussians outside the leading support, gives in place of (7.56)

$$\frac{\text{Var } F}{E_N^{\text{r}}} \sim \frac{1}{2}\int_{-1}^{1} f^2(x)\,dx + \frac{1}{2}\int_{-1}^{1} dx\, f(x)\int_{-1}^{1} dy\, f(y)\sqrt{2\pi N}\,\rho_{(2),\infty}^{T}(\sqrt{N}x, \sqrt{N}y).$$

From the explicit functional form (7.57) we observe that for large N

$$\sqrt{2\pi N}\,\rho_{(2),\infty}^{T}(\sqrt{N}x, \sqrt{N}y) \sim (1 - \sqrt{2})\delta(x - y).$$

This provides the result at least for piecewise continuous test function f. As noted in [231], following an idea in [370], taking advantage of the translation invariance of the limiting two-point truncated correlation function allows the integral instead to be computed in terms of Fourier transforms, and so further enlarging the class of f for its applicability. $\qquad\square$

In the case of the counting function for a determinantal point process, we know that a diverging variance is sufficient to conclude that the centred and rescaled linear statistics obey a central limit theorem. Here we see that for general linear statistics the variance diverges (this is in distinction to the case of smooth statistics for GinUE; recall Sect. 3.2), and furthermore the correlations are no longer determinantal but rather Pfaffian. Nonetheless, it was shown in [231, Th. 3.10] that the higher-order cumulants of F are of order $O(\sqrt{N})$ as $N \to \infty$. The conclusion is then that the statistic $(F - \langle F \rangle)/\sqrt{E_N^{\text{r}}}$ tends to a zero mean Gaussian, with variance given by the RHS of (7.58) with the factor E_N^{r} omitted.

7.8 Relationship to Diffusion Processes and Gap Probabilities

A Pfaffian point process with correlation kernel (7.38), distinct from the real eigenvalues of GinOE, has been known in the literature (at least implicitly) for a long time [427]. This point process is the annihilation process $A + A \to \varnothing$ on the real line, in which particles are initially non-interacting on the real line with density $\rho_0 = 1/\sqrt{2\pi}$, and evolving according to Brownian motion for all $t > 0$. Colliding particles are each annihilated, and the density is rescaled to have the initial value ρ_0. As explained in [244], assembling results in [427] gives that the correlation functions of this process are, in the limit $t \to \infty$ the Pfaffian point process with the same correlation kernel (7.38) as the real eigenvalues of GinOE matrices in the bulk scaling limit. This process was reconsidered in [528], in which the Pfaffian point process structure, and its correlation kernel, were made explicit. Our interest here is first to make note of some consequences of this co-incidence in relation to certain gap probabilities in the GinOE, and then to proceed to discuss special properties of other GOE gap probabilities.

For the first such consequence, knowledge about the correlations for another diffusion process—specifically the coalescence process $A + A \to A$—is required [80]. As for the annihilation process the particles are initially non-interacting on the real line with density $\rho_0 = \sqrt{2/\pi}$ (not $\sqrt{1/2\pi}$), and evolve according to Brownian motion for all $t > 0$. Now pairs of colliding particles merge to a single particle, and the density is rescaled to have the initial value ρ_0. Results from [244, 427, 528] give that that this is a Pfaffian point process with correlation kernel equal to (7.38), now multiplied by a factor of 2. From the theory of thinning a point process (recall Sect. 3.3) it follows that bulk-scaled GinOE real eigenvalues are statistically equivalent to the coalescence process with every particle deleted with probability $1/2$. Now denote by $E^{b,r}(\text{even}; J)$ the probability that a collection of intervals J contain an even number of particles for bulk-scaled GinOE real eigenvalues. Furthermore, we denote by $E^c(k; J)$ the probability that there are k particles in J for the coalescence process. The relationship between the two processes via deletion with probability half in the latter implies [244, 427]

$$E^{r,b}(\text{even}; J) = E^c(0; J) + \frac{1}{2}\sum_{k=1}^{\infty} E^c(k; J) = \frac{1}{2} + \frac{1}{2}E^c(0; J). \tag{7.59}$$

Furthermore, for a set $J = \cup_{j=1}^{m}(x_{2j-1}, x_{2j})$ of m disjoint intervals, $E^c(0; J)$ is given by [427]

$$E^c(0; J) = \text{Pf}\left[A_{ij}\right]_{i,j=1}^{2m}, \qquad A_{ij} = \text{erfc}\left(\frac{x_j - x_i}{\sqrt{2}}\right). \tag{7.60}$$

This leads to an explicit formula of $E^{r,b}(\text{even}; J)$. Thus in the simplest case of a single interval, it follows that

$$E^{r,b}(\text{even}; (0, s)) = \frac{1}{2}\left(1 + \text{erfc}\frac{s}{\sqrt{2}}\right). \tag{7.61}$$

The second and final consequences as noted in [244] relates to the small and large s expansions of $E^{b,r}(0; (0, s))$. As these had earlier been reported in the literature for the same gap probability in the case of the annihilation process [190], the fact that the statistical state of the latter coincides with that of bulk-scaled real GinOE eigenvalues gives the sought results.

Proposition 7.9 *For small s*

$$E^{r,b}(0; (0, s)) = 1 - \frac{s}{\sqrt{2\pi}} + \frac{s^3}{12\sqrt{2\pi}} - \frac{s^5}{80\sqrt{2\pi}} + \frac{s^6}{720\pi} + \cdots, \tag{7.62}$$

while for large s

$$E^{r,b}(0; (0, s)) = e^{-cs+C+o(1)},$$ (7.63)

where $c = \frac{1}{2\sqrt{2\pi}}\zeta(3/2) \approx 0.5211$, *and C is as in (7.12).*

Remark 7.6

1. We comment now on an application of $E^{r,b}$ to a certain topological transition relating to parametric energy levels associated with particular quantum dots, due to Beenakker et al. [77]. This appears in a setting of energy levels $\cdots > E_3 > E_2 > E_1 > 0$ and $|E_0| < E_1$. Thus each E_j is positive, while E_0 may be positive or negative, but no bigger in absolute value to E_1. Moreover, the energy levels are all functions of a parameter ϕ. The question of interest is the statistics of the values of ϕ such that $E_0(\phi) = 0$. As a theoretical model of the particular scattering problem (Andreev reflection) giving rise to these energy levels, the solution of this last equation is mapped to the real eigenvalues of the real, but non-symmetric, random matrix

$$M = (\mathbb{I}_{2N} - R)(\mathbb{I}_{2N} + R)^{-1}\mathbb{Z}_{2N},$$ (7.64)

where R is a $2N \times 2N$ Haar distributed real orthogonal matrix with determinant $+1$, and $Z_{2N} = \mathbb{I}_N \otimes \begin{bmatrix} 0 & 1 \\ -1 & 0 \end{bmatrix}$. Note that M has the skew symmetry $M^T = -Z_{2N}MZ_{2N}$, implying all eigenvalues of M are doubly degenerate. Although no exact solution relating to the ensemble $\{M\}$ is known, numerics indicate that the leading term for the number of real eigenvalues is $\sqrt{2N/\pi}$ as for GinOE, and most significantly for present purposes that after rescaling so that the mean density is $1/\sqrt{2\pi}$, these real eigenvalues have bulk spacing specified by the GinOE distribution $E^{r,b}$.

2. The ensemble of random matrices (7.64) is one example of a member of the same universality class for its real eigenvalues (at least according to numerical evidence) as GinOE. On the other hand, there are known ensembles for which the real eigenvalues appear (from numerical evidence) to have different statistics, for example

$$\begin{bmatrix} A & B \\ -B^T & C \end{bmatrix},$$

where A, B, C are square random matrices of the same size with A, C real symmetric from the GOE, and B from GinOE [536]. We make mention too of ensembles of non-Hermitian random matrices with complex entries which exhibit real eigenvalues; see [429, Sect. 8], [221] and [536]. An explicit example is random matrices of the form $I_{k,N-k}G$, where G is a member of the GUE, while $I_{k,N-k}$ is the diagonal matrix consisting of k entries equal to 1 and $N - k$ entries equal to -1. In [221] it is shown that for $\mu := k/N$ fixed as $N \to \infty$, there is a finite fraction of real eigenvalues provided $\mu \neq 1/2$, the density of which can be explicitly calculated. Let us also stress that the weakly non-symmetric regime [35, 38, 267] provides additional examples where the real eigenvalue statistics differ from those of GinOE.

Choosing $\hat{u}(x) = -\chi_{x \in (0,s)}$ in the generalised partition function $Z_N[1 + \hat{u}, 1]$ of Sect. 7.5 we see that $E^{\mathrm{r,b}}(0; (0, s)) = \lim_{N \to \infty} Z_N[1 + \hat{u}, 1]$ and thus from (7.29) we have a Fredholm Pfaffian formula for this probability. From the standard fact that a Pfaffian is the square root of the corresponding determinant (this can be deduced from the Pfaffian/determinant identity of Remark 7.4.2), the fact that the 2×2 entries of the operator defining the corresponding Fredholm determinant is a function of differences was used in [244] to give an alternative derivation of c. Moreover, this calculation was also carried out in the more general setting of the thinned point process, in which each real eigenvalue is deleted with probability $1 - \xi$; we know from Sect. 3.3 that its effect in the Fredholm Pfaffian formula is simply to multiply the kernel by ξ. Denoting the corresponding gap probability by $E^{\mathrm{r,b}}(0; (0, s); \xi)$, it was found

$$E^{\mathrm{r,b}}(0; (0, s); \xi) \underset{s \to \infty}{\sim} e^{-c(\xi)s + O(1)}, \quad c(\xi) = -\frac{1}{4\pi} \int_{-\infty}^{\infty} \log\left(1 - (2\xi - \xi^2)e^{-u^2/2}\right) du;$$
(7.65)

evaluating the integral as a series for $\xi = 1$ reclaims the value of c in (7.63). In [230] the theory of the asymptotics of Fredholm Pfaffians was further developed to derive the value of C independent of a certain continuity assumption implicit in the calculation of [190].

It was observed in [355] that the asymptotic result (7.63), with the interval of the LHS replaced by $(-s, s)$, and s in the exponent of the RHS replaced by $2s$, is suggestive in relation to the large N form of the probability $p_{N,N}^{\mathrm{GinOE}}$ that all the eigenvalues of an $N \times N$ matrix are real. First it is argued that the quantity $E^{\mathrm{r,b}}(0; (-s, s))$ can be replaced by the corresponding quantity for finite N, provided N is taken as large and $(-s, s)$ is contained in the leading interval of support $(-\sqrt{N}, \sqrt{N})$. But with $(-s, s)$ chosen as exactly this interval, up to corrections which vanish as N the finite N gap probability is the same quantity as $p_{N,0}^{\mathrm{GinOE}}$ for the probability of no real eigenvalues, and thus the prediction (7.11).

Further developing the viewpoint of the above paragraph, let $E_N^{\mathrm{r}}(0; (a, b))$ denote the probability that the interval (a, b) in the $N \times N$ GinOE is free of eigenvalues. From the definition of edge scaling, we see that

$$\lim_{N \to \infty} E_N^{\mathrm{r}}(0; (\sqrt{N} + s, \infty)) = E^{\mathrm{r,e}}(0; (s, \infty));$$
(7.66)

note that minus the derivative of the RHS with respect to s gives the PDF of the largest eigenvalue for $N \to \infty$. It turns out that the $s \to -\infty$ asymptotic of this quantity is closely related to (7.63) [230] (see also [69]),

$$E^{\mathrm{r,e}}(0; (s, \infty)) \underset{s \to -\infty}{\sim} e^{-c|s| - \frac{1}{2}\log 2 + C + O(s^{-1^+})},$$
(7.67)

where c, C are as in (7.63). Note that in both (7.63) and (7.67) a gap with no real eigenvalues is being created, which has size $|s|$ relative to the bulk region. This has been generalised to the thinned case in [70],

$$E^{r,e}(0; (s, \infty); \xi) \underset{s \to -\infty}{\sim} e^{-c(\xi)|s|+C(\xi)+o(1)},$$
(7.68)

where $c(\xi)$ is as in (7.65) and, with $\text{Li}_s(z) := \sum_{n=1}^{\infty} z^n/n^s$,

$$C(\xi) = \frac{1}{2} \log \frac{2}{2 - \xi} + \frac{1}{4\pi} \int_0^{2\xi - \xi^2} \left((\text{Li}_{1/2}(x))^2 - \frac{\pi x}{1 - x} \right) \frac{dx}{x}.$$

In relation to the diffusion processes that have featured in this section, we remark that for the annihilation process $A + A \to \varnothing$ with half line initial positions, an exact coincidence with the edge state Pfaffian point process of real Ginibre eigenvalues can be exhibited [285].

Knowledge that the edge-scaled real Ginibre eigenvalues form a Pfaffian point process with the explicit kernel determined by the formulas of Proposition 7.5 and the limit formula (7.40) imply a Fredholm Pfaffian formula for $E^{r,e}(0; (s, \infty); \xi)$ (see also [461, 470]). A recent result of Baik and Bothner [70] gives that a simpler Fredholm determinant formula holds true. For this denote by $\mathcal{S}_{(s,\infty)}$ the integral operator of (s, ∞) with kernel

$$S(x, y) = \frac{1}{2\sqrt{\pi}} e^{-(x+y)^2/4},$$

and write $\bar{\xi} := \xi(2 - \xi)$. One then has [70, Th 1.8]

$$E^{r,e}(0; (s, \infty); \xi) = \sqrt{\frac{1 - \sqrt{\bar{\xi}}}{2(2 - \xi)}} \det \left(\mathbb{I} + \sqrt{\bar{\xi}} \mathcal{S}_{(s,\infty)} \right) + \sqrt{\frac{1 + \sqrt{\bar{\xi}}}{2(2 - \xi)}} \det \left(\mathbb{I} - \sqrt{\bar{\xi}} \mathcal{S}_{(s,\infty)} \right).$$
(7.69)

Note that with $\xi = 1$ (no thinning) the coefficient in front of the first of these Fredholm determinants vanishes, and then [69, Th. 1.12]

$$E^{r,e}(0; (s, \infty)) = \det(\mathbb{I} - \mathcal{S}_{(s,\infty)}).$$
(7.70)

As emphasised in [70], both (7.69) and (7.70) are structurally identical to known Fredholm determinant formulas for the soft edge-scaled largest eigenvalue of GOE matrices; see [225] and [236] respectively. Moreover, it is shown in [69, 70] that associated with these gap probabilities are certain integrable systems, associated with the Zakharov–Shabat inverse scattering problem, which lead to alternative formulas again identical in structure to those known for the soft edge-scaled largest eigenvalue of GOE matrices in terms of transcendents from Painlevé II [195, 527].

For the Fredholm determinants in (7.70) and (7.69) the methods of Bornemann [101, 102] provide for fast, high precision numerical evaluations. This in turn allows for the determination of the values of statistical quantities associated with the corresponding distributions. Specifically, in relation to the distribution of the edge-scaled largest eigenvalue (no thinning), it is found that to five decimals the mean, variance,

skewness and kurtosis are equal to [69, Table 1] $-1.30319, 3.97536, -1.76969,$
5.14560, respectively.

Remark 7.7 Instead of the largest real eigenvalue, one can ask about the distribution
of the maximum of the real part of all GinOE eigenvalues, or the maximum modulus
of all the eigenvalues. Both are known. With global scaling $G \mapsto \frac{1}{\sqrt{N}} G$, the first
involves the extreme value scaled variable

$$1 + \sqrt{\frac{\gamma_N}{4N}} + \frac{x}{\sqrt{4N\gamma_N}}, \qquad \gamma_N = \frac{1}{2}\Big(\log N - 5 \log \log N - \log(2\pi^4)\Big),$$

while the second involves the same extreme value scaled variable as for GinUE,
(3.13). In these scalings, the extreme values obey the rescaled Gumbel law, with
cumulative distribution $\exp(-\frac{1}{2}e^{-t})$; see [35, 161] and [470] respectively. As already
remarked, the result for the largest real part is directly relevant to the stability of the
system of the matrix differential equation (1.11).

7.9 Elliptic GinOE

The elliptic GinOE interpolates between the asymmetric GinOE and the symmetric
Gaussian orthogonal ensemble (GOE). First we recall that a member of the GOE,
S say, can be constructed from a GinOE matrix G according to $S = \frac{1}{2}(G + G^T)$.
In addition to S, we introduce the antisymmetric real Gaussian matrix A with joint
distribution of its elements proportional to $e^{-\mathrm{Tr}\, A^2/2}$. For a parameter τ, $0 \le \tau < 1$,
and a scale factor b, following [401] a random matrix X is defined according to

$$X = \frac{1}{\sqrt{b}}(S + \sqrt{c}A), \qquad c = (1 - \tau)/(1 + \tau). \tag{7.71}$$

Up to the scale factor, when $\tau = 0$, X is a member of the GinOE, while for $\tau = 1$,
X is a member of the GOE.

Proposition 7.10 *The joint element distribution of X is*

$$A_{\tau,b} \exp\left(- \frac{b}{2(1-\tau)}\Big(\mathrm{Tr}\, XX^T - \tau\,\mathrm{Tr}\, X^2\Big)\right), \tag{7.72}$$

where

$$A_{\tau,b} = (\sqrt{c})^{-N(N-1)/2}(\sqrt{b})^{N^2}(2\pi)^{-N^2/2}.$$

Define the normalisation and the weight

$$C_N^e(\tau; b) = \frac{(\sqrt{b})^{N(N+1)/2}(\sqrt{1+\tau})^{N(N-1)/2}}{2^{N(N+1)/4}\prod_{l=1}^N \Gamma(l/2)},$$

$$\omega^e(z) = e^{-b|z|^2} e^{2by^2} \operatorname{erfc}\left(\sqrt{\frac{2b}{1-\tau}} y\right), \tag{7.73}$$

where $z = x + iy$. The probability density that there are k real eigenvalues for the random matrix X is then

$$C_N^e(\tau; b)\frac{2^{(N-k)/2}}{k!((N-k)/2)!}\prod_{s=1}^k (\omega^e(\lambda_s))^{1/2} \prod_{j=1}^{(N-k)/2} \omega^e(z_j)$$

$$\times \left|\Delta(\{\lambda_l\}_{l=1,\dots,k} \cup \{x_j \pm iy_j\}_{j=1,\dots,(N-k)/2})\right|. \tag{7.74}$$

Proof We see from (7.71) that

$$(dX) = 2^{N(N-1)/2}(\sqrt{c})^{N(N-1)/2}(\sqrt{b})^{-N^2}(dS)(dA).$$

Using this we can change variables in the joint element distribution of S and A to deduce that the joint element distribution of X is equal to (7.72).

Denote (1.10) by $P_{k,(N-k)/2}(\{\lambda_j\}_{j=1}^k; \{x_j \pm iy_j\}_{j=1}^{(N-k)/2})$, which corresponds to the eigenvalue PDF in the case $\tau = 0$, $b = 1$, conditioned so that there are exactly k real eigenvalues. For these matrices scale $X \mapsto \sqrt{b}X/(1 - \tau)^{1/2}$ to obtain that for matrices with PDF

$$A_{0,1}b^{N^2/2}(1 - \tau)^{-N^2/2}e^{-b\operatorname{Tr} XX^T/2(1-\tau)}, \tag{7.75}$$

the eigenvalue PDF is equal to

$$b^{N/2}(1 - \tau)^{-N/2}P_{k,(N-k)/2}\left(\{\sqrt{b}\lambda_j/(1 - \tau)^{1/2}\}_{j=1}^k; \right.$$
$$\left. \{\sqrt{b}x_j/(1 - \tau)^{1/2} \pm i\sqrt{b}y_j/(1 - \tau)^{1/2}\}_{j=1}^{(N-k)/2}\right).$$

Comparing (7.75) to (7.72) it follows that for matrices with element PDF (7.72) the eigenvalue PDF, conditioned so that there are exactly k eigenvalues, is equal to

$$\frac{A_{\tau,b}}{A_{0,1}}(1 - \tau)^{N(N-1)/2} \exp\left(\frac{\tau b}{2(1 - \tau)}\left(\sum_{j=1}^k \lambda_j^2 + 2\sum_{j=1}^{(N-k)/2}(x_j^2 - y_j^2)\right)\right)$$

$$\times P_{k,(N-k)/2}\left(\left\{\sqrt{\frac{b}{1 - \tau}}\lambda_j\right\}_{j=1}^k; \left\{\sqrt{\frac{b}{1 - \tau}}x_j \pm i\sqrt{\frac{b}{1 - \tau}}y_j\right\}_{j=1}^{(N-k)/2}\right).$$

This is (7.74). $\square$

We see that (7.74) with $b = 1/(1 + \tau)$ can, in the Coulomb gas picture of Sect. 7.2, be viewed as a modification of the Coulomb gas identified in relation to elliptic GinUE (see text below proof of Proposition 2.7), the modification being that there are imaginary terms and charges on the real line as for GinOE itself; recall Sect. 7.2. These modifications do not alter the interpretation of the origin of the one-body couplings in terms of the same neutralising background charge as for GinUE. The conclusion then is that after scaling $X \mapsto \frac{1}{\sqrt{N}} X$, and with $b = 1/(1 + \tau)$, the limiting eigenvalue density of elliptic GinOE is the constant $1/(\pi(1 - \tau^2))$, with the shape of the droplet being an ellipse specified by semi-axes $A = 1 + \tau$, $B = 1 - \tau$. Using different reasoning, this result was first deduced in [507].

Denote the analogue of (7.4) in the elliptic GinOE ensemble by $Z_{k,(N-k)/2}^{\mathrm{eGinOE}}[u, v]$, and use this to define $Z_N^{\mathrm{eGinOE}}[u, v]$ as specified above (7.7). Starting with (7.74), by following the strategy of the proof of Proposition 7.2, Pfaffian formulas for these quantities can be obtained. It suffices to record that for $Z_N^{\mathrm{eGinOE}}[u, v]$.

Proposition 7.11 *Let $\{p_{l-1}(x)\}_{l=1,\dots,N}$ be a set of monic polynomials of the indexed degree, and let $\alpha_{j,k}^{\mathrm{e}}[u]$, $\beta_{j,k}^{\mathrm{e}}[v]$ be defined as for $\alpha_{j,k}^{\mathrm{g}}[u]$, $\beta_{j,k}^{\mathrm{g}}[v]$ in Proposition 7.2 but with each ω^{g} replaced by ω^{e}. For k, N even we have*

$$Z_N^{\mathrm{eGinOE}}[u, v] = C_N^{\mathrm{e}}(\tau; b)\mathrm{Pf}\,[\alpha_{j,l}^{\mathrm{e}}[u] + \beta_{j,l}^{\mathrm{e}}[v]]_{j,l=1,\dots,N}. \tag{7.76}$$

Next, we would like to compute the skew-orthogonal polynomials associated with $(\alpha_{j,l}^{\mathrm{e}}[u] + \beta_{j,l}^{\mathrm{e}}[v])|_{u=v=1}$. While the formulas (7.18) for the latter still apply, according to the definition (7.71), for $\tau \neq 0$ the elements x_{jk} and x_{kj} in X are correlated, so the simple formula (7.19) is no longer valid. Nonetheless, an expansion of $\det(z\mathbb{I}_{2n} - G)$ can still be used to compute the averages required in (7.18)—see [253, proof of Proposition 11] for details of a closely related calculation—with the result

$$p_{2n}^{\mathrm{e}}(z) = C_{2n}(z),$$

$$p_{2n+1}^{\mathrm{e}}(z) = C_{2n+1}(z) - 2nC_{2n-1}(z) = -(1 + \tau)e^{\frac{z^2}{2(1+\tau)}}\frac{d}{dz}\left(e^{-\frac{z^2}{2(1+\tau)}}C_{2n}(z)\right),$$
$$\tag{7.77}$$

where $C_n(z) := (\tau/2)^{n/2} H_n(z/\sqrt{2\tau})$ and for convenience the specific choice of scale $b = 1/(1 + \tau)$ has been made. These polynomials were first identified in [267] without knowledge of (7.18). The corresponding normalisation is given by

$$r_n^{\mathrm{e}} = (2n)!2\sqrt{2\pi}(1 + \tau). \tag{7.78}$$

We can check that (7.77) and (7.78) reduce to the result of Proposition 7.4 in the limit $\tau \to 0^+$.

From knowledge of the skew-orthogonal polynomials, the entries in the correlation kernels specifying the Pfaffian point process are explicit. For example, in the case of the real–real correlations, these entries are specified in Proposition 7.5 and it is the quantity $S_N^r(x, y)$ therein which determines the other entries. It is possible to simplify the summation in its definition to obtain a structure identical to that in the $\tau = 0$ case (7.31), in which the first term is recognised from its appearance as the correlation kernel for elliptic GinUE (2.38) (up to proportionality, with w, z real, and with $N \mapsto N - 1$).

Proposition 7.12 *For elliptic GinOE with $b = 1/(1 + \tau)$ we have*

$$
S_N^{r,e}(x, y) = \frac{e^{-(x^2+y^2)/2(1+\tau)}}{\sqrt{2\pi}} \sum_{k=0}^{N-2} \frac{1}{k!} C_k(x) C_k(y) + \frac{e^{-y^2/2(1+\tau)}}{2\sqrt{2\pi}(1+\tau)} \frac{C_{N-1}(y)\Phi_{N-2}(x)}{(N-2)!}.
$$
(7.79)

Proof The derivative formula for p_{2n+1} in (7.77) implies a summation identity analogous to (7.32). We can furthermore check from (7.77) and (7.78) that analogous to (7.33) there is the further derivative formula

$$
\frac{2(k+1)}{r_{k+1}^e} p_{2k+2}(x) - \frac{1}{r_k^e} p_{2k}(x) = -\frac{1}{2\sqrt{2\pi}} \frac{1}{(2k+1)!} e^{x^2/2(1+\tau)} \frac{d}{dx} e^{-x^2/2(1+\tau)} C_{2k+1}(x),
$$

which implies a summation identity analogous to (7.34). $\square$

The result (7.79) can be used to show that both the bulk-scaled, and edge-scaled real-real correlation functions of elliptic GinOE are identical to that for the original GinOE, with the change of scale in the latter $\lambda_j \mapsto \lambda_j/\sqrt{1 - \tau^2}, z_j \mapsto z_j/\sqrt{1 - \tau^2}$ [267]. This implies that, structurally, the formula (7.54) for the variance of the number of real eigenvalues with N large, remains true. Another application of (7.79) is in relation to the weakly non-symmetric limit, obtained by the scaling $\tau \mapsto 1 - \alpha^2/N$ before the computation of the large N limit [267]. Thus

$$
\lim_{N\to\infty} \frac{\pi}{\sqrt{N}} S_N^{r,e}\left(\frac{\pi x}{\sqrt{N}}, \frac{\pi y}{\sqrt{N}}\right)\Big|_{\tau=1-\alpha^2/N} = \int_0^1 e^{-\alpha^2 u^2} \cos \pi u(x - y)\, du,
$$

which implies in particular that the correlations now decay algebraically, in contrast to the Gaussian decay exhibited by (7.37). See also [35] for the scaling limit at the edge of the spectrum.

Setting $x = y$ in (7.79) gives the eigenvalue density. From this, with τ fixed, it was shown in [124, 267] that the expected number of real eigenvalues $E_N^{r,e}$ satisfies the large N asymptotic expansion

$$
E_N^{r,e} - \frac{1}{2} \underset{N\to\infty}{\sim} \sqrt{\frac{2}{\pi} \frac{1+\tau}{1-\tau}} N\left(1 + \frac{\tau - 3}{8(1-\tau)} \frac{1}{N} + \frac{5\tau^2 - 14\tau - 3}{128(1-\tau)^2} \frac{1}{N^2} + \cdots\right),
$$
(7.80)

which recovers (7.14) for $\tau = 0$, cf. [115]. On the other hand, in the scaling $\tau \mapsto 1 - \alpha^2/N$, one finds

$$E_N^{\mathrm{r,e}} \underset{N\to\infty}{\sim} c(\alpha)N + O(1), \qquad c(\alpha) := e^{-\alpha^2/2}\left(I_0\left(\frac{\alpha^2}{2}\right) + I_1\left(\frac{\alpha^2}{2}\right)\right); \qquad (7.81)$$

see [124, Th. 2.1]. (We mention that the function $c(\alpha)$ also appears in several different contexts; cf. (3.16) and Prop. 6.3.) Furthermore, as an analogue of Proposition 7.7, it was shown in [124, Th. 2.3] that in the weakly non-symmetric regime the variance $(\sigma_N^{\mathrm{r,e}})^2$ is again proportional to $E_N^{\mathrm{r,e}}$ as

$$(\sigma_N^{\mathrm{r,e}})^2 \sim 2\left(1 - \frac{c(\sqrt{2}\alpha)}{c(\alpha)}\right)E_N^{\mathrm{r,e}}; \qquad (7.82)$$

note that as $\alpha \to \infty$ the ratio in (7.82) tends to $2 - \sqrt{2}$ in (7.54). The global density of real eigenvalues for τ fixed tends to be uniform in the interval $(-1-\tau, 1+\tau)$; see [267]. In contrast, in the regime $\tau \mapsto 1 - \alpha^2/N$, it was proposed by Efetov [214] using the supersymmetry method that for $x \in (-2, 2)$,

$$\lim_{N\to\infty} \rho_N^{\mathrm{r,e}}(\sqrt{N}x)\Big|_{\tau=1-\alpha^2/N} = \frac{1}{c(\alpha)}\frac{1}{2\alpha\sqrt{\pi}}\mathrm{erf}\left(\frac{\alpha}{2}\sqrt{4-x^2}\right). \qquad (7.83)$$

This convergence was recently shown in [124] by properly analysing the double scaling limits of (7.79). As expected, the limiting density in (7.83) interpolates between the uniform density ($\alpha \to \infty$) with the semi-circle law ($\alpha \to 0$). A point of interest is that Efetov was motivated by an application of almost symmetric random matrices to the study of vortices in disordered superconductors with columnar defects [320].

To provide more details of the statistics of the real eigenvalues, we again consider the generating function $Z_N^{\mathrm{e}}(\zeta) := \sum_{k=0}^{N/2} \zeta^k p_{2k,N}^{\mathrm{eGinOE}}$ of the probabilities $p_{2k,N}^{\mathrm{eGinOE}}$ of having exactly $2k$ real eigenvalues. Then by using the explicit form of the skew-orthogonal polynomials (7.77), it was shown in [126, Proposition 2.1] that $Z_N^{\mathrm{e}}(\zeta)$ can be written as

$$Z_N^{\mathrm{e}}(\zeta) = \det\left[\delta_{j,k} + \frac{(\zeta-1)}{\sqrt{2\pi}}M_n^{(\tau)}(j,k)\right]_{j,k=1,\dots,N/2}, \qquad (7.84)$$

where

$$M_n^{(\tau)}(j,k) = \frac{1}{\sqrt{\Gamma(2j-1)\Gamma(2k-1)}}\int_{\mathbb{R}} e^{-\frac{x^2}{1+\tau}}C_{2j-2}(x)C_{2k-2}(x)\,dx$$

$$= \left(\frac{1+\tau}{1-\tau}\right)^{\frac{1}{2}}\frac{\Gamma(j+k-\frac{3}{2})\,_2F_1(k-j+\frac{1}{2}, j-k+\frac{1}{2}; -j-k+\frac{5}{2}; -\frac{\tau}{1-\tau})}{\sqrt{\Gamma(2j-1)\Gamma(2k-1)}}.$$

$$(7.85)$$

It can be easily observed that for $\tau = 0$, the determinantal formula (7.84) recovers (7.10). As an application of (7.84), it was proven in [126] that with τ fixed,

$$\lim_{N \to \infty} \frac{\log p_{0,N}^{\text{eGinOE}}}{\sqrt{N}} = -\sqrt{\frac{1+\tau}{1-\tau}} \frac{\zeta(3/2)}{\sqrt{2\pi}}, \tag{7.86}$$

generalising the leading-order asymptotic of (7.11), whereas with the scaling $\tau \mapsto 1 - \alpha^2/N$,

$$\lim_{N \to \infty} \frac{\log p_{0,N}^{\text{eGinOE}}}{N} \leq \sum_{m=1}^{\infty} \frac{c(\sqrt{m}\,\alpha)}{2m} = \frac{2}{\pi} \int_0^1 \log\left(1 - e^{-\alpha^2 s^2}\right) \sqrt{1 - s^2}\, ds, \tag{7.87}$$

where $c(\alpha)$ is given in (7.81). In [251], the formula (7.84) was further applied to prove the local central limit theorem of $p_{2k,N}^{\text{eGinOE}}$, cf. Proposition 7.3.

Remark 7.8
1. It can be checked from the formula for $p_{2n}^{\text{e}}(z)$ and the first formula for $p_{2n}^{\text{e}}(z)$ in (7.77), together with the formula (7.78) for the normalisation that the analogue of (7.43), with K_{N-1} now the kernel for elliptic GinUE (2.38), is also valid for $S_N^{\text{r,e}}(x, y)$.
2. Denote the probability density (7.74) by $P_{N,k}(\{\lambda_j\}_{j=1}^k; \{x_j \pm iy_j\}_{j=1}^{(N-k)/2}; \tau, b)$. It follows from (7.73) and (7.74) that

$$P_{N,N}(\{\lambda_j\}_{j=1}^N; \tau, b) = (1+\tau)^{N(N-1)/4} P_{N,N}(\{\lambda_j\}_{j=1}^N; \tau = 0, b = 1).$$

Hence upon integration and use of Proposition 7.1, $p_{N,N}^{\text{eGinOE}} = ((1+\tau)/2)^{N(N-1)/4}$. Note in particular that this is equal to unity for $\tau = 1$, as is consistent with $\tau = 1$ corresponding to real symmetric matrices.
3. A variation of the construction of elliptic GinOE matrices (7.71) is to consider asymmetric random matrices $S + A_0$, where $S \in$ GOE, while A_0 is a fixed real antisymmetric matrix. An analysis of this setting in [279] reclaimed the result of [214] for the density of both the real (as given by (7.83)) and complex eigenvalues, although this was shown to break down if A_0 was of finite rank.

7.10   An Application of Elliptic GinOE to Equilibria Counting

In this section, following Fyodorov and Khoruzhenko [274], it will be shown that generalisations of the considerations underpinning the random differential equation (1.11) as formulated in [431] relate to elliptic GinOE. The starting point (see the review [44] for an extended description) is the coupled nonlinear differential equations

$$\frac{dX_i(t)}{dt} = f_i(\mathbf{X}(t)), \qquad i = 1, \ldots, N. \tag{7.88}$$

Here the f_i are not known explicitly. An equilibrium point $\mathbf{X}^*$ is when $f_i(\mathbf{X}^*) = 0$. Stability around the fixed point is probed in terms of the Jacobian $M_{ij} = \frac{\partial f_i(\mathbf{X})}{\partial X_j}|_{\mathbf{X}^*}$.

May [431], on consideration of the applied setting within ecology, took the diagonal elements to have mean -1. Taking the off-diagonal elements to have mean zero and all elements to have variance α^2 then leads to (1.11).

The work [274] takes a global approach to the study of equilibria in (7.88), with the RHS written with the addition of $-\mu X_i(t)$. This allows the question: "what is the probability that a randomly chosen equilibrium is stable?" to be probed. To proceed further, the f_i are decomposed into the sum of a gradient (curl-free) and solenoidal (divergence free) components,

$$
f_i(\mathbf{X}) = -\frac{\partial V(\mathbf{X})}{\partial X_i} + \frac{1}{\sqrt{N}} \sum_{j=1}^{N} \frac{\partial A_{ij}(\mathbf{X})}{\partial X_i}.
$$

The matrix $A(\mathbf{X})$ is anti-symmetric. Both $V(\mathbf{X})$ and $A(\mathbf{X})$ are assumed to be statistically independent, isotropic, centred, homogeneous random Gaussian fields with covariances

$$
\langle V(\mathbf{X})V(\mathbf{Y}) \rangle = v^2 \Gamma_V(|\mathbf{X} - \mathbf{Y}|^2),
$$
$$
\langle A_{ij}(\mathbf{X})A_{nm}(\mathbf{Y}) \rangle = a^2 \Gamma_A(|\mathbf{X} - \mathbf{Y}|^2)(\delta_{i,n}\delta_{j,m} - \delta_{i,m}\delta_{j,n}),
$$

and with $\Gamma_V''(0) = \Gamma_A''(0) = 1$ as normalisations.

The primary observable considered in [274] is the expected number $\langle \mathcal{N} \rangle$ of equilibrium points implied by this model. Through use of the Kac–Rice formula, this is reduced to the form of a random matrix average

$$
\langle \mathcal{N} \rangle = \left\langle \left| \det \left[\delta_{i,j}(1 + \xi\sqrt{\tau}/m) - \frac{1}{m\sqrt{N}} J_{ij} \right]_{i,j=1}^{N} \right| \right\rangle,
$$
$$
m = \frac{\mu}{\sqrt{4N(v^2 + a^2)}}, \qquad \tau = \frac{v^2}{v^2 + a^2}, \tag{7.89}
$$

where ξ is distributed as $N[0, 1/\sqrt{N}]$ and the J_{ij} are entries of a zero mean Gaussian random matrix with correlations (for large N)

$$
\langle J_{ij} J_{kl} \rangle = \left(\delta_{i,k}\delta_{j,l} + \tau(\delta_{i,j}\delta_{k,l} + \delta_{i,l}\delta_{j,k}) \right). \tag{7.90}
$$

In the case that $\tau = 0$ the entries of $[J_{ij}]$ are seen to be statistically independent, which relates to May's setting. On the other hand, when $\tau = 1$, (7.90) specifies $[J_{ij}]$ as proportional to a real symmetric GOE matrix. We refer to [420] for an extensive discussion of May's model when $\tau = 1$ in connection to the universality of the Tracy–Widom distribution. For general $0 < \tau < 1$, $[J_{ij}]$ defines an elliptic GinOE matrix as specified in Sect. 7.9.

In the theory of the real eigenvalues of GinOE matrices, note has been made of the formula (7.42) relating the average of the absolute value of the characteristic polynomial to the density. Such a formula holds equally as well for the elliptic GinOE, and gives [274, Eq. (12)]

$$\langle \mathcal{N} \rangle = \frac{C_N}{m^N} \int_{-\infty}^{\infty} e^{-NS(\lambda)} \rho_{(1),N+1}^{\mathrm{r,e}}(\lambda \sqrt{N}) \, d\lambda, \qquad (7.91)$$

where

$$C_N = \frac{2\sqrt{1 + 1/\tau}(N-1)!}{N^{(N-1)/2}(N-2)!!}, \qquad S(\lambda) = (\lambda - m)^2/(2\tau) - \lambda^2/(2(1+\tau)).$$

The minimum of $S(\lambda)$ occurs at $\lambda^* = m(1 + \tau)$. On the other hand, we know that in the global variable $\lambda\sqrt{N}$ the support of the real eigenvalue for elliptic GinOE is $|\lambda| < 1 + \tau$. Hence λ^* lies inside the support for $m < 1$, and outside otherwise. Moreover, inside the support $\rho_{(1),N+1}^{\mathrm{r,e}}(\lambda\sqrt{N})$ has the large N value $1/\sqrt{2\pi(1 - \tau^2)}$. It follows that for $m < 1$ [274, Eq. (14)]

$$\langle \mathcal{N} \rangle = \sqrt{\frac{2(1 + \tau)}{(1 - \tau)}} e^{N((1/2)(m^2-1)-\log m)}, \qquad (7.92)$$

which grows exponentially in N. Analysis in the case of $m > 1$ requires knowledge of $\rho_{(1),N+1}^{\mathrm{r,e}}(\lambda\sqrt{N})$ in the large deviation regime outside of the support. This is derived in [274] starting from the finite N expression (7.79) with $x = y$, which in turn is used to deduce that then $\langle \mathcal{N} \rangle \to 1$. Moreover, in [274, Eq. (16)] a formula involving (7.41) was given which interpolates between (7.92) and the regime of a single equilibrium for m scaled about unity.

Chapter 8
Analogues of GinUE Statistical Properties for GinOE

A fundamental property of GinOE is that away from the real axis its statistical properties coincide with GinUE. In keeping with this, the limiting variance formula is found to the be the same as for GinUE. Also, the Coulomb gas interpretation of the eigenvalue PDF for GinOE implies, that as for GinOE, screening sum rules linking correlation functions hold true. The singular values and eigenvectors both permit studies analogous to those for GinUE, although in the case of the eigenvectors, there is now an advantage in considering the joint probability distribution of the position of the eigenvalue (assumed real) and the invariant overlap.

8.1 Bulk and Edge Correlations

The bulk correlation kernels for the real–real, real–complex and complex–complex correlations are given by (7.38) and the formulas of Remark 7.4.3. However, with regard to the latter two, it needs to be clarified that here the term bulk is used in the sense of being away from the spectrum boundary, but still in the vicinity of the real axis. Suppose we take the limits $v, y \to \infty$ with $v - y$ fixed in $\mathcal{K}_\infty^c$ of Remark 7.4.3. Then from the large s form of $\mathrm{erfc}(\sqrt{2}s)$ we calculate

$$\lim_{\substack{v,y\to\infty \\ |v-y| \text{ fixed}}} \mathcal{K}_\infty^c(w, z) = \frac{1}{\pi} \begin{bmatrix} 0 & e^{-(|w|^2+|z|^2)/2}e^{w\bar{z}} \\ -e^{-(|w|^2+|z|^2)/2}e^{\bar{w}z} & 0 \end{bmatrix}. \tag{8.1}$$

The Pfaffian of this kernel as required according to the first displayed formula of Remark 7.4.3 with $k_1 = 0, k_2 = k$ to compute the k-point complex–complex correlation therefore reduces to a determinant. Moreover, the correlation kernel in the determinant is precisely that for bulk-scaled GinUE (2.18).

S.-S. Byun and P. J. Forrester, *Progress on the Study of the Ginibre Ensembles*, KIAS Springer Series in Mathematics 3, https://doi.org/10.1007/978-981-97-5173-0_8

The edge scaling of the finite N complex–complex correlation kernel (7.45) is readily calculated from (7.46). Thus in the neighbourhood of the spectrum edge of the real eigenvalues one finds [104, Cor. 9]

$$\lim_{N\to\infty} S_N^c(\sqrt{N}+w, \sqrt{N}+z) = \frac{i(\bar{z}-w)}{\sqrt{2\pi}}\left(\text{erfc}(\sqrt{2}u)e^{2u^2}\text{erfc}(\sqrt{2}y)e^{2y^2}\right)^{1/2}e^{-\frac{1}{2}(w-\bar{z})^2}.$$

(8.2)

Now taking $v, y \to \infty$ with $v - y$ fixed in $\mathcal{K}_\infty^c$ gives for the analogue of (8.1)

$$\lim_{\substack{v,y\to\infty \\ |v-y|\,\text{fixed}}} \mathcal{K}_\infty^{c,e}(w, z) = \begin{bmatrix} 0 & K_\infty^e(w, z) \\ -K_\infty^e(\bar{w}, \bar{z}) & 0 \end{bmatrix}.$$

(8.3)

Here $K_\infty^{c,e}$ denotes the correlation kernel for edge-scaled GinUE as given in (2.21) with $v = 1$. Again the corresponding Pfaffian as determines the edge complex–complex correlations reduces to a determinant, with the latter that for edge-scaled GinUE with origin $(\sqrt{N}, 0)$.

For real random matrices of identical and independently distributed, zero mean and fixed standard deviation entries, a universality result for their edge statistics agreeing with those for GinOE has been established in [159].

8.2 Global Density and Fluctuations

We know that with global scaling $z_j \mapsto \sqrt{N}z_j$, the density for both the GinOE and GinUE obeys the circular law (2.17). In the GinUE case convergence to this limit can be proved by computing the radial moments $\langle \frac{1}{N}\sum_{j=1}^N |r_j|^p\rangle$ $(p = 1, 2, \dots)$; recall Sect. 4.2 and (4.16) with $\beta = 2$. Instead of the radial moments, more natural in the case of GinOE are averages of the eigenvalue power sums $\sum_{j=1}^N z_j^k$. We have

$$\left\langle \sum_{j=1}^N z_j^k \right\rangle = \int_{-\infty}^\infty x^k \rho_{(1),N}^r(x)\,dx + \int_{\mathbb{R}_+^2} ((x+iy)^k + (x-iy)^k)\rho_{(1),N}^c((x,y))\,dxdy.$$

(8.4)

Only for k even is this nonzero, so we write $k = 2p$ with p a non-negative integer. Note that for the GinUE, this average vanishes for all positive integers p, due to rotation invariance. Although the exact functional forms of $\rho_{(1),N}^r(x)$ (7.41) and of $\rho_{(1),N}^c((x,y))$ (7.47) are known, evaluation of (8.4) in a structured form following from direct integration does not seem possible. However, a less direct approach in which the average of a general Schur polynomial in GinOE eigenvalues is first computed, implies the sort structured evaluation [508, Sect. 4] (see also [268])

$$\left\langle \sum_{j=1}^N z_j^{2p} \right\rangle = \prod_{j=1}^p (N + 2p - 2j).$$

(8.5)

Changing variables $z_j \mapsto \sqrt{N}z_j$, and making an ansatz of expansions in $1/\sqrt{N}$,

$$\rho^c_{(1),N}(\sqrt{N}(x, y)) = \frac{1}{\pi}\chi_{|z|<1} + \frac{1}{\sqrt{N}}\mu^c_{(1)}((x, y)) + \cdots,$$

$$\rho^r_{(1),N}(\sqrt{N}x) = \frac{1}{\sqrt{2\pi}}\chi_{|x|<1} + \cdots,$$

the equating of powers of N with (8.5) substituted on the LHS allows deductions about the functional form in the ansatz to be made.

Proposition 8.1 *We have*

$$\mu^c_{(1)}((x, y)) = -\frac{1}{\sqrt{2\pi}}\delta(y)\chi_{|x|<1}, \tag{8.6}$$

which is valid up to the possible addition of a rotationally invariant functional form which integrates to zero.

Proof After making the substitutions, we see that the highest power on the LHS is N^p, while on the RHS the highest power is N^{p+1}, with next highest power of order $N^{p+1/2}$. Thus we begin by equating the term of order N^{p+1} to zero, which gives $\frac{1}{\pi}\int_{|z|<1} z^{2p} d^2z = 0$. This is identically true since p is a positive integer. Equating the terms of order $N^{p+1/2}$ to zero gives

$$\frac{1}{\sqrt{2\pi}}\int_{-1}^{1} x^{2p}\,dx + \int_{\mathbb{C}} z^{2p}\mu^c_{(1)}((x, y))\,d^2z = 0,$$

from which we deduce (8.6). $\qquad\square$

We remark that the result in (8.6) has, in the Coulomb gas picture, the interpretation of perfect screening of the charge density on the unit interval by the charges in the disk. With the next term in the $N^{-1/2}$ expansion of $\rho^r_{(1),N}(\sqrt{N}x)$ denoted by $\mu^r_{(1)}(x)$, the exact functional form (7.41) of $\rho^{r,e}_{(1),\infty}(X)$ can be used for its evaluation. Thus

$$\int_{-\infty}^{\infty} x^{2p}\mu^r_{(1)}(x)\,dx = 2\lim_{N\to\infty} N^{-p}\int_0^{\infty} x^{2p}\left(\rho^r_{(1),N}(x) - \frac{1}{\sqrt{2\pi}}\chi_{|x|<\sqrt{N}}\right)dx$$

$$=2\int_{-\infty}^{\infty}\left(\rho^{r,e}_{(1),\infty}(x) - \frac{1}{\sqrt{2\pi}}\chi_{x<0}\right)dx = \frac{1}{2}, \tag{8.7}$$

where the second equality follows by shifting the origin to the right spectrum edge $x = \sqrt{N}$ according to the change of variables $x \mapsto \sqrt{N} + x$, and the final equality is the explicit evaluation of the integral using (7.41). This implies

$$\mu^r_{(1)}(x) = \frac{1}{4}\Big(\delta(x - 1) + \delta(x + 1)\Big). \tag{8.8}$$

Let the order $1/N$ term in the expansion of $\rho^c_{(1),N}(\sqrt{N}(x, y))$ be denoted by $v^c_{(1)}((x, y))$. Equating terms of order N^p on both sides of (8.5) and using too (8.8) gives

$$\int_{\mathbb{C}} z^{2p} v^c_{(1)}((x, y)) \, d^2 z = \frac{1}{2}. \tag{8.9}$$

However, unlike the circumstance for (8.7) this does not allow us to deduce $v^c_{(1)}((x, y))$. Instead, returning to (7.47), with the knowledge that the ratio of the incomplete gamma function to the gamma function therein approaches unity exponentially fast in N for $|z| < 1$ (see below (4.17) for references), simple asymptotics associated with $\mathrm{erfc}(\sqrt{2N}y)$ give the expansion [158, Remark 2.6]

$$v^c_{(1)}((x, y)) = \frac{1}{\pi} - \frac{1}{4\pi N y^2} + \mathrm{O}\Big(\frac{1}{N^{3/2}}\Big). \tag{8.10}$$

While an order $1/N$ correction is clearly distinguished, in general this cannot be integrated in the full unit disk due to the order $1/y^2$ singularity as the real axis is approached. In fact it has been proved by Cipolloni, Erdős and Schröder [158, Th. 2.2] that for a large class of test functions f,

$$\Big\langle \frac{1}{N} \sum_{j=1}^{N} f(z_j)\Big\rangle^{\mathrm{g}} - \frac{1}{\pi} \int_{|z|<1} f(z) \, d^2 z$$

$$= \frac{1}{N}\Big(\frac{1}{8\pi} \int_{|z|<1} \nabla^2 f(z) \, d^2 z + \frac{1}{4\pi} \int_{|z|<1} \frac{f(x) - f(z)}{y^2} \, d^2 z + \frac{1}{2\pi} \int_{-1}^{1} \frac{f(x)}{\sqrt{1-x^2}} \, dx$$

$$- \frac{1}{2\pi} \int_{0}^{2\pi} f(e^{i\theta}) \, d\theta + \frac{1}{4}\Big(f(x-1) + f(x+1)\Big)\Big) + \mathrm{o}(N^{-1}), \tag{8.11}$$

where the superscript "g" on the average indicates the use of global scaling. Here $f(x) = f(z)|_{y=0}$ and similarly the interpretation of $f(e^{i\theta})$. This shows that the correction (8.6) completely cancels the expected term from the density of real eigenvalues at order $N^{-1/2}$. The final term is identified with the real density correction (8.8). For f with support off the real axis, the contribution from (8.11) is recognised. However, if this is not the case, a formula for $v^c_{(1)}((x, y))$ cannot be read off. It can be verified that with $f(z) = z^{2p}$, (8.11) is consistent with the leading N form of (8.5) [158, Remark 2.7]. Also, we can check that the RHS vanishes for f a constant, as it must.

A fundamental question is the variance associated with the linear statistic $\sum_{j=1}^{N} f(z_j)$, and furthermore its distribution. In the case of GinUE, we know that the behaviour is different depending on the smoothness of f. On the other hand, we have already seen for the real eigenvalues of GinOE that the fluctuations are proportional to $\sqrt{N}$ independent of this property. However, for smooth f it is known that when averaging over all the eigenvalues, this effect is suppressed, and again the limiting variance is of order unity [158, 370, 453].

Proposition 8.2 *Let f, g be smooth real or complex-valued functions in the plane, and subject to a constraint on their growth. Define the Fourier components of their Fourier expansion on the unit circle in the plane as in Proposition 3.5. Define too $f^s = \frac{1}{2}(f(x, y) + f(x, -y))$ and similarly the meaning of g^s. We have*

$$
\lim_{N \to \infty} \mathrm{Cov}^{\mathrm{GinOE}}\left(\sum_{j=1}^N f(\mathbf{r}_j/\sqrt{N}), \sum_{j=1}^N \bar{g}(\mathbf{r}_j/\sqrt{N}) \right)
$$

$$
= \frac{1}{2\pi} \int_{|\mathbf{r}|<1} \nabla f^s \cdot \nabla \bar{g}^s \, dx dy + \sum_{n=-\infty}^{\infty} |n| f_n^s \bar{g}_{-n}^s; \quad (8.12)
$$

cf. (3.21).

Proof (Comments only) With $X \in \mathrm{GinOE}$, define the $2N \times 2N$ Hermitian matrix

$$
H^z = \begin{bmatrix} 0_{N \times N} & X - z\mathbb{I}_N \\ X^\dagger - \bar{z}\mathbb{I}_N & 0_{N \times N} \end{bmatrix},
$$

and write $G^z(w) = (H^z - w\mathbb{I}_N)^{-1}$ for the resolvent of H^z. The starting point of the calculations of [158] is the formula

$$
\sum_{j=1}^N f(z_j) = -\frac{1}{4\pi} \int_{\mathbb{C}} \nabla^2 f(z) \int_0^\infty \mathrm{Im} \, \mathrm{Tr} \, G^z(i\eta) \, d\eta d^2 z,
$$

where $\{z_j\}$ are the eigenvalues of X (both real and complex). Hence the task is reduced to first finding the limiting covariance for the traces of the resolvents. Aiding this is the fact that for large N, $G^z(w)$ becomes approximately deterministic, with its limit expressed in terms of the solution of the scalar equation

$$
-\frac{1}{m^z} = w + m^z - \frac{|z|^2}{w + m^z}, \quad \eta \mathrm{Im} \, m^z > 0, \quad \eta = \mathrm{Im} \, w \neq 0,
$$

which is a special case of the matrix Dyson equation; see [218] for an introduction to the latter. □

Remark 8.1 1. It is elementary to compute that for $G \in \mathrm{GinOE}$,

$$
\mathrm{Var}^g(\mathrm{Tr} \, G) := \langle (\mathrm{Tr} \, G)^2 \rangle^g - (\langle \mathrm{Tr} \, G \rangle^g)^2 = 1. \quad (8.13)
$$

Since $\mathrm{Tr} \, G = \sum_{j=1}^N z_j$, this corresponds to the circumstance that $f^s = g^s = x$ on the RHS of (8.12), and furthermore $f_n^s = g_n^s = \frac{1}{2}$ for $n = \pm 1$, $f_n^s = g_n^s = 0$ otherwise. We verify agreement between the value obtained from the general formula (8.12) and (8.13).

2. In the setting of Proposition 8.2, and after centring by subtracting from the linear

statistic its mean, the results of [158, 370, 453] give that for $N \to \infty$ a central limit theorem holds, whereby the distribution is a mean zero Gaussian with variance as implied by (8.12).

3. For the counting function $\sum_{j=1}^{N} \chi_{z_j/\sqrt{N} \in \mathcal{D}}$, where the set $\mathcal{D}$ contained entirely within the upper half disk $|z| < 1$, $\mathrm{Im}\, z > 0$, the recent work [297] makes use of (a refinement of) (8.1) to establish that (3.7) is true for the complex eigenvalues of the GinOE. The variance therein is that implied by (3.9) and thus also as for the GinUE.

8.3 Sum Rules

In this section we view the eigenvalues from the Coulomb gas perspective of Sect. 7.2. Suppose we fix a real eigenvalue at the origin. Requiring that the corresponding total charge be cancelled by a redistribution of all the other charges implies the sum rule

$$2 \int_{\mathbb{C}_+} \rho_{(2),\infty}^{\mathrm{c,b},T}(0, z)\, d^2 z + \int_{-\infty}^{\infty} \rho_{(2),\infty}^{\mathrm{r,b},T}(0, y)\, dy = -\rho^{\mathrm{r}}. \tag{8.14}$$

If instead the charge was fixed in the upper half plane at a point z_0, then (8.14) would need to be modified to read

$$2 \int_{\mathbb{C}_+} \rho_{(2),\infty}^{\mathrm{c,b},T}(z_0, z)\, d^2 z + \int_{-\infty}^{\infty} \rho_{(2),\infty}^{(\mathrm{c,r}),\mathrm{b},T}(z_0, x)\, dx = -2\rho_{(1),\infty}^{\mathrm{c,b}}(z_0). \tag{8.15}$$

As for in the one-component case (recall Sect. 4.3) the sum rules (8.14) and (8.15) can be generalised to involve higher point correlation functions. In stating the general result it is convenient to make use of the truncated correlations associated with the latter; see e.g. [237, Sect. 5.1.1] for their definition.

Proposition 8.3 *We have*

$$\int_{-\infty}^{\infty} \rho_{(k_1+1,k_2),\infty}^{T}(\{x_j\}_{j=1,\dots,k_1} \cup \{y\}; \{z_j\}_{j=1,\dots,k_2})\, dy$$

$$+2 \int_{\mathbb{R}_+^2} \rho_{(k_1,k_2+1),\infty}^{T}(\{x_j\}_{j=1,\dots,k_1}; \{z_j\}_{j=1,\dots,k_1} \cup \{z\})\, d^2 z$$

$$= -(k_1 + 2k_2)\rho_{(k_1,k_2),\infty}^{T}(\{x_j\}_{j=1,\dots,k_1}; \{z_j\}_{j=1,\dots,k_2}). \tag{8.16}$$

Proof Underlying this sum rule are the matrix correlation kernel identities

$$\int_{-\infty}^{\infty} \mathcal{K}^{\mathrm{r}}(x, u)\mathcal{K}^{\mathrm{r}}(u, y)\, du + 2 \int_{\mathbb{C}_+} \mathcal{K}^{\mathrm{r,c}}(x, z)\mathcal{K}^{\mathrm{c,r}}(z, y)\, d^2 z$$

$$= \mathcal{K}^{\mathrm{r}}(x, y)\begin{bmatrix} 0 & 0 \\ 0 & 1 \end{bmatrix} + \begin{bmatrix} 1 & 0 \\ 0 & 0 \end{bmatrix}\mathcal{K}^{\mathrm{r}}(x, y),$$

$$\int_{-\infty}^{\infty} \mathcal{K}^{c,r}(z, y)\mathcal{K}^{r,c}(y, v)\, dy + 2\int_{\mathbb{C}_+} \mathcal{K}^c(z, w)\mathcal{K}^c(w, v)\, d^2w = 2\mathcal{K}^c(z, v),$$

$$\int_{-\infty}^{\infty} \mathcal{K}^{c,r}(z, y)\mathcal{K}^r(y, x)\, dy + 2\int_{\mathbb{C}_+} \mathcal{K}^c(z, w)\mathcal{K}^{c,r}(w, x)\, d^2w = \mathcal{K}^{c,r}(z, x),$$

$$\int_{-\infty}^{\infty} \mathcal{K}^r(x, y)\mathcal{K}^{r,c}(y, z)\, dy + 2\int_{\mathbb{C}_+} \mathcal{K}^{r,c}(x, w)\mathcal{K}^c(w, z)\, d^2w = \mathcal{K}^{r,c}(x, z).$$

Further details are given in [265, Sect. 4.4]. We remark that identities of this sort in relation to the correlation kernel for a random matrix ensemble with orthogonal symmetry first appeared in [208, 419]. □

As discussed in Sect. 4.3, the fast decay of the correlations implies that not only the total charge in the screening cloud, but also its integer moments should vanish. We will illustrate this in relation to (8.15). First, in forming the moments we should interpret $2\rho_{(2),\infty}^{c,b,T}(z_0, z)$ as $\rho_{(2),\infty}^{c,b,T}(z_0, z) + \rho_{(2),\infty}^{c,b,T}(z_0, \bar{z})$, thus making the contribution of the image explicit. Similarly, we should interpret $-2\rho_{(1),\infty}^c(z_0)$ as the integral over z of $-\delta(z_0 - z)\rho_{(1),\infty}^{c,b}(z) - \delta(z_0 - \bar{z})\rho_{(1),\infty}^{c,b}(\bar{z})$. Weighting the integrands by $(z - z_0)^k$ then gives the k-th moments. We can see that this is identically zero for k odd. Our interest then is in the sum rule implied by the $k = 2p$ even case.

Proposition 8.4 *For p a non-negative integer we have*

$$2\int_{\mathbb{R}_+^2} w^{2p} \rho_{(0,2)}^T(z, w)\, d^2w + \int_{-\infty}^{\infty} x^{2p} \rho_{(1,1)}^T(z, x)\, dx = -2z^{2p} \rho_{(1),\infty}^{c,b}(z). \quad (8.17)$$

Proof Multiplying by $\alpha^p / p!$, for $|\alpha| < 1$ a parameter and summing over p replaces the corresponding terms by exponentials. This transformed identity can be checked to be a corollary of an appropriately exponential weighted version of the second of the sum rules listed in the proof of Proposition 8.3 with $v = z$,

$$\int_{-\infty}^{\infty} e^{\alpha y^2}\mathcal{K}^{c,r}(z, y)\mathcal{K}^{r,c}(y, z)\, dy$$

$$+ 2\int_{\mathbb{C}_+} e^{\alpha w^2}\mathcal{K}^c(z, w)\mathcal{K}^c(w, z)\, d^2w = 2e^{\alpha z^2}\mathcal{K}^c(z, z).$$

This can be checked component-wise; see [265, proof of Prop. 4.8]. □

8.4 Singular Values

Matrices X from GinOE being real, the squared singular values are the eigenvalues of $X^T X$. With the meaning of GinOE extended to include rectangular $p \times N$ matrices, the random matrices $X^T X$ have an interpretation in multivariate statistics. Thus one interprets each column as a particular trait that is being measured from a population of size N so that X is regarded as a data matrix. Moreover, let the distribution of each of the traits be a standard Gaussian, which serves as a structureless base case. Up to a simple normalisation factor, $X^T X$ is the matrix of sample covariances between the traits. Because of its statistical interest, this class of random matrix attracted attention in one of the earliest works relating to random matrix theory [534].

The result of [534] was to establish that the Jacobian for the change of variables $A = X^T X$ is proportional to $(\det A)^{(p-N-1)/2}$. Around a decade later, the Jacobian for the change of variables from a real symmetric matrix to its eigenvalues and eigenvectors was calculated [327]. This exhibited a factorised form, with Jacobian $\prod_{1 \le j < k \le N} |\lambda_k - \lambda_j|$ relating only to the eigenvalues. Denoting the squared singular values of a $p \times N$ rectangular GinOE matrix by $\{s_j\}_{j=1}^N$, it then follows that the corresponding joint PDF is proportional to

$$\prod_{j=1}^N s_j^{(p-N-1)/2} e^{-s_j/2} \prod_{1 \le j < k \le N} |s_k - s_j|. \tag{8.18}$$

Introducing the global scaling $X^T X \mapsto \frac{1}{N} X^T X$, and with p scaling with N according to $p = \alpha N$, $\alpha > 1$, it is a well-known result [288] that the smallest and largest squared singular values tend almost surely to the values $(\sqrt{\alpha \mp 1} + 1)^2$. Hence the ratio of the largest to the smallest singular value (i.e. the condition number κ_N of X) tends to a constant. However, with $p = N$ (or more generally $p = N + p_1$ with p_1 independent of N), while the largest squared singular value tends to 4, the smallest now tends to zero at the rate $1/N^2$. Specifically in the square case $p = N$, it has been shown by Edelman [210] that this implies κ_N/N has a limiting distribution with the heavy tailed PDF

$$\frac{2x + 4}{x^3} e^{-2/x - 2/x^3}. \tag{8.19}$$

In the rectangular case, for any $N \ge 2$, a result of [153] gives the bound

$$\langle \log \kappa_N \rangle < \frac{N}{|p - N| + 1} + 2.258,$$

which apart from the last two digits of the constant, is the same as that known for rectangular GinUE matrices (recall Sect. 6.3).

Shifting a global-scaled square GinOE matrix $\tilde{X} = \frac{1}{\sqrt{N}} X$ by defining

$$Y = -z\mathbb{I}_N + \tilde{X} \tag{8.20}$$

leads to a transition effect for the corresponding smallest singular value in the neighbourhood of $|z| = 1$. Thus for $|z| > 1$ and independent of N, the smallest singular value is bounded away from 0 as $N \to \infty$. The transition region $|z| < 1 + c/N^{1/2}$ is studied in [157], with a distinction between z complex and $z \approx \pm 1$ quantified.

For X a square GinOE matrix, we turn our attention to the distribution of $(\det X)^2$, which is well known in multivariate statistics [442].

Proposition 8.5 *For X a GinOE matrix*

$$(\det X)^2 \overset{\mathrm{d}}{=} \prod_{j=1}^{N} \chi_j^2. \tag{8.21}$$

Proof Either of the strategies used to establish (6.13) can be used. Specifically, to make use of (8.18) (with $p = N$), we use the fact that $(\det X)^2 = \prod_{l=1}^{N} s_l$ and hence that the Mellin transform of the distribution of $(\det X)^2$ is given by

$$\frac{1}{C_N} \int_0^\infty ds_1 \cdots \int_0^\infty ds_N \prod_{j=1}^{N} s_j^{s-3/2} e^{-s_j/2} \prod_{1 \le j < k \le N} |s_k - s_j|^2$$

$$= \prod_{j=1}^{N} \frac{2^{s-1} \Gamma(s + j/2)}{\Gamma(1 + j/2)},$$

where C_N is the same multiple integral in the case $s = 1$. The gamma function evaluation follows from the Laguerre weight Selberg integral as given in, e.g. [237, Proposition 4.7.3]. It follows from this that $(\det X)^2 \overset{\mathrm{d}}{=} \prod_{j=1}^{N} F_j$, where F_j is independent with the Mellin transform of its PDF given by $\frac{2^s \Gamma(s+j/2)}{\Gamma(1+j/2)}$. One verifies that thus F_j is equal to χ_j^2, as required.   □

There is an interpretation of $(\det X)^2$ in integral geometry. The idea is to regard the columns of X as vectors in $\mathbb{R}^N$. Then $|\det X|$ is equal to the volume of the parallelotope generated by these vectors. For X a GinUE matrix, the parallelotope is random, and said to be Gaussian. Then (8.21) gives the distribution of the squared volume. For large N, (8.21) can be used to establish that the distribution of $\log(\det X)^2$ has, to leading-order mean equal to $-N$, variance $2 \log N$, and after recentring and rescaling satisfies a central limit theorem [428, 439].

Let Y_k denote the shifted GinOE matrix (8.20) restricted to the first k columns. For z real the positive definite matrix $W_k = Y_k^T Y_k$ is an example of a non-central Wishart matrix, well known in mathematical statistics [442]. An exact evaluation of $\langle (\det W_k)^\alpha \rangle$ in terms of a generalised hypergeometric function of k variables has been given in [169]; see [269, Sect. 4] for a discussion of the implications of this result in relation to the Lyapunov spectrum for products of shifted GinOE matrices.

Remark 8.2 3. The squared singular values as specified by the PDF (8.18) specify the classical Laguerre orthogonal ensemble and form a Pfaffian point process, see e.g. [237, Chaps. 3 and 6]. However, there is no known analogous result for elliptic GinOE. Also, with regards to the squared singular values of the various extensions of GinOE to be considered below, in particular the spherical model and truncated real orthogonal matrices, the classical Jacobi orthogonal ensemble results, which is a

Pfaffian point process. However, no such structure is known for the squared singular values of products of GinOE matrices.

8.5 Real GinOE Eigenvector Statistics

The basic question of interest is, as for GinUE, the statistical properties of the scaled invariant overlaps of the left and right eigenvectors, $O_{ij} := \langle \ell_i, \ell_j \rangle \langle r_i, r_j \rangle$. In the diagonal case, the approach of Fyodorov [272] was to study the joint PDF

$$
\mathcal{P}^{\mathrm{r}}(t, \lambda) := \Big\langle \sum_{j=1}^{N} \delta(\tilde{O}_{jj} - 1 - t)\delta(\lambda - \lambda_j) \Big\rangle, \qquad \tilde{O}_{jj} := O_{jj}/N,
$$

and where λ is restricted to real values (thus the superscript "r"). This was shown to permit an exact evaluation generalising the GinOE eigenvalue density formula for $\rho_{(1),N}^{\mathrm{r}}(\lambda)$, (7.31) with $x = y$ [272, Th.2.1 and Eq. (2.5)].

Proposition 8.6 *We have*

$$
\mathcal{P}^{\mathrm{r}}(t, \lambda) = \frac{1}{2\sqrt{2\pi}} e^{\frac{\lambda^2}{2}\frac{1}{1+t}} \frac{1}{t(1+t)}\left(\frac{t}{1+t}\right)^{(N-1)/2}
$$
$$
\times \left(\frac{e^{-\lambda^2}\lambda^{2(N-1)}}{(N-2)!} + \frac{1}{(N-2)!}\Gamma(N-1; \lambda^2)\Big((N-1) - \lambda^2\frac{t}{1+t}\Big)\right).
$$
(8.22)

Proof (Comments only) The first step is to introduce the Laplace transform $\int_0^\infty e^{-pt}\mathcal{P}^{\mathrm{r}}(t, \lambda)\,dt$. Next, it was shown that this Laplace transform can be expressed in terms of a GinOE average involving ratios of characteristic polynomials

$$
\left\langle \frac{\det((zI_N - G)(\bar{z}I_N - G^\dagger))}{(\det(2pI_N + (zI_N - G)(\bar{z}I_N - G^\dagger))^{1/2}} \right\rangle.
$$

Supersymmetric integration techniques are then used to evaluate this average. □

We remark that an extension of this proposition to the elliptic GinOE has been given in [278, Th. 2.1]. From the definition $\rho_{(1),N}^{\mathrm{r}}(\lambda) = \int_0^\infty \mathcal{P}^{\mathrm{r}}(t, \lambda)\,dt$, which is readily verified. Bulk and edge scaling limits follow [272, Eqs. (2.7) and (2.10)].

Corollary 8.1 *We have*

$$
P^{\mathrm{r,b}}(s, x) := \lim_{N\to\infty} N\mathcal{P}^{\mathrm{r}}(Ns, \sqrt{N}x) = \frac{1}{2\sqrt{2\pi}} \frac{e^{-\frac{(1-x^2)}{2s}}}{s^2}(1 - x^2)\chi_{|x|<1} \qquad (8.23)
$$

and

$$P^{\mathrm{r,e}}(s, x) := \lim_{N \to \infty} \sqrt{N} \mathcal{P}^{\mathrm{r}}(\sqrt{N}s, \sqrt{N} + \delta)$$

$$= \frac{1}{2\sqrt{2\pi}} \frac{1}{\sigma^2} e^{-\frac{1}{4\sigma^2} + \frac{\delta}{\sigma}} \left(\frac{1}{\sqrt{2\pi}} e^{-2\delta^2} + \left(\frac{1}{\sigma} - 2\delta \right) \mathrm{erfc}(\sqrt{2}\delta) \right). \quad (8.24)$$

Here one has the sum rules

$$\int_0^\infty P^{\mathrm{r,b}}(s, x)\, ds = \frac{1}{\sqrt{2\pi}} \chi_{|x|<1}, \qquad \int_0^\infty P^{\mathrm{r,b}}(s, x)\, ds = \rho_{(1),\infty}^{\mathrm{r,e}}(x)$$

reclaiming the global and edge-scaled densities (7.36) and (7.41). The result (8.23) is the real analogue of the result for GinUE (6.24). Note that the tail is now proportional to $1/s^2$, so only the zeroth integer moment in s (which gives the global density) is defined. In particular, the mean value with respect to s, which for GinUE is given by the result of Mehlig and Chalker (6.26), formally diverges, telling us that for GinOE, O_{ii} averaged over the full spectrum is no longer proportional to N. If instead O_{ii} is conditioned on a single eigenvalue away from the real away from the real axis, numerical experiments from [160, Fig. 3] indicate that O_{ii} is proportional to N. Hence eigenvalues on (or close to) the real axis are responsible for this breaking down in general.

Chapter 9
Further Extensions to GinOE

As for GinUE, GinOE permits a number of related ensembles which share many of its special properties. One of these, already introduced in the previous chapter, is elliptic GinOE. Considered in this chapter are the induced GinOE, which relates to a polar decomposition involving a rectangular generalisation of a GinOE matrix and a random Haar real orthogonal matrix; the real spherical ensemble of matrices $G_1^{-1} G_2$, with both G_1, G_2 GinOE matrices; the sub-block of a Haar real orthogonal matrix; and products of GinOE matrices. Common structures and statistical features are emphasised, with again the occurrence of real eigenvalues responsible for topics of study beyond those for GinUE and its extensions.

9.1 Common Structures

Recall that for a given parameter $\tau \in [0, 1)$, the elliptic GinOE is defined in terms of (anti-)symmetric GOE matrices by (7.71), follows the distribution (7.72), and recovers the GinOE when $\tau = 0$. Comparison between the development of the theory of the eigenvalue statistics for GinOE and elliptic GinOE shows a number of common structures:

(S1) The joint eigenvalue PDF has the functional form (1.10) for a certain weight function $\omega(z)$, and a certain normalisation C_N; cf. (7.74). As a consequence the summed generalised partition function $Z_N[u, v]$ has the Pfaffian form (7.7) involving the same weights and normalisation.

(S2) The functional form for the joint eigenvalue PDF in the sector that all eigenvalues are real leads to a closed form expression for the probability that all eigenvalues are real.

(S3) Associated with $Z_N[u, v]$ in the case $u = v = 1$ is a particular skew inner product, while associated with the latter are a family of skew-orthogonal polynomials. These skew-orthogonal polynomials admit the matrix theoretic form

© The Author(s) 2025
S.-S. Byun and P. J. Forrester, *Progress on the Study of the Ginibre Ensembles*, KIAS Springer Series in Mathematics 3, https://doi.org/10.1007/978-981-97-5173-0_9

(7.18), and their normalisations r_{j-1} can be computed from C_N. Moreover, the odd degree skew-orthogonal polynomials can be written in a derivative form involving the weight $(\omega(x))^{1/2}$.

(S4) The general m-point correlation function between real–real, complex–complex and real–complex eigenvalues is of the form of a Pfaffian point process and as such are determined by particular 2×2 correlation kernels. The latter in turn are in fact fully determined by their upper off-diagonal entry, denoted by $S_N(x, y)$. In the real–real case, due to the derivative formula for the odd degree skew-orthogonal polynomials, and a further derivative formula for a certain linear combination of successive even degree skew-orthogonal polynomials, the normalisations and the weight $(\omega(x))^{1/2}$, the summation can be written in a simplified form involving a rank one perturbation of the kernel known in the case of the GinUE and elliptic GinUE. Another form valid in both cases is (7.43), while its analogue in the complex–complex case (7.46) is similarly valid in both cases.

(S5) The structural form of the large N asymptotic formula (7.54) for the variance of the number of real eigenvalues is valid, or more generally the formula (7.58) for the variance of a linear statistic.

(S6) The global density of the complex eigenvalues is the same as found for GinUE (the circular law) and elliptic GinUE.

As we further develop the theory of ensembles related to GinOE below, we will see that most of the properties listed above again hold true.

9.2   Induced GinOE

Let G be an $n \times N$ $(n \geq N)$ random matrix with independent real standard Gaussian entries. Let R be a Haar distributed real orthogonal matrix (for this notion see [194]). In the case that G is square, it is easy to check that $(G^T G)^{1/2} R$ has the same element distribution as G, and thus is an element of GinOE; see e.g. Proposition 2.10 in the case that G is an element of GinUE, and R is a Haar complex unitary matrix. In the rectangular case, the element distribution of $\tilde{G} = (G^T G)^{1/2} R$ is the same as the element distribution of G multiplied by the factor $(\det G^T G)^{(n-N)/2}$ [228]; cf. Proposition 2.11. The essential fact required to establish this is that the factor is the Jacobian for the change of variables $A = (G^T G)^{1/2}$, which itself can be established in two steps: first change variables $G^\dagger G = B$, then change variables $B = A^2$. The Jacobian in both these steps can be read off from the change of variables required in the theory of real Wishart matrices; see [442, Ch. 2].

Of relevance is the explicit form of the proportionality coefficient.

Proposition 9.1 *Set $L := n - N$. The element distribution on $N \times N$ real random matrices $\tilde{G}$ specified by*

$$\frac{1}{C_{L,N}}(\det \tilde{G}^T \tilde{G})^{L/2} e^{-\operatorname{Tr} \tilde{G}^T \tilde{G}/2}, \quad C_{L,N} = \pi^{N^2/2} 2^{N^2/2+NL/2} \prod_{j=1}^{N} \frac{\Gamma((j+L)/2)}{\Gamma(j/2)}$$

is correctly normalised.

Proof According to the QR decomposition, we can write $\tilde{G} = QR$, where Q is an $N \times N$ real orthogonal matrix, and R is an $N \times N$ real upper triangular matrix with positive diagonal entries. With this change of variables, it is known that [442, Th. 2.1.13 with $n = m = N$],

$$(d\tilde{G}) = \prod_{j=1}^{N} r_{jj}^{N-j}(dR)(Q^T dQ),$$

where $(Q^T dQ)$ is the Haar measure on real orthogonal matrices. Hence

$$\int (\det \tilde{G}^\dagger \tilde{G})^{L/2} e^{-\frac{1}{2}\operatorname{Tr} \tilde{G}^\dagger \tilde{G}} (d\tilde{G})$$

$$= \left(\prod_{j=1}^{N} \int_0^\infty r^{L+j-1} e^{-\frac{1}{2}r^2} dr\right)\left(\int_{-\infty}^\infty e^{-\frac{1}{2}r^2} dr\right)^{N(N-1)/2} \int (Q^T dQ).$$

Computing the integrals over t, and using the known result for the volume of the orthogonal group [442, Corollary 2.1.16]

$$\int (Q^T dQ) = \frac{2^N \pi^{N(N+1)/4}}{\prod_{l=1}^{N} \Gamma(l/2)}, \tag{9.1}$$

we can identify the RHS with $C_{n,N}$. $\qquad \square$

For random matrices $\{\tilde{G}\}$—referred to as induced GinOE matrices—the element distribution of Proposition 9.1 tells us that the eigenvalue PDF in the sector of k real eigenvalues is again given by (1.10), multiplied by $\frac{C_{0,N}}{C_{L,N}} \prod_{l=1}^{k} |\lambda_l|^L \prod_{l=1}^{(N-k)/2} |z_l|^L$. With $Z_N^{\text{iGinOE}}[u, v]$ denoting the analogue of $Z_N[u, v]$ as specified above (7.7), we see that (7.7) again holds true, but with the modification that there is an extra factor of $C_{0,N}/C_{L,N}$, and that $\alpha_{j,k}, \beta_{j,k}$ are replaced by $\alpha_{j,k}^i, \beta_{j,k}^i$, defined as in Proposition 7.2, but with extra factors of $(xy)^L$ and $(x^2 + y^2)^L$ in the integrands respectively. Most crucially, the formulas (7.19) for the skew-orthogonal polynomials corresponding to $(\alpha_{j,k}^i + \beta_{j,k}^i)|_{u=v=1}$ are again valid with $\langle \operatorname{Tr} G^2 \rangle = (2j + L)$ (we replace the index n in (7.19) by j to avoid conflict with the n appearing in the definition of the induced GinOE, and specifically the parameter L). Moreover, with $C_j(z) = z^j$ they permit the structural form

$$p_{2j}^i(z) = C_{2j}(z), \quad p_{2j+1}^i(z) = -z^{-L} e^{z^2/2} \frac{d}{dz}\left(z^L e^{-z^2/2} C_{2j}(z)\right) \tag{9.2}$$

cf. (7.77), and the derivation of (7.21) implies

$$r^i_{j-1} = 2\sqrt{2\pi}\,\Gamma(L + 2j - 1). \tag{9.3}$$

Taking into consideration these modifications, we can check that the determinant formula (7.10) remains valid for $Z^i_N(\zeta)$, except that L needs to be added to the argument of each of the gamma functions. As a consequence, for the expected number of real eigenvalues, we obtain the formula

$$E^{r,i}_N = \sqrt{\frac{2}{\pi}} \sum_{k=1}^{N/2} \frac{\Gamma(L + 2k - 3/2)}{\Gamma(L + 2k - 1)}, \tag{9.4}$$

cf. the first equality in (7.13). From the determinant formula the validity of the analogue of the local central limit theorem Proposition 7.3 can also be checked. Another consequence of (9.2) and (9.3), substituted in the formula for S^r_N given in (7.24), together with the validity of a derivative formula analogous to (7.33), is the simplified summation form

$$S^{r,i}_N(x, y) = \frac{e^{-(x^2+y^2)/2}}{\sqrt{2\pi}} \sum_{k=0}^{N-2} \frac{(xy)^{k+L}}{(k + L)!} + \frac{y^{L+N-1}e^{-y^2/2}}{2\sqrt{2\pi}} \frac{\Phi_{N-2}(x)}{(L + N - 2)!}, \tag{9.5}$$

which reduces to (7.31) when $L = 0$.

In studying the large N limit, the most interesting circumstance is to simultaneously set $L = N\alpha$, ($\alpha > 0$). It follows from (9.4) that to leading-order the expected number of real eigenvalues is then $\sqrt{2N/\pi}(\sqrt{\alpha + 1} - \sqrt{\alpha})$. Setting $x = y$ in (9.5) gives the eigenvalue density. Further introducing the global scaling $x \mapsto \sqrt{N}x$ we can check that the final term goes to zero, while up to proportionality the summation can be identified with $K^{iG}_{N-1}(x, x)$ as given by (2.46). From the known global-scaled limit of the latter (2.47) it follows that

$$\lim_{N \to \infty} \rho^{r,i}_{(1),N}(\sqrt{N}x) = \frac{1}{\sqrt{2\pi}}\left(\chi_{|x|<\sqrt{\alpha+1}} - \chi_{|x|<\sqrt{\alpha}}\right). \tag{9.6}$$

Note that this can be used to provide an alternative derivation of the leading form of the expected number of real eigenvalues as stated above. Bulk and edge scalings in this limit exhibit the same functional forms as for GinOE, given by (7.37) and (7.40) respectively. The former of these implies that the analogue of (7.54) remains valid. Also, using (7.44), the analogue of (7.46) can be obtained for $S^{c,i}_N(w, z)$, thus giving an expression in terms of the induced GinUE kernel $K^{iG}_{N-1}(w, z)$, with in fact precisely the same prefactor. A consequence is that the global scaling limit of the complex eigenvalues coincides with that of the induced GinUE (2.47),

$$\lim_{N \to \infty} \rho^{c,i}_{(1),N}(\sqrt{N}x) = \frac{1}{\pi}\left(\chi_{|z|<\sqrt{\alpha+1}} - \chi_{|z|<\sqrt{\alpha}}\right). \tag{9.7}$$

9.3 Spherical GinOE

For (X_1, X_2) a pair of $N \times N$ matrices, the generalised eigenvalues are defined as the solution of the equation $\det(X_1 - \lambda X_2) = 0$. We see that for X_2 invertible, the generalised eigenvalues then coincide with the eigenvalues of $X_2^{-1} X_1$. Here our interest is in the case that both X_1, X_2 are GinOE matrices—then the ensemble specified by $\{X_2^{-1} X_1\}$ is referred to as the spherical GinOE.

Already in the pioneering paper on this topic, it was identified that statistical properties of real eigenvalues relate to integral geometry [212]. To see this, the matrix pair (X_1, X_2) is viewed as two vectors in $\mathbb{R}^{N^2}$. The plane spanned by these two vectors intersects the sphere S^{N^2-1} to define a great circle. Introducing $X = c(X_1 - \lambda X_2)$, with c such that $\mathrm{Tr}(X^T X) = 1$, we see that real generalised eigenvalues correspond to intersections of the great circle with a unit vector corresponding to a singular matrix of unit Frobenius norm. In fact this viewpoint was used to deduce that the expected number of real eigenvalues, $E_N^{r,s}$ say, is given by

$$E_N^{r,s} = \frac{\sqrt{\pi}\,\Gamma((N+1)/2)}{\Gamma(N/2)} \underset{N\to\infty}{\sim} \sqrt{\frac{\pi N}{2}}\left(1 - \frac{1}{4N} + \cdots\right); \qquad (9.8)$$

cf. (7.13), (7.14). With $E_N^{r,s}$ determined, it was also noted in [212] that the density of real eigenvalues can be deduced. To see this, set $\lambda = \tan\theta$ so the equation determining the generalised eigenvalues can be written $\det(\cos\theta X_1 - \sin\theta X_2) = 0$. Since X_1, X_2, being GinOE matrices, are invariant with respect to rotations it follows that $(\cos\theta, \sin\theta)$ must be uniformly distributed on the unit circle and so λ must have a Cauchy distribution. Consequently,

$$\rho_{(1),N}^{\mathrm{sGinOE}}(\lambda) = \frac{1}{\pi}\frac{E_N^{r,s}}{1+\lambda^2}. \qquad (9.9)$$

Subsequent to [212] it has been established that the eigenvalues of the spherical GinOE form a Pfaffian point process, with (9.8) and (9.9) corollaries of this general structure [265]. We require the normalisation

$$C_N^{\nu s} = \Gamma\left(\nu - \frac{N}{2} + \frac{1}{2}\right)^N \prod_{j=1}^{N} \frac{1}{\Gamma(j/2)\Gamma(\nu - N + j/2)} \qquad (9.10)$$

and weight

$$\omega^{\nu s}(z) = \frac{2}{\sqrt{\pi}}\frac{\Gamma(\nu - \frac{N}{2} + 1)}{\Gamma(\nu - \frac{N}{2} + \frac{1}{2})}\frac{1}{|1+z^2|^{2\nu-N+1}}\int_{2y/(1+|z|^2)}^{\infty}\frac{dt}{(1+t^2)^{\nu-N/2+1}}, \qquad (9.11)$$

where $z = x + iy$. With $z = x$ real, this reduces to $\omega^{\nu s}(x) = (1+x^2)^{-2\nu+N-1}$.

Proposition 9.2 *In the above notation and that of (1.10), the joint eigenvalue PDF for k real eigenvalues $\{\lambda_l\}_{l=1,\ldots,k}$ and the $(N-k)/2$ complex eigenvalues $\{z_j := x_j + iy_j\}_{j=1,\ldots,(N-k)/2}$ in the upper half plane for the spherical GinOE is*

$$C_N^{vs} \frac{2^{(N-k)/2}}{k!((N-k)/2)!} \prod_{s=1}^{k} (\omega^{vs}(\lambda_s))^{1/2} \prod_{j=1}^{(N-k)/2} \omega^{vs}(z_j)$$

$$\times \left| \Delta(\{\lambda_l\}_{l=1,\ldots,k} \cup \{x_j \pm iy_j\}_{j=1,\ldots,(N-k)/2}) \right|, \qquad (9.12)$$

with $v = N$.

Proof To derive (9.12), aspects of the derivation of the eigenvalue PDF for the spherical GinUE, and the GinOE, are required. In order, these are:

(i) Determine the analogue of the joint element distribution (2.49). Using the same method, one obtains that for $Y = A^{-1}B$ with A, B GinOE matrices, the joint element distribution is

$$A_{N,v} \det(\mathbb{I}_N + Y^T Y)^{-v}\Big|_{v=N}, \quad A_{N,v} = \pi^{-N^2/2} \prod_{j=1}^{N} \frac{\Gamma((2v-N+j)/2)}{\Gamma((2v-2N+j)/2)}.$$
$$(9.13)$$

This is an example of the matrix t-distribution [170]. There is some advantage in keeping the variable v general throughout the remainder of the calculation. In particular, knowledge of the corresponding functional forms becomes useful in the computation of the skew-orthogonal polynomials to be undertaken subsequently.

(ii) Introduce the real Schur decomposition $Y = QRQ^T$, conditioned on k real eigenvalues, with the notations of Sect. 7.1. Integrate out the strictly upper triangular entries of R using a modification of the technique used for spherical GinUE Sect. 2.5. This results in the conditional PDF

$$C_N^{vs} \left(\frac{\Gamma(v - \frac{N}{2} + 1)}{\sqrt{\pi}\,\Gamma(v - \frac{N}{2} + \frac{1}{2})} \right)^{(N-k)/2} \left| \tilde{\Delta}(\{\lambda_l\}_{l=1,\ldots,k} \cup \{x_j \pm iy_j\}_{j=1,\ldots,(N-k)/2}) \right|$$

$$\times \prod_{s=1}^{k} \frac{1}{(1+\lambda_s^2)^{v-(N-1)/2}} \prod_{s=1}^{(N-k)/2} \frac{2|b_s - c_s|}{\det(\mathbb{I}_2 + R_s R_s^T)^{v-N/2+1}}, \qquad (9.14)$$

where C_N^{vs} is given by (9.10). The quantity $\tilde{\Delta}$ is defined as for Δ in (1.10) except that the difference factor between each $z_j := x_j + iy_j$ and its complex conjugate is omitted, so that

$$\tilde{\Delta}(\{\lambda_l\}_{l=1,\ldots,k} \cup \{x_j \pm iy_j\}_{j=1,\ldots,(N-k)/2}) = \prod_{1 \le p < q \le k} (\lambda_q - \lambda_p)$$

$$\times \prod_{p=1}^{k} \prod_{l=1}^{(N-k)/2} (\lambda_p - z_l)(\lambda_p - \bar{z}_l) \prod_{1 \le l_1 < l_2 \le (N-k)/2} |z_{l_2} - z_{l_1}|^2 |z_{l_2} - \bar{z}_{l_1}|^2.$$

Each R_s is a 2×2 matrix on the part of the diagonal of the block diagonal matrix R corresponding to the complex eigenvalues.

(iii) The final step is to change variables from $\{b_j, c_j\}$ to $\{y_j, \delta_j\}$ and to integrate over δ_j as for GinOE; recall Sect. 7.1. Now setting $\nu = N$, after some minor manipulation (9.12) results. □

Let $\alpha_{j,k}^{vs}[u]$, $\beta_{j,k}^{vs}[v]$ be defined as for $\alpha_{j,k}^{g}[u]$, $\beta_{j,k}^{g}[v]$ in Proposition 7.2 but with each ω^{g} replaced by ω^{vs} as defined in (9.11). The analogue of (7.7) can then be checked to be

$$Z_N^{vs}[u, v] = C_N^{vs} \operatorname{Pf}[\alpha_{j,k}^{vs} + \beta_{j,k}^{vs}]_{j,k=1,\dots,N}. \tag{9.15}$$

Further progress relies on the polynomials $\{p_{l-1}(x)\}$ having the skew-orthogonality property (7.17) with respect to the skew inner product corresponding to $\gamma_{j,k}^{vs} := \alpha_{j,k}^{vs}|_{u=1} + \beta_{j,k}^{vs}|_{v=1}$. We see from (9.11) that with the replacement $\nu \mapsto (\nu + N)/2$ the weights in the definitions of $\alpha_{j,k}^{vs}$, $\beta_{j,k}^{vs}$ are independent of N. On the other hand, there is a known construction of an $N \times N$ matrix with element distribution (9.13) and this value of ν. Thus form the product $W^{-1/2}B$ with B an $N \times N$ Haar distributed orthogonal matrix and W the Wishart matrix $W = \tilde{G}^T \tilde{G}$ with $\tilde{G}$ a rectangular $\nu \times N$ GinOE matrix—generally the exponent is $1/2$ of the sum of the row and column size in $\tilde{G}$; see e.g. [237, Exercises 3.6 q.3]. It is immediate how to modify the construction to a $2n \times 2n$ matrix with the corresponding weight similarly independent of n. This can then be used in the formulas of Proposition 7.4 to determine the skew-orthogonal polynomials, first derived using evaluations of the averages in (7.18) deduced from symmetric function theory [227, 240].

Proposition 9.3 *Consider the ensemble specified by $N \times N$ matrices with the element distribution (9.13) modified so that $\nu \mapsto (\nu + N)/2$, and the corresponding skew inner product implied by $\gamma_{j,k}^{vs}$. We have*

$$p_{2n}^{vs}(z) = z^{2n},$$

$$p_{2n+1}^{vs}(z) = z^{2n+1} - \frac{2n}{\nu - 2n - 1} z^{2n-1} = -\frac{(1+z^2)^{(\nu+1)/2}}{\nu - 2n - 1} \frac{d}{dz}\left((1+z^2)^{-(\nu-1)/2} z^{2n}\right).$$

Furthermore,

$$r_n^{vs} = \frac{\pi 2^{2-\nu} \Gamma(2n+1)\Gamma(\nu - 2n - 1)}{\Gamma((\nu+1)/2)^2}.$$

Proof From the theory of the previous paragraph, the matrix G in (7.18) can be constructed as the matrix product $W^{-1/2}B$ with B a $2n \times 2n$ GinOE matrix and W a real Wishart matrix constructed from a rectangular $\nu \times 2n$ GinOE matrix. Thus this gives a $2n \times 2n$ random matrix with the same eigenvalue weight (9.11) made independent of N by $\nu \mapsto (\nu + N)/2$. Because the matrix product is isotropic, the distribution of each element unchanged by negation, and moreover the joint first moment of distinct pairs of elements are uncorrelated. As a result the skew-orthogonal polynomials are again given by the simple formula (7.19). For the quantity $\langle \mathrm{Tr}\, G^2 \rangle$ therein, we have

$$\langle \mathrm{Tr}\, G^2 \rangle = \langle \mathrm{Tr}\, B^2 \rangle \langle \mathrm{Tr}\, W^{-1} \rangle = 2n \left(\frac{1}{\nu - 2n - 1} \right).$$

Here we have used the elementary result that for B a $2n \times 2n$ GinOE matrix, $\langle \mathrm{Tr}\, B^2 \rangle = 2n$, and the fact that for a Wishart matrix W constructed from a rectangular $\nu \times 2n$ GinOE matrix, $\langle \mathrm{Tr}\, W^{-1} \rangle = 1/(\nu - 2n - 1)$; see [429, Appendix A], [473].

The formula for the normalisation $r_n^{\nu s}$ again relies on the weights for $\nu \mapsto (\nu + N)/2$ being independent of N, together with the facts that for $u = v = 1$, $Z_N^{\nu s \mathrm{GinOE}}[u, v]$ in (9.15) has the value unity by its construction, while by skew-orthogonality the Pfaffian equals $\prod_{l=0}^{N/2-1} r_l^{\nu s}$. An important point for the calculation is that the product in (9.10) can be written in the form

$$\prod_{l=1}^{N} \frac{1}{\Gamma(l/2)\Gamma(\nu - N + l/2)} \bigg|_{\nu \mapsto (\nu+N)/2} = \prod_{l=1}^{N/2} \frac{1}{\pi \, 2^{2-\nu} \Gamma(2l-1)\Gamma(\nu - 2l + 1)}$$

so that the only dependence on N is in the upper limit of the product. $\square$

Using the derivative formula for $p_{2n+1}^{\nu s}(z)$ from Proposition 9.3 in the definition above (9.15) of $\alpha_{j,k}^{\nu s}$ shows

$$\alpha_{2j-1,2k}^{\nu s} \bigg|_{\substack{\nu \mapsto (\nu+N)/2 \\ u=v=1}} = \frac{2}{\nu - 2k + 1} \frac{\Gamma(j + k - 3/2)\Gamma(3/2 - j - k + \nu)}{\Gamma(\nu)} =: \frac{\tilde{\alpha}_{j,k}^{\nu s}}{\nu - 2k + 1}.$$

With $\tilde{r}_{j-1}^{\nu s} := (\nu - 2j + 1) r_{j-1}^{\nu s}$ It follows that the analogue of (7.10) is the formula

$$Z_N^{\nu s}(\zeta) \bigg|_{\nu \mapsto (\nu+N)/2} = \det \left[\delta_{j,k} + \frac{\zeta - 1}{(\tilde{r}_{j-1}^{\nu s} \tilde{r}_{k-1}^{\nu s})^{1/2}} \tilde{\alpha}_{j,k}^{\nu s} \right]_{j,k=1,\ldots,N/2}. \tag{9.16}$$

This can be used to deduce the analogue of (7.54) and the local central limit theorem Proposition 7.3 (the case $\nu = N$ of the local central limit theorem is known from the earlier study [265, Proposition 3.5], based instead on (9.22) below). The derivative formula for $p_{2n+1}^{\nu s}(z)$, supplemented by the further derivative relation

$$\frac{2(k+1)}{(\nu - 2k - 3) r_{k+1}^{\nu s}} p_{2k+2}(z) - \frac{1}{r_k^{\nu s}} p_{2k}(z)$$

$$= -\frac{\Gamma((\nu+1)/2))^2}{\pi \, 2^{2-\nu} \Gamma(2k+2)\Gamma(\nu - 2k - 1)} (1 + z^2)^{(\nu+1)/2} \frac{d}{dz} \frac{z^{2k+1}}{(1 + z^2)^{(\nu-1)/2}},$$

gives for the analogue of the summation identity in the first line of (7.31).

Proposition 9.4 *We have*

$$
S_N^{\mathrm{r,vs}}(x,y)\Big|_{\nu\mapsto(\nu+N)/2} = \frac{\Gamma((\nu+1)/2)^2}{\pi\,2^{1-\nu}}(1+y^2)^{-(\nu+1)/2}
$$

$$
\times\left((1+x^2)^{-(\nu-1)/2}\sum_{k=0}^{N-2}\frac{(xy)^k}{k!(\nu-k-1)!} + \frac{1}{2\Gamma(N-1)\Gamma(\nu-N+1)}y^{N-1}\Phi_{N-2}(x)\right).
$$

$$(9.17)$$

In the case $\nu = N$ *this further simplifies to give* [265]

$$
S_N^{\mathrm{r,vs}}(x,y)\Big|_{\nu=N} = \frac{\Gamma((N+1)/2)^2}{\pi\,2^{1-N}\Gamma(N)}(1+x^2)^{-(N-1)/2}(1+y^2)^{-(N+1)/2}(1+xy)^{N-1}.
$$

$$(9.18)$$

For $\nu = N + p$ for any fixed $p > 0$ the bulk scaling limit involves the scalings $x \mapsto X/\sqrt{N}$. With this assumed, the result (7.37) first derived for GinOE itself, is reclaimed. We note there is no edge for spherical GinOE. In (9.17), setting $\nu = N\alpha_2$, $\alpha_2 \geq 1$ as is consistent with the notation below (2.54), setting $x = y$ and taking the large N limit gives for the global density of real eigenvalues [227, Th. 4.2.14]

$$
\lim_{N\to\infty}\frac{1}{\sqrt{N}}\rho_{(1),N}^{\mathrm{r,vs}}(x) = \sqrt{\frac{\alpha_2}{2\pi}}\frac{1}{1+x^2}\chi_{|x|<\sqrt{1/(\alpha_2-1)}}.
$$

$$(9.19)$$

In relation to the complex eigenvalues, the key formula is (7.44), further simplified to the form analogous to (7.46) with this involving the (generalised) GinUE spherical ensemble kernel. As is consistent with (9.19) we first suppose that in (9.13) the replacement $\nu \mapsto (N+\nu)/2$ has been made, and then set $\nu = \alpha_2 N$, $\alpha_2 > 1$. The corresponding GinUE spherical ensemble kernel is then given by (2.56) with $M = N$, $n = \nu$. From this one finds the same functional form as obtained in the GinUE case (2.57)

$$
\lim_{N\to\infty}\frac{1}{N}\rho_{(1),N}^{\mathrm{c,vs}}(x) = \frac{\alpha_2}{\pi}\frac{1}{(1+|z|^2)^2}\chi_{|z|<\sqrt{1/(\alpha_2-1)}};
$$

$$(9.20)$$

see [227, Th. 4.2.14]. One observes that the global real density (9.19) is proportional to the square root of the global complex density (9.20) continued to the real line. The proportionality constant is $1/\sqrt{2}$. This inter-relation (apart from the value of the proportionality) has been predicted by Tarnowski [516] to hold for a wide class of asymmetric real random matrices. For example, it holds for the GinOE itself (compare (7.36) and (7.49)), and the induced GinOE (compare (9.6) and (9.7)).

Remark 9.1
1. Setting $x = y$ in (9.18) reclaims (9.9). Also, since the bulk GinUE correlations are the limit of bulk scaling, the analogue of the asymptotic variance formula (7.54) holds [265, stated below (14)].
2. In [265] an approach to studying the eigenvalue distribution for the spherical GinOE (i.e. the case $\nu = N$ of the above) making use of a stereographic projection

to the sphere was detailed. In this approach it was found that the corresponding skew-orthogonal polynomials have the skew-orthogonality property with respect to the analogues of $\alpha_{j,k}^s$ and $\beta_{j,k}^s$ individually. Consequently, the generating function for the probability of k real eigenvalues, Z_N^s, was shown to be given in the product form

$$
Z_N^s(\zeta) = \frac{(-1)^{(N/2)(N/2-1)/2}}{2^{N(N-1)/2}}
$$

$$
\times \, \Gamma((N+1)/2)^{N/2} \Gamma(N/2+1)^{N/2} \prod_{s=1}^{N} \frac{1}{\Gamma(s/2)^2} \prod_{l=0}^{N/2-1} (\zeta \alpha_l + \beta_l), \quad (9.21)
$$

where

$$
\alpha_l = \frac{2\pi}{N-1-4l} \frac{\Gamma((N+1)/2)}{\Gamma(N/2+1)},
$$

$$
\beta_l = \frac{2\sqrt{\pi}}{N-1-4l} \left(2^N \frac{\Gamma(2l+1)\Gamma(N-2l)}{\Gamma(N+1)} - \sqrt{\pi} \frac{\Gamma((N+1)/2)}{\Gamma(N/2+1)} \right); \quad (9.22)
$$

cf. the determinant formula (9.16) with $\nu = N$.

3. Again in the case $\nu = N$, the matrix element $S_N^{c,s}$ as specified by (7.44) has the explicit form [265, Proposition 4.3]

$$
S_N^{c,s}(w, z) = \frac{N(N-1)}{2^{N+1} \pi r_w r_z} \left[\int_{\frac{r_w^{-1} - r_w}{2}}^{\infty} \frac{dt}{\left(1+t^2\right)^{N/2+1}} \right]^{1/2} \left[\int_{\frac{r_z^{-1} - r_z}{2}}^{\infty} \frac{dt}{\left(1+t^2\right)^{N/2+1}} \right]^{1/2}
$$

$$
\times \left(\frac{e^{i(\theta_z - \theta_w)/2}}{(r_w r_z)^{1/2}} + \frac{e^{-i(\theta_z - \theta_w)/2}}{(r_w r_z)^{-1/2}} \right)^{N-2} \left(\frac{e^{i(\theta_z - \theta_w)/2}}{(r_w r_z)^{1/2}} - \frac{e^{-i(\theta_z - \theta_w)/2}}{(r_w r_z)^{-1/2}} \right),
$$

where $w, z := r_w e^{i\theta_w}, r_z e^{i\theta_z}$. Used here are coordinates inside the unit disk, which result from the conformal mapping of the half plane $Z = i(1-z)/(1+z)$. Setting $w = z = re^{i\theta}$ in this gives for the transformed complex density [265, Eq. (17)]

$$
\rho_{(1),N}^{c,s}(z) = \frac{N(N-1)}{2^{N+1} \pi r^2} \left(\frac{1}{r} + r \right)^{N-2} \left(\frac{1}{r} - r \right) \int_{\frac{r^{-1}-r}{2}}^{\infty} \frac{dt}{(1+t^2)^{N/2+1}}.
$$

We calculate from this the limiting form

$$
\lim_{N \to \infty} \frac{1}{N} \rho_{(1),N}^{c,s}(z) = \frac{1}{\pi} \frac{1}{(1+r^2)^2}, \quad (9.23)
$$

which is in fact invariant upon the inverse of the applied conformal mapping and so applies too in the original coordinates. Note that this is consistent with (9.20) in the case $\alpha_2 = 1$.

9.4 Truncations of Haar Real Orthogonal Matrices

Starting with an $(n + N) \times (n + N)$ Haar distributed real orthogonal matrix (see [194] for a discussion of the origin of this class of random matrices), we consider an $N \times N$ sub-block A_N say and seek the corresponding eigenvalue PDF [227, 363, 432]. As for truncations of Haar distributed unitary matrices, the eigenvalues must be contained inside the unit disk $|z| < 1$.

Proposition 9.5 *With $z = x + iy$, introduce the weights*

$$
\omega^{\mathrm{t}}(z) = \begin{cases} \frac{n(n-1)}{2\pi} |1 - z^2|^{n-2} \int_{2|y|/|1-z^2|}^{1} (1 - u^2)^{(n-3)/2} \, du, & n \neq 1 \\ \frac{1}{\pi} |1 - z^2|^{-1}, & n = 1. \end{cases} \tag{9.24}
$$

After denoting (9.1) as $\mathrm{vol}(O(N))$, *define the normalisation*

$$
C_{N,n}^{\mathrm{t}} = \frac{\mathrm{vol}(O(n))\mathrm{vol}(O(N))}{\mathrm{vol}(O(n+N))} \left(\frac{(2\pi)^n}{n!} \right)^{N/2} = \left(\frac{2^n}{n!} \right)^{N/2} \prod_{j=1}^{N} \frac{\Gamma((n+j)/2)}{\Gamma(j/2)}.
$$

The joint eigenvalue PDF for k real eigenvalues $\{\lambda_l\}_{l=1,\dots,k}$ and the $(N-k)/2$ complex eigenvalues $\{z_j := x_j + iy_j\}_{j=1,\dots,(N-k)/2}$ in the upper half plane for the an $N \times N$ truncation of a Haar distributed $O(n+N)$ matrix is then equal to

$$
C_{N,n}^{\mathrm{t}} \frac{2^{(N-k)/2}}{k!((N-k)/2)!} \prod_{s=1}^{k} (\omega^{\mathrm{t}}(\lambda_s))^{1/2} \prod_{s=1}^{(N-k)/2} \omega^{\mathrm{t}}(z_s)
$$

$$
\times \left| \Delta(\{\lambda_l\}_{l=1,\dots,k} \cup \{x_j \pm iy_j\}_{j=1,\dots,(N-k)/2}) \right|. \tag{9.25}
$$

Proof Following the working given to derive the analogous result for the GinUE in Sect. 2.6, which is based on [17, Appendix B], the calculation can be carried out according to a number of specific steps.

(i) Make use of a matrix integral form of the matrix delta function constraint relating to the top $N \times N$ block A_N of the $(n + N) \times (n + N)$ real orthogonal matrix, to deduce that the element PDF of an $N \times N$ sub-block A_N is up to normalisation equal to (2.60), now with the integral over the space of $N \times N$ real symmetric matrices $\{H_N\}$.

(ii) Introduce the real Schur decomposition $A_N = QRQ^T$ conditioned on the number of real eigenvalues, and integrate over the off (block diagonal) entries of R therein with the columns as variables. This requires carrying out the analogue of the steps used to deduce (2.60). However, there is the complication that in the columns corresponding to the complex eigenvalues, each column has a block structure. From this viewpoint, we then need to proceed analogous to step (ii) of the proof of Proposition 9.2, the details of which are given in [265]. With

$n \neq 1$ (this case needs to be considered separately) the conditional eigenvalue PDF

$$C_{N,n}^{vt} \left(\frac{n(n-1)}{2\pi} \right)^{N/2} \left| \tilde{\Delta}(\{\lambda_l\}_{l=1,\dots,k} \cup \{x_j \pm iy_j\}_{j=1,\dots,(N-k)/2}) \right|$$

$$\times \prod_{s=1}^{k} \left(\frac{\sqrt{\pi}\Gamma((n-1)/2)}{\Gamma(n/2)} |1 - \lambda_s^2|^{n-2} \right)^{1/2}$$

$$\times \prod_{s=1}^{(N-k)/2} 2|b_s - c_s| \det(\mathbb{I}_2 - R_s R_s^T)^{(n-3)/2} \tag{9.26}$$

results.

(iii) As for the derivations of the eigenvalue PDF detailed for the GinOE and spherical GinOE, the final step is to change variables from $\{b_j, c_j\}$ to $\{y_j, \delta_j\}$ and to integrate over δ_j.   □

Next, we let $\alpha_{j,k}^t[u]$, $\beta_{j,k}^t[v]$ be defined as for $\alpha_{j,k}^g[u]$, $\beta_{j,k}^g[v]$ in Proposition 7.2, but now with weight ω^t as defined in (9.24). For the analogue of (7.7) we then have

$$Z_N^t[u, v] = C_{N,n}^t \, \mathrm{Pf}[\alpha_{j,k}^t + \beta_{j,k}^t]_{j,k=1,\dots,N}. \tag{9.27}$$

We now come to the skew-orthogonal polynomials associated with $\gamma_{j,k}^t := \alpha_{j,k}^t|_{u=1} + \beta_{j,k}^t|_{v=1}$. For this the formulas of Proposition 7.4 can be used.

Proposition 9.6 *Consider the ensemble specified by $N \times N$ random matrices obtained by deleting n rows and columns of an $(n + N) \times (n + N)$ Haar distributed orthogonal matrix. For the skew-orthogonal polynomials associated with $\gamma_{j,k}^t$ we have*

$$p_{2j}^t(z) = z^{2j},$$

$$p_{2j+1}^t(z) = z^{2j+1} - \frac{2j}{n+2j} z^{2j-1} = -\frac{(1-z^2)^{-n/2+1}}{n+2j} \frac{d}{dz}\left((1-z^2)^{n/2} z^{2j} \right).$$

Furthermore,

$$r_j^t = \frac{n!(2j)!}{(n+2j)!}.$$

Proof The matrix G in (7.18) (with n replaced by j to avoid a conflict in notation) is constructed as the $2j \times 2j$ sub-block of an $(n + 2j) \times (n + 2j)$ Haar distributed orthogonal matrix. The requirements of the applicability of the formulas (7.19), namely that the distribution of each element is unchanged by negation, and that the joint first moment of distinct pairs of elements are uncorrelated, is therefore valid. The square of an element of G is then equal in distribution to $a_1^2/(a_1^2 + \cdots + a_{n+2j}^2)$, where each a_j is a standard Gaussian, and is thus a beta $\mathrm{B}[1/2, (n + 2j - 1)/2]$

random variable. Its mean is $1/(n + 2j)$ and so $\operatorname{Tr} \langle G^2 \rangle = 2j/(n + 2j)$, which so specifies p_{2j+1}. This reasoning is a special case of a calculation first given in [256]. The formula for the normalisation r_j^t follows from setting $u = v = 1$ in (9.27), and noting that the LHS then has the interpretation of a sum over probabilities of eigenvalues being real and so is unity, while by the skew-orthogonality property the RHS equals $C_{N,n}^t \prod_{l=0}^{N/2-1} r_l^t$. □

Next, we use the derivative formula for $p_{2j+1}^i(z)$ of Proposition 9.6, and the definition of $\alpha_{j,k}^t$ in Proposition 7.2 with ω^g replaced by ω^t to compute

$$\alpha_{2j-1,2k}^t \Big|_{u=v=1} = \frac{1}{(n+2j)\sqrt{\pi}} \frac{\Gamma((n+1)/2)\Gamma(n+1)}{\Gamma(n/2)} \frac{\Gamma(-3/2+j+k)}{\Gamma(-3/2+j+k+n)} =: \frac{\tilde{\alpha}_{j,k}^t}{n+2j}. \tag{9.28}$$

Hence for the generating function for the number of real eigenvalues we obtain a formula structurally identical to (9.16), but with $\tilde{\alpha}_{j,k}^{vs}$ replaced by $\tilde{\alpha}_{j,k}^t$ and $\tilde{r}_{j-1}^{vs}$ replaced by $(n + 2j)r_{j-1}^t$. This in turn allows the analogues of (7.54) and the local central limit theorem Proposition 7.3 to be deduced.

In relation to the real–real correlation kernel S_N^r from Proposition 7.5, we have seen that supplementing the derivative formula for $p_{2j+1}^t(z)$ by a further derivative relation involving the normalisations of the skew-orthogonal polynomials, allows for a simplification [238].

Proposition 9.7 *We have*

$$S_N^{r,t}(x, y) = \frac{1}{\pi} \frac{\Gamma(1/2)\Gamma((n+1)/2)}{\Gamma(n/2)} (1 - x^2)^{n/2-1}(1 - y^2)^{n/2} \sum_{j=0}^{N-2} \frac{(n+j-1)!}{(n-1)!j!} (xy)^j$$

$$- \frac{1}{r_{N/2-1}} \Phi_{N-2}(y)\omega^t(x)x^{N-1}. \tag{9.29}$$

Proof The required additional derivative formula is

$$\frac{2(j+1)}{(n+2j+2)r_{j+1}^t} p_{2j+2}(z) - \frac{1}{r_j^t} p_{2j}(z)$$

$$= -\frac{(n+2j)!}{n!(2j+1)!} (1 - z^2)^{(-n/2+1)} \frac{d}{dz} \left((1 - z^2)^{n/2} z^{2j+1} \right).$$

Introducing the incomplete beta function $I_x(a, b)$ and beta integral $B(x, y)$ in standard notation, one observes that (9.29) admits an expression in terms of these functions. Specifically, with $x = y$ as corresponds to the density [227, 363, 432],

$$S_N^{r,t}(x, x) = \frac{1}{B(n/2, 1/2)} \frac{I_{1-x^2}(n+1, N-1)}{1 - x^2}$$

$$+ \frac{(1 - x^2)^{(n-2)/2}|x|^{N-1}}{B(N/2, n/2)} I_{x^2}((N-1)/2, (n+2)/2). \tag{9.30}$$

From this form the scaled large N, n asymptotics with $\alpha = N/(N+n)$ can be computed as [363]

$$\lim_{N \to \infty} \frac{1}{\sqrt{N}} \rho_{(1)}^{r,t}(x) = \sqrt{\frac{1-\alpha}{2\pi\alpha}} \frac{1}{1-x^2} \chi_{-\sqrt{\alpha} < x < \sqrt{\alpha}}. \tag{9.31}$$

A corollary of this, obtained by integrating the RHS, is the asymptotic formula for the number of real eigenvalues

$$E_N^{r,t}\Big|_{\alpha=N/(N+n)} \sim \sqrt{\frac{2N(1-\alpha)}{\pi\alpha}} \operatorname{Arctanh} \sqrt{\alpha}. \tag{9.32}$$

The corresponding bulk and edge scaling of $S_N^{r,t}(x, y)$ in this setting reclaims the GinOE results (7.37) and (7.40). Hence we know that the analogue of the asymptotic formula (7.54) for the variance of the number of real eigenvalues will hold true.

For finite N the formula (7.44) upon a simplification analogous to (7.46) provides a simple to analyse expression for the complex density. From this one can calculate [363]

$$\lim_{N \to \infty} \frac{1}{N} \rho_{(1),N}^{c,t}\Big|_{\alpha=N/(N+n)} = \frac{1-\alpha}{\pi\alpha} \frac{1}{(1-|z|^2)^2} \chi_{|z| < \sqrt{\alpha}}. \tag{9.33}$$

Here one observes that the global real density (9.31) is proportional to the square root of the global complex density (9.33) continued to the real line, with proportionality $1/\sqrt{2}$ in keeping with [516]. We note too that the functional form (9.33) is identical to that given in (2.63) for the eigenvalue density of truncated unitary random matrices in the same global scaling limit, in accordance with common feature (S6) of Sect. 9.1.

With n fixed, and thus $\alpha = 1$, the asymptotic formula (9.32) breaks down. In keeping with this is that for n fixed the $N \to \infty$ limit of (9.29) is well defined without any scaling of x and y. Thus

$$\lim_{N \to \infty} S^{r,t}(x, y) = \frac{1}{\pi} \frac{\Gamma(1/2)\Gamma((n+1)/2)}{\Gamma(n/2)} \frac{(1-x^2)^{n/2-1}(1-y^2)^{n/2}}{(1-xy)^n}. \tag{9.34}$$

Also, there is an edge scaling distinct from (7.40). Thus from (9.30) we have [363]

$$\lim_{N \to \infty} \frac{1}{N} \rho_{(1)}^{r,t}\left(1 - \frac{x}{N}\right) = \tilde{\rho}_{(1)}^{e,t}(x), \tag{9.35}$$

where, with $\bar{\gamma}(n, x) := (x^n \Gamma(n))^{-1} \gamma(n, x)$,

$$\rho_{(1)}^{e,t}(x) = \frac{x^{n/2-1} e^{-x}}{2\Gamma(n/2)} \left(1 - x^{n/2+1} \bar{\gamma}(n/2+1, x) + \frac{(2x)^n}{B(n/2, 1/2)} \bar{\gamma}(n+1, 2x)\right).$$

This exhibits the large x tail behaviour $1/(B(n/2, 1/2)x)$ which is in keeping with the large N form of the expected number of real eigenvalues for n fixed [363]

$$E_N^{\mathrm{r,t}} \sim \frac{\log N}{B(n/2, 1/2)}. \tag{9.36}$$

The limit $N \to \infty$ with $n = 1$ is of special interest due to an interpretation in terms of the zeros of a certain random power series [238, 373].

Proposition 9.8 *Consider the random power series $f(z) = \sum_{p=0}^{\infty} a_p z^p$, where each a_p is an independently distributed standard real Gaussian random variable. We have that the distribution of the zeros of $f(z)$ coincides with the limiting distribution of the eigenvalues of an $N \times N$ sub-block of $(N + 1) \times (N + 1)$ Haar distributed real orthogonal matrices for $N \to \infty$.*

Proof Let U_{N+1} denote the Haar distributed orthogonal matrix. Set $A_{N+1} = \mathrm{diag}\,(a, 1, \ldots, 1)$, $a > 0$. Then the matrix $U_{N+1} A_{N+1}$ has the same eigenvalues as $A_{N+1}^{1/2} U_{N+1} A_{N+1}^{1/2}$. In the limit $a \to 0$ these eigenvalues are 0 and the eigenvalues of the $N \times N$ sub-block of U_{N+1} obtained by deleting the first row and column. Introducing $\mathbb{I}'_{N+1} = \mathrm{diag}\,(0, 1, \ldots, 1)$ and $\mathbf{e}_1 = (1, 0, \ldots, 0)$, manipulations including use of (7.26) reduce the characteristic equation for the nonzero eigenvalues of $U_{N+1} A_{N+1}$ in the limit $a \to 0$ to the secular equation

$$\mathbf{e}_1^T (\mathbb{I}_{N+1} - \lambda U_{N+1}^\dagger \mathbb{I}'_{N+1})^{-1} U_{N+1}^\dagger \mathbf{e}_1 = 0;$$

see [255, proof of Proposition 1]. Since for $|a| < 1$ the eigenvalues also have modulus less than one, the inverse can be expanded according to the geometric series, with coefficient of λ^k equal to $\mathbf{e}_1^T (U_{N+1}^\dagger \mathbb{I}'_{N+1})^k U_{N+1}^\dagger \mathbf{e}_1$. Noting that for large N this coefficient has the form $\mathbf{e}_1^T (U_{N+1}^\dagger \mathbb{I}'_{N+1})^k \mathbf{e}_1 (1 + \mathrm{O}(k/N))$, then applying the known result [373]

$$\{\sqrt{N}(\mathbf{e}_1^T (U_{N+1}^\dagger \mathbb{I}'_{N+1})^k \mathbf{e}_1\}_{k=0,1,\ldots} \overset{\mathrm{d}}{=} \{a_k\}_{k=0,1,\ldots},$$

where each a_k is an independent standard real Gaussian, implies the statement of the proposition. $\square$

The random power series in this result can itself be viewed as the $N \to \infty$ limit of the random polynomial $p_N(z) = \sum_{p=0}^{N} a_k z^k$ with standard real Gaussian coefficients. The density of the real zeros of this random polynomial, $\rho_{(1),N}^{\mathrm{r,K}}(x)$ say, was first investigated by Kac [349], making use of what is now referred to as the Kac–Rice formula; for an introduction see [447]. It was found

$$\rho_{(1),N}^{\mathrm{r,K}}(x) = \frac{1}{\pi} \left(\frac{1}{(1 - x^2)^2} - \frac{(N + 1)^2 x^{2N}}{(1 - x^{2N+2})^2} \right)^{1/2}.$$

Taking the limit $N \to \infty$ with $|x| < 1$ reclaims the functional form (9.34) with $x = y$ in the case $n = 1$, as is consistent with this being the limiting density for the random matrix problem of Proposition 9.8. The higher point correlation functions for the real zeros of the random polynomial are, according to the Kac–Rice formalism, structurally given by a k-dimensional Gaussian integral with a particular covariance matrix, weighted by a product of the absolute value of the integration variables [94]. In addition, this structure is maintained upon taking the $N \to \infty$ limit. On the other hand, the result of Proposition 9.8, together with Proposition 7.5 as it applies to truncations of Haar real orthogonal matrices, tells us that these same correlations can be written as a Pfaffian; see also [430]. This is also true of for the correlations between the complex zeros, where the Kac–Rice formalism leads to an expression in terms of a so-called Hafnian (a Hafnian relates to a Pfaffian as a permanent does to a determinant) [465].

We have remarked that the generating function for the number of real eigenvalues is given by a formula structurally identical to (9.16), but with $\tilde{\alpha}_{j,k}^{\text{vs}}$ replaced by $\tilde{\alpha}_{j,k}^{\text{t}}$ defined in (9.28) and $\tilde{r}_{j-1}^{\text{vs}}$ replaced by $(n + 2j)r_{j-1}^{\text{t}}$. Specifically, with $n = 1$ this shows [462]

$$Z_N^{\text{t}}(\zeta)\Big|_{n=1} = \det\left[\delta_{j,k} + \frac{\zeta - 1}{\pi(j + k - 3/2)}\right]_{j,k=1,\ldots,N/2} \underset{\substack{\zeta=0 \\ N\to\infty}}{\sim} N^{-3/8}, \qquad (9.37)$$

where the asymptotic result, applying when $\zeta = 0$, relates to the probability of no real eigenvalues; for more on the asymptotics see [230, 287]. The work [462] (see also [137, 186, 187, 197, 487, 488]) relates the exponent in the asymptotic formula, interpreted in terms of the probability of no real zeros for random Kac polynomials, to the so-called persistence exponent for two-dimensional ($d = 2$) diffusion.

The diffusion in question is of a scalar field $\phi(\mathbf{x}, t)$, which at time $t = 0$ is a zero mean Gaussian random field with short-range (delta function) correlations. The persistence is the probability that, for a system of linear size L and with the origin $\mathbf{0}$ in the bulk, $\phi(\mathbf{0}, t)$ does not change sign from its initial value. (We refer the reader to [109] for a comprehensive review on the persistence in extended many-body nonequilibrium systems.) The persistence exponent $\theta(d)$ quantifies the expected large L asymptotic form that the probability decays as $L^{-2\theta(d)}$ for $t \gg L^d$. Through a common inter-relation with a Gaussian stationary process with (in logarithmic time) covariance $\text{sech}(T/2)$ [186, 487, 488] it is predicted that $\theta(2)$ is equal to $1/4$ of the exponent in the analogue of the asymptotic formula (9.37) for Kac random polynomials, which in turn is twice the exponent in (9.37). The leads to the conclusion that $\theta(2) = 3/16$.

9.5   Products of GinOE

As for products of GinUE matrices, the most general case of products of compatibly sized rectangular matrices G_i can be reduced to products of $N \times N$ square matrices $\tilde{G}_i$, now with element distribution proportional to

$$|\det \tilde{G}_i \tilde{G}_i^T|^{\nu_i/2} e^{-\text{Tr}\,\tilde{G}_i \tilde{G}_i^T/2},$$

again with $v_i > 0$ equal to the difference in the number of rows in G_i and the number of rows in $G_i (= N)$ [333]. The eigenvalue probability density function, conditioned on the number of real eigenvalues, can be computed for a general number M of matrices in the product [254, Proposition 8].

Proposition 9.9 *Define the real weight*

$$w_r^v(\lambda) = \prod_{j=1}^{m} \left[\int_{\mathbb{R}} d\lambda^{(j)} \left(\frac{\lambda^{(j)}}{2} \right)^{v_j/2} e^{-\frac{1}{2}(\lambda^{(j)})^2} \right] \delta(\lambda - \lambda^{(1)} \cdots \lambda^{(M)}) \qquad (9.38)$$

and the complex weight

$$w_c^v(x, y) = 2\pi \int_{\mathbb{R}} d\delta \frac{|\delta|}{\sqrt{\delta^2 + 4y^2}} W^v \left(\begin{bmatrix} \mu_+ & 0 \\ 0 & \mu_- \end{bmatrix} \right),$$

$$\mu_{\pm} = \frac{1}{2} \left(\pm |\delta| + [\delta^2 + 4(x^2 + y^2)]^{1/2} \right) \qquad (9.39)$$

with

$$W^v(G) = \prod_{l=1}^{M} \left[\int_{\mathbb{R}^{2 \times 2}} (dX^{(l)}) \det \left(\frac{X^{(l)} X^{(l)T}}{2} \right)^{v_l/2} \frac{e^{-\frac{1}{2} \mathrm{Tr}\, X^{(l)} X^{(l)T}}}{\sqrt{2\pi^3}} \right]$$
$$\times \delta(X - X^{(1)} \cdots X^{(M)}). \qquad (9.40)$$

Define too the normalisation

$$Z_{N,k}^{M,v} = 2^{MN(N+1)/4} \prod_{l=1}^{M} \prod_{j=1}^{N} \Gamma \left(\frac{j + v_l}{2} \right). \qquad (9.41)$$

With this notation, we have that the joint eigenvalue eigenvalue PDF of the random product matrix $\tilde{G}_1 \cdots \tilde{G}_M$, conditioned on there being k eigenvalues, is given by the functional form (1.10) with the normalisation C_N^g replaced by $1/Z_{N,k}^{M,v}$, $(\omega^g(\lambda_s))^{1/2}$ replaced by $w_r^v(\lambda_s)$ and $\omega^g(z_j)$ replaced by $w_c^v(x_j, y_j)$.

Proof Enabling this calculation is the real matrix version of the periodic Schur form (2.71), where now each Z_i is a block diagonal matrix of the structure of R in the real Schur decomposition of Sect. 7.1, and is thus conditioned on k real eigenvalues. In keeping with the notation of Sect. 7.1, we write $\tilde{G}_l = Q_l(D_l + R_l)Q_{l+1}, (l = 1, \ldots, M)$, where each D_l is block diagonal, and each R_l is strictly block upper triangular. We denote the scalar diagonal elements of D_l by $\{\lambda_s^{(l)}\}_{s=1}^{k}$ and the 2×2 block diagonal entries by $\{X_s^{(l)}\}_{s=1}^{k}$.

The first k diagonal entries are scalars, $\{\lambda_i := \lambda_i^{(1)} \cdots \lambda_t^{(l)}\}_{t=1}^{k}$, while the latter $(N - k)/2$ entries are 2×2 matrices, $\{G_s := G_s^{(1)} \cdots G_s^{(l)}\}_{s=k}^{(N+k)/2}$. With this notation, the Jacobian for the above given change of variables reads [330, Proposition A.26], [254]

$$\prod_{l=1}^{M} (d\tilde{G}_l) = \left| \tilde{\Delta}(\{\lambda_l\}_{l=1,\dots,k} \cup \{x_j \pm iy_j\}_{j=1,\dots,(N-k)/2}) \right|$$

$$\times \prod_{l=1}^{M} (dR_l)(Q_l^T dQ_l) \prod_{l=1}^{M} \left(\prod_{j=1}^{k} d\lambda_j^{(l)} \prod_{s=k+1}^{(N+k)/2} dX_s^{(l)} \right), \quad (9.42)$$

where $\lambda_s := \prod_{j=1}^{M} \lambda_s^{(j)}$ and $X_s := \prod_{j=1}^{M} X_s^{(j)}$, while $\tilde{\Delta}$ is as in (1.10) but with the difference between each pair $x_s \pm iy_s$ (which are the eigenvalues of X_s) omitted. Making the change of variables in the element PDF of the matrix product gives

$$\prod_{l=1}^{M} e^{-\frac{1}{2} \operatorname{Tr} \tilde{G}_l \tilde{G}_l^T} = \prod_{l=1}^{M} e^{-\frac{1}{2} \sum_{s=1}^{k} (\lambda_s^{(l)})^2 - \frac{1}{2} \sum_{s=k+1}^{(N+k)/2} \operatorname{Tr} X_s^{(l)} (X_s^{(l)})^T} e^{-\frac{1}{2} \sum_{i<j} (r_{ij}^{(l)})^2}.$$

Hence, after integrating over $\{r_{ij}^{(l)}\}$, we obtain that up to proportionality the eigenvalue PDF is equal to

$$\prod_{j=1}^{k} \delta(\lambda_j - \lambda_j^{(1)} \cdots \lambda_j^{(M)}) \, d\lambda_j \prod_{s=k+1}^{(N+k)/2} \delta(G_s - G_s^{(1)} \cdots G_s^{(M)}) \, (dG_s)$$

$$\times \frac{1}{Z_{k,N}^{M,\nu}} \prod_{l=1}^{M} \left[\prod_{j=1}^{k} \left(e^{-\frac{1}{2}(\lambda_j^{(l)})^2} d\lambda_j^{(l)} \right) \prod_{s=k+1}^{(N+k)/2} \left(\frac{e^{-\frac{1}{2} \operatorname{Tr} X_s^{(l)} X_s^{(l)T}}}{\sqrt{2\pi^3}} (dX_s^{(l)}) \right) \right]. \quad (9.43)$$

The final step is to change variables from the off-diagonal elements of the X_s, parameterised as in Sect. 7.1, to the quantities y_j (the imaginary part of j-th complex eigenvalue) and δ_j; recall the final paragraph in Sect. 7.1.   □

Remark 9.2 The real weight (9.38) can be written in terms of the Meijer G-function according to

$$\omega_{\mathrm{r}}^{\nu}(\lambda) = G_{0,M}^{M,0} \left(\begin{array}{c} - \\ \frac{\nu_1}{2}, \dots, \frac{\nu_M}{2} \end{array} \middle| \frac{\lambda^2}{2^M} \right);$$

cf. (2.74). With $M = 2$, $\nu_1 = \nu_2 = 0$ this simplifies to

$$\omega_{\mathrm{r}}^{\nu}(\lambda) = 2K_0(|\lambda|). \quad (9.44)$$

For general $M \geq 2$ no special function evaluation of the matrix integral defining $\omega_{\mathrm{c}}^{\nu}$ is known. An exception is $M = 2$, $\nu_1 = \nu_2 = 0$, for which [38, 254]

$$\omega_{\mathrm{c}}^{\nu}(x, y) = 4 \int_0^{\infty} \frac{1}{t} \exp\left(-2(x^2 - y^2)t - \frac{1}{4t} \right) K_0(2(x^2 + y^2)t) \operatorname{erfc}(2\sqrt{t}y) \, dt. \quad (9.45)$$

Comparing (9.44) and (9.45) makes it clear that unlike the structure highlighted in point (S3) of Sect. 9.1, it is no longer true that $\omega_{\mathrm{r}}(x) = (\omega_{\mathrm{c}}(x, 0))^{1/2}$.

We turn our attention now to the calculation of the skew-orthogonal polynomials associated with $\gamma^p_{j,k} := \alpha^p_{j,k}|_{u=1} + \beta^p_{j,k}|_{v=1}$. Here we define $\alpha^p_{j,k}[u]$, $\beta^p_{j,k}[v]$ as in Proposition 7.2, but with $(\omega^g(x)\omega^g(y))^{1/2}$ replaced by $w^v_r(x)w^v_r(y)$ and $\omega^g(z)$ replaced by $w^v_c(x,y)$. Through the use of (7.19), a simple calculation gives that the skew-orthogonal polynomials are given by [254, Proposition 9]

$$p^{v,M}_{2n}(z) = z^{2n}, \qquad p^{v,M}_{2n+1}(z) = z^{2j+1} - z^{2j-1}\prod_{k=1}^{M}(2j + v_k), \tag{9.46}$$

with normalisation

$$h^M_{j-1} = \prod_{k=1}^{M} \frac{2\sqrt{2\pi}}{2^{v_k}}\Gamma(2j + v_k - 1). \tag{9.47}$$

However, there is no longer a derivative formula for $p^{v,M}_{2n+1}(z)$ analogous to that for $p^g_{2n+1}(z)$ in (7.20). Associated with the latter is that the structural feature (S2) of earlier cases for the probability of all eigenvalues being real is no longer true; see [254, Eq. (5.2)] for a determinant formula involving particular Meijer G-functions. Asymptotics of the latter allows for an effect first noticed in [385]—that the probability of all eigenvalues being real tends to 1 as $M \to \infty$ for large classes of product random matrices—to be proven in the Gaussian case [241]. Subsequent references on this topic include [314, 467, 468].

In keeping with the effect of the probability of all eigenvalues being real increasing as M increases, is the result of Simm [501, Th. 1.1] for the expected number of real eigenvalues, $E^{v,M}_N$ say. Here the point to draw attention to is the factor of $\sqrt{M}$ relative to the $M = 1$ case.

Proposition 9.10 *We have*

$$E^{v,M}_N \underset{N\to\infty}{\sim} \sqrt{\frac{2NM}{\pi}} + \mathrm{O}(\log N). \tag{9.48}$$

Proof (Sketch) The starting point for this result, obtained with the specialisation each $v_i = 0$, is the exact expression for the density $\rho^{r,M}_{(1),N}(x)$ implied by knowledge of the skew-orthogonal polynomials, written in terms of the corresponding product GinUE correlation kernel (2.78) according to the structural form (7.43) with $x = y$. The implied sum is then integrated term-by-term, which results in a sum of particular Meijer G-functions. The summand is now in a form suitable for asymptotic analysis in the large N limit. $\qquad\square$

The different behaviour of $E^{v,M}_N$ for N fixed and $M \to \infty$, relative to M fixed and $N \to \infty$, makes it natural to inquire about the asymptotic behaviour of $E^{v,M}_N$ in the circumstance that $M = \alpha N$. This has been determined in [19, Th. 1.1] as being given by

$$\lim_{N\to\infty} \frac{E_N^{\nu,M}}{N}\bigg|_{M=\alpha N} = \left(1+\frac{\alpha}{4}\right)\mathrm{erf}\left(\sqrt{\frac{\alpha}{8}}\right) - \frac{\alpha}{4} + \sqrt{\frac{\alpha}{2\pi}}e^{-\frac{\alpha}{8}}, \qquad (9.49)$$

which can be demonstrated to provide an interpolation between the previously found behaviours.

Remark 9.3

1. A further application of $\rho_{(1),N}^{\mathrm{r},M}(x)$ expressed in the structural form (7.43) is to the calculation of the global scaling limit. Thus in [501, Th. 1.2] this was used to establish the validity of the corresponding product GinUE result (2.79).

2. As noted in Remark 2.6.4, there is interest in a product of M random matrices in the limit $M \to \infty$ from a dynamical systems perspective. For GinOE matrices, the explicit Lyapunov spectrum was computed long ago by Newman [314]. This exact result in the case of the largest Lyapunov exponent was generalised in [269, 356] to allow for the GinOE matrices to be left multiplied by a positive definite matrix. A direct study of the stability exponents associated with the modulus of the eigenvalues—these becoming all real as $M \to \infty$—was undertaken in [331].

3. The averaged absolute value of the product of GinOE matrices, which analogous to (7.42) relates to the density of the real eigenvalues, has appeared in a counting problem for equilibria in an analysis of a discrete analogue of the random nonlinear differential equations (7.88) [332].

4. The spectral density of the product of M independent elliptic GinOE matrices has been studied in [454]. The limiting global density is the M-th power of the circular law, independent of the parameter τ. Note in particular that this result applies to the global density of products of GOE matrices.

5. The products of M truncated orthogonal matrices in Sect. 9.4 have also been studied in the literature [256, 261, 407]. In particular, for strong non-orthogonality, it was shown in [407] that the leading-order asymptotic of the expected number of real eigenvalues is of the form (9.32) multiplied by $\sqrt{M}$; cf. (9.48). On the other hand, for weak non-orthogonality, the asymptotic form (9.36) with $B(n/2, 1/2)$ replaced by $B(Mn/2, 1/2)$ holds.

The structural form (7.43) for $S_N^{\mathrm{r},M}(x, y)$ also has consequences for the analysis of the two-point correlation and associated fluctuation formula. In particular, it is used in [231] to establish that the structural formula (7.58) for the variance of a linear statistic as $N \to \infty$ holds true independent of M. For the particular linear statistic corresponding to the counting function for the number of real eigenvalues, this formula was shown to break down in the case that M simultaneously tends to ∞ with N. Rather then $(\sigma_N^{\mathrm{r},M})^2/E_N^{\mathrm{r},M}|_{M=\alpha N}$ tends to an α dependent quantity; see [19, Th. 1.2].

Chapter 10
Statistical Properties of GinSE and Elliptic GinSE

Eigenvalues of GinSE matrices occur in complex conjugate pairs as for complex GinOE eigenvalues, but now with zero probability of an eigenvalue being real. Also in common with complex GinOE eigenvalues is that the GinSE eigenvalues form a Pfaffian point process, and that their PDF admits a Coulomb gas interpretation with image terms. While the determination of the required skew-orthogonal polynomials is straightforward, the particular form gives a new challenge in relation to analysing the various scaling limits. This is overcome by making use of a certain inhomogeneous partial differential equation satisfied by the pre-kernel. Such an approach carries over to the analysis of a weakly non-Hermitian regime. Also contained in this chapter is the large N analysis of the partition function and gap probability associated with GinSE eigenvalues, and results relating to the singular values and eigenvectors. Furthermore, a study is undertaken of an elliptic generalisation, which interpolates between GinSE and Hermitian GSE matrices.

10.1 Eigenvalue PDF

Using a similarity transformation, the symplectic Ginibre matrix G can be identified via the matrix-valued version of the quaternion realisation (1.1) as

$$\begin{bmatrix} A & B \\ -\bar{B} & \bar{A} \end{bmatrix} \in \mathbb{C}^{2N \times 2N}, \tag{10.1}$$

where A, B are independent copies of GinUE. Due to this form, the matrix G has $2N$ eigenvalues that come in complex conjugate pairs $\pm z_j$, where we take $\operatorname{Im} z_j > 0$. Importantly from the viewpoint of calculating the Jacobians associated with a change of variables involving eigenvalues, the eigenvectors of complex conjugate eigenvalues are related by a linear transformation. This can be encoded by forming a block Schur decomposition $G = UZU^\dagger$ with U a $2N \times 2N$ unitary matrix with

© The Author(s) 2025
S.-S. Byun and P. J. Forrester, *Progress on the Study of the Ginibre Ensembles*, KIAS
Springer Series in Mathematics 3, https://doi.org/10.1007/978-981-97-5173-0_10

2×2 block entries of the form (1.1)—a conjugation equivalent symplectic unitary matrix—and Z a block triangular matrix, with diagonal blocks

$$\left\{ \begin{bmatrix} 0 & z_j \\ \bar{z}_j & 0 \end{bmatrix} \right\}_{j=1}^{N}$$

containing the eigenvalues of G, and off-diagonal elements having the quaternion form (1.1); see [331, Proposition A.25]. As in the proof of Proposition 2.1, the Jacobian calculation is carried out by forming the wedge product of the matrix of differentials of $U^\dagger dGU$, in the order of the block indices (j, k) with j decreasing from N to 1, and k increasing from 1 to N. This gives the eigenvalue dependent factor

$$\prod_{j=1}^{N} e^{-2W(z_j, \bar{z}_j)} |z_j - \bar{z}_j|^2 \prod_{1 \le j < k \le N} |z_k - z_j|^2 |z_k - \bar{z}_j|^2,$$

and moreover (after some working) the product of differentials can be shown to factorise as in (2.6). The eigenvalue PDF now follows by integrating over the independent off-diagonal entries of Z. The resulting functional form (1.8) is formally the same as the GinUE eigenvalue PDF (1.7) with $N \mapsto 2N$, and where z_{j+N} is identified with $\bar{z}_j$, but with terms involving only differences of $\{\bar{z}_j\}$ ignored (due to dependencies in the quaternion structure these are not associated with independent differentials). As previously mentioned, the eigenvalue PDF (1.8) of GinSE was derived already in the original work [293] of Ginibre starting from the eigenvalue–eigenvector decomposition. For the details of the above outlined method using the quaternionic Schur decomposition, see [331, proof of Proposition A.25].

In general, an eigenvalue PDF of the non-Hermitian random matrices in the same symmetry class of the GinSE is of the form

$$\frac{1}{N! Z_N^{\mathbb{H}}(W)} \prod_{j=1}^{N} e^{-2W(z_j, \bar{z}_j)} |z_j - \bar{z}_j|^2 \prod_{1 \le j < k \le N} |z_k - z_j|^2 |z_k - \bar{z}_j|^2, \quad \operatorname{Im} z_j > 0,$$

(10.2)

where $Z_N^{\mathbb{H}} \equiv Z_N^{\mathbb{H}}(W)$ is the partition function, which turns (10.2) into a probability measure. Here W can be an arbitrary real function such that $Z_N^{\mathbb{H}}$ exists, which furthermore is assumed to satisfy the complex conjugation symmetry $W(z, \bar{z}) = W(\bar{z}, z)$. The ensemble of the form (10.2) is called the planar symplectic ensemble [22].

10.2   Coulomb Gas Perspective

The eigenvalue PDF (10.2) can be rewritten in terms of the Hamiltonian

$$H_{\mathbb{H}}(z_1, \ldots, z_N) := \sum_{1 \le j < l \le N} \log \frac{1}{|z_j - z_l|^2 |z_j - \bar{z}_l|^2}$$

$$+ \sum_{j=1}^{N} \left(2W(z_j, \bar{z}_j) + \log \frac{1}{|z_j - \bar{z}_j|^2} \right) \tag{10.3}$$

as

$$\frac{1}{N! \, Z_N^{\mathbb{H}}(W)} \, e^{-H_{\mathbb{H}}(z_1, \ldots, z_N)}. \tag{10.4}$$

Analogous to the discussion of Sect. 7.2, this can be regarded as a two-dimensional Coulomb gas [245] in the upper half plane $\mathbb{H}$ with image charges in the lower half plane; cf. [367]. As such, this is an image system counterpart of the eigenvalue PDF of the normal matrix model, which by way of comparison is given by

$$\frac{1}{N! \, Z_N^{\mathbb{C}}(W)} \, e^{-H_{\mathbb{C}}(z_1, \ldots, z_N)}, \tag{10.5}$$

where

$$H_{\mathbb{C}}(z_1, \ldots, z_N) := \sum_{1 \le j < l \le N} \log \frac{1}{|z_j - z_l|^2} + \sum_{j=1}^{N} W(z_j, \bar{z}_j);$$

see Sect. 5 and references therein.

The Coulomb gas interpretation (10.4) allows one to describe the limiting spectral distribution using logarithmic potential theory. In the scaling $W(z, \bar{z}) = N Q(z)$, chosen to make the interaction and the potential term in (10.3) of the same order, one can observe that the continuum limit of the Hamiltonian (10.3) divided by $2N^2$ is given by

$$\begin{aligned} I_Q[\mu] &:= \frac{1}{4} \int_{\mathbb{C}^2} \log \frac{1}{|z - w||z - \bar{w}|} \, d\mu(z) \, d\mu(w) + \int_{\mathbb{C}} Q \, d\mu \\ &= \int_{\mathbb{C}^2} \log \frac{1}{|z - w|} \, d\mu(z) \, d\mu(w) + \int_{\mathbb{C}} Q \, d\mu. \end{aligned} \tag{10.6}$$

Here, we have used the complex conjugation symmetry $Q(z) = Q(\bar{z})$ for the second line. Thus it is natural to expect that the empirical measure μ_Q of (10.4) converges to Frostman's equilibrium measure, a unique probability measure minimising the energy functional (10.6). This convergence was shown by Benaych-Georges and Chapon [82] for a general potential Q. In particular it shows that the limiting spectral distributions of (10.4) and (10.5) are identical Sect. 5.2. In particular, for $Q(z) = |z|^2$, this gives rise to the universal appearance of the circular law for the GinUE and GinSE.

With regards to quantitative features, let us first recall that under minor assumptions on Q, the equilibrium measure μ_Q is absolutely continuous and takes the form

$$d\mu_Q(z) = \frac{\partial_z \partial_{\bar{z}} Q(z)}{\pi} \chi_{z \in S_Q} \, d^2z, \tag{10.7}$$

where the compact set S_Q is called the droplet. For a radially symmetric potential $q(r) = Q(|z| = r)$ which is strictly subharmonic in $\mathbb{C}$, the droplet is of the form $S_Q = \{R_1 \le |z| \le R_2\}$, where the pair of constant (R_1, R_2) is characterised by

$$R_1 q'(R_1) = 0, \quad R_2 q'(R_2) = 2, \tag{10.8}$$

see [479, Sect. IV.6]. In particular, for $Q(z) = |z|^2$, it follows that $\partial_z \partial_{\bar{z}} Q(z) = 1$ and $R_0 = 0$, $R_1 = 1$, which coincides with the circular law of the GinSE.

Remark 10.1 Underlying the image charge viewpoint of (10.3) is the pair potential (7.2), which we know is the solution of the two-dimensional Poisson equation in Neumann boundary conditions along the x-axis. If instead we take the point $\mathbf{r}$ to be inside a disk of radius R and require Neumann boundary conditions on the boundary of the disk, the pair potential is [237, Eq. (15.188) with $\epsilon = 0$]

$$\phi(\mathbf{r}, \mathbf{r}') = -\log\left(|z - z'||R - zz'/R|\right).$$

Imposing a smeared out charge neutral background of density $1/\pi$ (and hence taking $R = \sqrt{N}$) the corresponding charge neutral Boltzmann factor is [237, Eq. (15.190)]

$$A_{N,\beta} e^{-\beta \sum_{j=1}^{N} |z_j|^2/2} \prod_{1 \le j < k \le N} |z_k - z_j|^\beta |1 - z_j \bar{z}_k/N|^\beta \prod_{j=1}^{N} (1 - |z_j|^2/N)^{\beta/2}, \tag{10.9}$$

where $A_{N,\beta} = e^{-\beta N^2((1/4) \log N - 3/8)}$. Here $\beta > 0$ is the inverse temperature, with the case $\beta = 2$ being the disk analogue of (10.3), although this viewpoint gives the self-energy term of exponent 1 rather than 2 as in (1.8); see [245, Sect. 2.1] for more on this point.

10.3   Skew-Orthogonal Polynomials

We define the skew-symmetric form $\langle \cdot, \cdot \rangle_{s,S}$ by

$$\langle f, g \rangle_{s,S} := \int_{\mathbb{C}} \left(f(z) g(\bar{z}) - g(z) f(\bar{z}) \right) (z - \bar{z}) e^{-2W(z,\bar{z})} \, d^2 z, \tag{10.10}$$

where in keeping the notation of (7.16) the subscripts indicate skew and GinSE. Note the similarity with the second term in (7.16). As discussed in Sect. 7.4, a family $\{q_m\}_{m \ge 0}$ of monic polynomials q_m of degree m is said to be a family of skew-orthogonal polynomials if the following skew-orthogonality conditions hold: for all $k, l \in \mathbb{N}$

$$\langle q_{2k}, q_{2l}\rangle_{s,S} = \langle q_{2k+1}, q_{2l+1}\rangle_{s,S} = 0,$$
$$\langle q_{2k}, q_{2l+1}\rangle_{s,S} = -\langle q_{2l+1}, q_{2k}\rangle_{s,S} = r_k\,\delta_{k,l}. \tag{10.11}$$

We mention that the matrix averages formulas (7.18) are again valid; see [353]. In distinction to ensembles based on GinOE, there are also alternative methods which in fact have a broader scope, so these instead will be discussed below.

Proposition 10.1 *For a radially symmetric potential* $W(z, \bar{z}) = \omega(|z|)$, *let*

$$h_k = 2\pi \int_0^\infty r^{2k+1} e^{-2\omega(r)}\,dr \tag{10.12}$$

be the squared orthogonal norm. Then

$$q_{2k+1}(z) = z^{2k+1}, \qquad q_{2k}(z) = z^{2k} + \sum_{l=0}^{k-1} z^{2l} \prod_{j=0}^{k-l-1} \frac{h_{2l+2j+2}}{h_{2l+2j+1}} \tag{10.13}$$

forms a family of skew-orthogonal polynomials. Furthermore, the skew-norm is given by $r_k = 2h_{2k+1}$.

Proof Since the monomials $\{z^k\}$ are orthogonal polynomials with respect to a rotationally symmetric weight function, it follows that

$$\langle z^k, z^l\rangle_{s,S} = \int_{\mathbb{C}} \left(z^{k+1}\bar{z}^l - z^{l+1}\bar{z}^k - z^k\bar{z}^{l+1} + z^l\bar{z}^{k+1}\right) e^{-2\omega(|z|)}\,d^2z$$
$$= 2\delta_{k+1,l}h_{k+1} - 2\delta_{l+1,k}h_k.$$

Note here that the indices in the Kronecker delta differ by one, which in turn immediately leads to $\langle q_{2k+1}, q_{2l+1}\rangle_{s,S} = 0$. Furthermore, it follows that $\langle q_{2k}, q_{2l+1}\rangle_{s,S} = 0$ if $l > k$. Let us write

$$a_l = \prod_{j=0}^{k-l-1} \frac{h_{2l+2j+2}}{h_{2l+2j+1}}. \tag{10.14}$$

Then we have

$$\langle q_{2k}, q_{2k+1}\rangle_{s,S} = \left\langle z^{2k} + \sum_{l=0}^{k-1} a_l z^{2l}, z^{2k+1}\right\rangle_{s,S} = 2h_{2k+1}.$$

On the other hand, for the case $k < l$, after straightforward computations, we obtain

$$\langle q_{2k+1}|q_{2l}\rangle_{s,S} = 2(a_k h_{2k+1} - a_{k+1}h_{2k+2}) = 0.$$

Therefore, we have shown that $\langle q_{2k+1}, q_{2l+1}\rangle_{s,S} = r_k\delta_{k,j}$. The other cases follow from similar computations with minor modifications. $\qquad\square$

As an example of Proposition 10.1, for $W^g(z, \bar{z}) = |z|^2$, the associated skew-orthogonal polynomials q_k^g are given by

$$q_{2k+1}^g(z) = z^{2k+1}, \qquad q_{2k}^g(z) = \sum_{l=0}^{k} \frac{k!}{l!} z^{2l}, \qquad r_k^g = \frac{(2k+1)!}{2^{2k+1}} \pi. \qquad (10.15)$$

These are a special case of the skew-orthogonal polynomials obtained in [353] for the elliptic GinSE; see the sentence below (10.23). We also refer to [237, Exercises 15.9 q.2] and [240, Proposition 1] for a derivation of (10.15).

The crux of Proposition 10.1 is that one can construct the skew-orthogonal polynomials using the associated (monic) orthogonal polynomials p_j with respect to the same weight, i.e.

$$\int_{\mathbb{C}} p_j(z)\overline{p_k(z)} e^{-2W(z,\bar{z})} \, d^2z = h_k \delta_{j,k}. \qquad (10.16)$$

A setting beyond the radially symmetric case in which we can construct the skew-orthogonal polynomials is when the associated orthogonal polynomials satisfy a three-term recurrence relation.

Proposition 10.2 *Suppose that the sequence of monic orthogonal polynomials* (p_j) *satisfies the three-term recurrence relation*

$$zp_k(z) = p_{k+1}(z) + b_k p_k(z) + c_k p_{k-1}(z), \qquad b_k, c_k \in \mathbb{R}. \qquad (10.17)$$

Then

$$q_{2k+1}(z) = p_{2k+1}(z),$$

$$q_{2k}(z) = \sum_{l=0}^{k} a_l z^l, \qquad a_l := \prod_{j=0}^{k-l-1} \frac{h_{2l+2j+2} - c_{2l+2j+2} h_{2l+2j+1}}{h_{2l+2j+1} - c_{2l+2j+1} h_{2l+2j}} \qquad (10.18)$$

satisfies (10.11) with $r_k = 2(h_{2k+1} - c_{2k+1}h_{2k})$. *Conversely, if the skew-orthogonal polynomials have the form (10.18), then the three-term recurrence (10.17) holds.*

Proof (Sketch) The essential idea of the proof has already been given in the proof of Proposition 10.1 above. The notable difference is that while computing the skew-symmetric form $\langle q_k, q_l \rangle_{s,S}$, we expand the term

$$\left(q_k(z)\overline{q_l(z)} - \overline{q_k(z)}q_l(z) \right)(z - \bar{z}) \qquad (10.19)$$

in the integrand using the three-term recurrence relation (10.17). $\square$

The idea of constructing skew-orthogonal polynomials in Proposition 10.2 first appeared in [353], where the Hermite polynomials were considered. Later, this was extended to the Laguerre polynomials in [11]. The general statement in Proposition 10.2 was given in [26]. As an example, we consider the elliptic GinSE potential

$$W^{\mathrm{e}}(z, \bar{z}) = \frac{1}{1 - \tau^2}(|z|^2 - \tau \,\mathrm{Re}\, z^2), \qquad \tau \in [0, 1). \tag{10.20}$$

The associated monic orthogonal polynomials and norms are given by

$$p_k^{\mathrm{e}}(z) = \left(\frac{\tau}{4}\right)^{k/2} H_k\left(\frac{z}{\sqrt{\tau}}\right), \qquad h_k^{\mathrm{e}} = \sqrt{1 - \tau^2}\, \frac{k!}{2^{k+1}}\pi, \tag{10.21}$$

see e.g. [24, Lem. 7]. Then by using the recurrence relation

$$z p_k^{\mathrm{e}}(z) = p_{k+1}^{\mathrm{e}}(z) + \frac{\tau}{2} k\, p_{k-1}^{\mathrm{e}}(z) \tag{10.22}$$

and Proposition 10.2, we have

$$q_{2k+1}^{\mathrm{e}}(z) = p_{2k+1}^{\mathrm{e}}(z), \qquad q_{2k}^{\mathrm{e}}(z) = \sum_{l=0}^{k} \frac{k!}{l!} p_{2l}^{\mathrm{e}}(z),$$
$$r_k^{\mathrm{e}} = (1 - \tau)\sqrt{1 - \tau^2}\, \frac{(2k + 1)!}{2^{2k+1}}\pi. \tag{10.23}$$

Note that (10.15) can be recovered by taking the $\tau \to 0$ limit of (10.23).

10.4 Correlation Functions and Sum Rules

The k-point correlation function $\rho_{(k),N}^{\mathrm{s}}$ of the ensemble (10.4) is given by

$$\rho_{(k),N}^{\mathrm{s}}(z_1, \ldots, z_k) := \frac{N!}{(N-k)!} \frac{1}{Z_N^{\mathbb{H}}(W)} \int_{\mathbb{C}^{N-k}} e^{-H_{\mathbb{H}}(z_1, \ldots, z_N)} \prod_{j=k+1}^{N} d^2 z_j. \tag{10.24}$$

As an analogue of Proposition 7.5, a Pfaffian formula for $\rho_{(k),N}^{\mathrm{s}}$ follows [353].

Proposition 10.3 *Let q_j be the skew-orthogonal polynomials as specified in (10.11). Using these polynomials, define*

$$\kappa_N^{\mathrm{s}}(z, w) = \sum_{k=0}^{N-1} \frac{q_{2k+1}(z)q_{2k}(w) - q_{2k}(z)q_{2k+1}(w)}{r_k}. \tag{10.25}$$

Then we have

$$\rho_{(k),N}^{\mathrm{s}}(z_1, \ldots, z_k) = \prod_{j=1}^{k} (\bar{z}_j - z_j) \mathrm{Pf}\left[\mathcal{K}_N^{\mathrm{s}}(z_j, z_l)\right]_{j,l=1,\ldots,k}, \tag{10.26}$$

where

$$\mathcal{K}_N^s(z, w) = e^{-W(z,\bar{z})-W(w,\bar{w})} \begin{bmatrix} \kappa_N^s(z, w) & \kappa_N^s(z, \bar{w}) \\ \kappa_N^s(\bar{z}, w) & \kappa_N^s(\bar{z}, \bar{w}) \end{bmatrix}. \tag{10.27}$$

Using Proposition 10.3 and (10.15), it follows that the matrix entry $\kappa_N^g(z, w)$—referred to as the pre-kernel—of the GinSE is given by

$$\kappa_N^g(z, w) = \frac{\sqrt{2}}{\pi} \left(\sum_{k=0}^{N-1} \frac{(\sqrt{2}z)^{2k+1}}{(2k+1)!!} \sum_{l=0}^{k} \frac{(\sqrt{2}w)^{2l}}{(2l)!!} - \sum_{k=0}^{N-1} \frac{(\sqrt{2}w)^{2k+1}}{(2k+1)!!} \sum_{l=0}^{k} \frac{(\sqrt{2}z)^{2l}}{(2l)!!} \right). \tag{10.28}$$

One strategy for analysing the double summation is to derive a suitable differential equation for the kernel. This idea essentially goes back to an early work [437] of Mehta and Srivastava.

As an analogue of Proposition 7.6, the following holds true; see [22].

Proposition 10.4 *Letting* $\widehat{\kappa}_N^g(z, w) := e^{-2zw}\kappa_N^g(z, w)$, *we have*

$$\partial_z \widehat{\kappa}_N^g(z, w) = 2(z - w)\widehat{\kappa}_N^g(z, w) + \frac{2}{\pi}\frac{\Gamma(2N; 2zw)}{(2N-1)!} - \frac{1}{\pi}(2z)^{2N}e^{-z^2}\frac{\Gamma(N; w^2)}{(2N-1)!}. \tag{10.29}$$

Proof Note that

$$\partial_z \sum_{k=0}^{N-1} \frac{(\sqrt{2}z)^{2k+1}}{(2k+1)!!} \sum_{l=0}^{k} \frac{(\sqrt{2}w)^{2l}}{(2l)!!} = \sqrt{2}\sum_{k=0}^{N-1} \frac{(\sqrt{2}z)^{2k}}{(2k-1)!!} \sum_{l=0}^{k} \frac{(\sqrt{2}w)^{2l}}{(2l)!!}$$

$$= \sqrt{2}\sum_{k=1}^{N-1} \frac{(\sqrt{2}z)^{2k}}{(2k-1)!!} \sum_{l=0}^{k-1} \frac{(\sqrt{2}w)^{2l}}{(2l)!!} + \sqrt{2}\sum_{k=0}^{N-1} \frac{(\sqrt{2}z)^{2k}}{(2k-1)!!} \frac{(\sqrt{2}w)^{2k}}{(2k)!!}.$$

Rearranging the terms, we have

$$\partial_z \sum_{k=0}^{N-1} \frac{(\sqrt{2}z)^{2k+1}}{(2k+1)!!} \sum_{l=0}^{k} \frac{(\sqrt{2}w)^{2l}}{(2l)!!} = 2z\sum_{k=0}^{N-1} \frac{(\sqrt{2}z)^{2k+1}}{(2k+1)!!} \sum_{l=0}^{k} \frac{(\sqrt{2}w)^{2l}}{(2l)!!}$$

$$+ \sqrt{2}\sum_{k=0}^{N-1} \frac{(\sqrt{2}z)^{2k}}{(2k-1)!!} \frac{(\sqrt{2}w)^{2k}}{(2k)!!} - \sqrt{2}\frac{(\sqrt{2}z)^{2N}}{(2N-1)!!} \sum_{l=0}^{N-1} \frac{(\sqrt{2}w)^{2l}}{(2l)!!}.$$

Similarly, we have

$$\partial_z \sum_{k=0}^{N-1} \frac{(\sqrt{2}w)^{2k+1}}{(2k+1)!!} \sum_{l=0}^{k} \frac{(\sqrt{2}z)^{2l}}{(2l)!!}$$

$$= 2z \sum_{k=0}^{N-1} \frac{(\sqrt{2}w)^{2k+1}}{(2k+1)!!} \sum_{l=0}^{k} \frac{(\sqrt{2}z)^{2l}}{(2l)!!} - 2z \sum_{k=0}^{N-1} \frac{(\sqrt{2}w)^{2k+1}}{(2k+1)!!} \frac{(\sqrt{2}z)^{2k}}{(2k)!!}.$$

Combining above identities, we conclude (10.29). □

As a consequence of Proposition 10.4, the uniform asymptotic expansion (3.31) can be used to derive a linear inhomogeneous differential equation of order one satisfied by the limiting correlation kernels. Then the anti-symmetry of the limiting pre-kernels characterises a unique solution, which in turn determines the limiting kernels; see [22, 118].

For the bulk case, we have

$$\rho_{(k),\infty}^{s,b}(z_1, \ldots, z_k)$$

$$:= \lim_{N\to\infty} \rho_{(k),N}^{s}(z_1, \ldots, z_k) = \prod_{j=1}^{k} (\bar{z}_j - z_j) \mathrm{Pf}\, [\mathcal{K}_\infty^{s,b}(z_j, z_l)]_{j,l=1,\ldots,k}, \quad (10.30)$$

where $\mathcal{K}_\infty^{s,b}$ is of the form

$$\mathcal{K}_\infty^{s,b}(z, w) := e^{-|z|^2 - |w|^2} \begin{bmatrix} \kappa_\infty^{s,b}(z, w) & \kappa_\infty^{s,b}(z, \bar{w}) \\ \kappa_\infty^{s,b}(\bar{z}, w) & \kappa_\infty^{s,b}(\bar{z}, \bar{w}) \end{bmatrix}, \quad \kappa_\infty^{s,b}(z, w) := \frac{e^{z^2+w^2}}{\sqrt{\pi}} \mathrm{erf}(z - w).$$

$$(10.31)$$

In particular, this gives the bulk-scaled density

$$\rho_{(1),\infty}^{s,b}(x + iy) = \frac{4}{\pi} y F(2y), \quad (10.32)$$

where $F(z) := e^{-z^2} \int_0^z e^{t^2}\, dt$ is Dawson's integral function. As $y \to 0^+$, this tends to zero with leading term $\frac{8y^2}{\pi}$, while for $y \to \infty$ this limits to $\frac{1}{\pi}$ which is the interior bulk density away from the real axis. There is a peak in this profile, as is consistent with the sum rule (10.39) below. It can be checked from (10.31) that the truncated two-point correlation implied by (10.30) decays like a Gaussian in all directions. The limiting pre-kernel (10.31) first appeared in the second edition of Mehta's book [436] using a different approach; cf. [416, Lem. 3.5.2]. Later, this was rederived by Kanzieper in [353] using a similar method described above.

Generally, if each $\mathrm{Im}(z_j)$, $(j = 1, \ldots, k)$ in (10.32) is taken to infinity, but with the relative distance between each z_i still finite, the RHS simplies to the Pfaffian of the anti-symmetrix matrix (8.1), as for GinOE. We know that this in turn reduces to the determinantal formula implied by Proposition 2.3 for bulk-scaled GinUE.

Contrary to the bulk case, the analogous result for the edge case appeared in the literature [22] only recently. The same Pfaffian structure was found

$$\lim_{N\to\infty} \rho_{(k),N}^{s}(\sqrt{N} + z_1, \ldots, \sqrt{N} + z_k) = \prod_{j=1}^{k} (\bar{z}_j - z_j) \mathrm{Pf}\, [\mathcal{K}_\infty^{s,e}(z_j, z_l)]_{j,l=1,\ldots,k},$$

$$(10.33)$$

with $\mathcal{K}^{s,e}_\infty$ as for $\mathcal{K}^{s,b}_\infty$ in (10.33) but with $\kappa^{s,b}_\infty$ replaced by $\kappa^{s,e}_\infty$ throughout, where

$$\kappa^{s,e}_\infty(z, w) := \frac{e^{2zw}}{\pi} \int_{-\infty}^0 e^{-t^2} \sinh(2t(w - z))\mathrm{erfc}(z + w - t)\, dt. \qquad (10.34)$$

Setting $w = \bar{z}$, this gives for the density

$$\rho^{s,e}_{(1),\infty}(x + iy) = -\frac{2y}{\pi} \int_{-\infty}^0 e^{-s^2} \sin(4sy)\mathrm{erfc}(2x - s)\, ds. \qquad (10.35)$$

Note here that unlike the bulk case, the edge scaling limit does not have the translation invariance along the horizontal x-direction. We also remark that

$$\rho^{s,e}_{(1),\infty}(x + iy) \sim \frac{\mathrm{erfc}(2x)}{2\pi} + \frac{e^{-4x^2}}{8\pi^{3/2} y^2} + \cdots, \qquad y \to \infty. \qquad (10.36)$$

The leading term of the RHS of (10.36) as a function of x coincides with the limiting edge density of the GinUE up to scaling. As in Sect. 8.1, such a limiting relation holds too for the general k-point function, which in particular exhibits a deformation from a Pfaffian to a determinant; see [22, Cor. 2.2].

A structural feature, highlighted in [22, 119], is that the kernels (10.31) and (10.34) can be expressed in a unified way

$$\kappa^s_\infty(z, w) := \frac{e^{z^2+w^2}}{\sqrt{\pi}} \int_E W(f_w, f_z)(u)\, du, \qquad (10.37)$$

where $W(f, g) := fg' - gf'$ is the Wronskian, and

$$f_z(u) := \frac{\mathrm{erfc}(\sqrt{2}(z - u))}{2}, \qquad E := \begin{cases} (-\infty, \infty) & \text{for the bulk case,} \\ (-\infty, 0) & \text{for the edge case.} \end{cases} \qquad (10.38)$$

As in Sect. 7.5, the expression (10.37) allows one to observe the bulk limiting form from the edge limiting form with $z, w \to -\infty$.

We now discuss sum rules for the limiting densities (10.32) and (10.35); cf. Propositions 4.3 and 4.4.

Proposition 10.5 *We have*

$$\int_{-\infty}^\infty \left(\rho^{s,b}_{(1),\infty}(x + iy) - \frac{1}{\pi} \right) dy = 0 \qquad (10.39)$$

and

$$\int_{(-\infty,\infty)^2} \left(\rho^{s,e}_{(1),\infty}(x + iy) - \rho^{s,b}_{(1),\infty}(x + iy)\chi_{x<0} \right) dx\, dy = -\frac{1}{8}. \qquad (10.40)$$

Proof The first identity (10.39) immediately follows from the property of Dawson's
integral $F(2y)/2 = \int (1 - 4y\,F(2y))\,dy$. For the second identity (10.40), letting D_R
be a disk of radius $R > 0$, we first observe that by the change of variables,

$$
\int_{D_R} \left(\rho^{s,b}_{(1),\infty}(x + iy)\chi_{x<0} - \rho^{s,e}_{(1),\infty}(x + iy) \right) dx\,dy
$$

$$
- \int_{D_R} \left(\rho^{s,b}_{(1),\infty}(x + iy) - \frac{1}{\pi} \right)\chi_{x<0}\,dx\,dy
$$

$$
= \int_{D_R} \left(\frac{\chi_{x<0}}{\pi} - \rho^{s,e}_{(1),\infty}(x + iy) \right) dx\,dy = \int_{D_R} \left(\frac{1}{2\pi} - \rho^{s,e}_{(1),\infty}(x + iy) \right) dy\,dx
$$

$$
= \frac{1}{\pi} \int_{D_R} \left(\frac{1}{2} + 2y \int_{-\infty}^{0} e^{-s^2} \sin(4sy)\mathrm{erfc}(2x - s)\,ds \right) dy\,dx
$$

$$
= \frac{1}{\pi} \int_{D_R} \left(\frac{1}{2} - 2y \int_{0}^{\infty} e^{-s^2} \sin(4sy)\big(2 - \mathrm{erfc}(2x - s)\big)\,ds \right) dy\,dx.
$$

By adding the last two expressions and using

$$
4y \int_{0}^{\infty} e^{-s^2} \sin(4sy) = 4y\,F(2y) = \pi \rho^{s,e}_{(1),\infty}(x + iy), \tag{10.41}
$$

it follows that

$$
\int_{D_R} \left(\rho^{s,b}_{(1),\infty}(x + iy)\chi_{x<0} - \rho^{s,e}_{(1),\infty}(x + iy) \right) dx\,dy
$$

$$
= \frac{1}{\pi} \int_{D_R} y \left(\int_{-\infty}^{\infty} e^{-s^2} \sin(4sy)\mathrm{erfc}(2x - s)\,ds \right) dy\,dx.
$$

Furthermore, by using $\int_{-\infty}^{\infty} e^{-s^2} \sin(4sy)\,ds = 0$ and the principal value integral

$$
\lim_{R\to\infty} \int_{-R}^{R} \mathrm{erf}(s - 2x)\,dx = -s, \tag{10.42}
$$

we obtain

$$
- \lim_{R\to\infty} \frac{1}{\pi} \int_{D_R} y \left(\int_{-\infty}^{\infty} e^{-s^2} \sin(4sy)\mathrm{erf}(s - 2x)\,ds \right) dy\,dx
$$

$$
= \frac{1}{\pi} \int_{-\infty}^{\infty} y \left(\int_{-\infty}^{\infty} e^{-s^2} \sin(4sy)s\,ds \right) dy = \frac{2}{\sqrt{\pi}} \int_{-\infty}^{\infty} y^2 e^{-4y^2}\,dy = \frac{1}{8},
$$

which leads to (10.40).   □

Note that compared to the analogous identity (4.12) for the GinUE edge scaling
limit, the RHS of (10.40) takes on the nonzero value $-1/8$. Curiously, the companion

identity to (4.12), namely Proposition 4.4 with $\beta = 2$, which relates to the dipole moment of the edge density profile takes on the nonzero value $-1/8\pi$.

We remark that the bulk multipole screening sum rule analogous to (8.17)

$$\int_{\mathbb{C}_+} w^{2p} \rho^{s,b\,T}_{(2),\infty}(z, w)\, d^2 w = -z^{2p} \rho^{s,b}_{(1),\infty}(z), \quad p \in \mathbb{Z}_{\geq 0}, \tag{10.43}$$

has been established in [244].

10.5   Elliptic GinSE

The elliptic GinSE is defined in a similar way to Sect. 7.9. Namely, for a parameter $\tau, 0 \leq \tau < 1$, set

$$X = \sqrt{1+\tau}\, S + \sqrt{1-\tau}\, A. \tag{10.44}$$

Here S is a member of Gaussian symplectic ensemble (GSE) of Hermitian matrices (see e.g. [237, Sect. 1.3.2]), whereas A is a member of anti-symmetric GSE. Its eigenvalue PDF is of the form (10.4) with the elliptic GinSE potential (10.20) previously mentioned.

As discussed in Sect. 10.2, the global-scaled eigenvalues $z_j \to \sqrt{N} z_j$ distribute in a way to minimise the energy (10.6) with the potential (10.20). This minimisation problem can be exactly solved, which gives that the limiting spectrum is given by the ellipse

$$\left\{ (x, y) \in \mathbb{R}^2 : \left(\frac{x}{1+\tau} \right)^2 + \left(\frac{y}{1-\tau} \right)^2 \leq 1 \right\}; \tag{10.45}$$

see e.g. Sect. 2.3 and [114]. As in the elliptic GinU/OE, this is the elliptic law for the elliptic GinSE.

Turning to the correlation functions, by combining (10.25) and (10.23), one can show that the associated pre-kernel $\kappa^e_N(z, w)$ is evaluated as

$$\kappa^e_N(z, w) = \frac{\sqrt{2}}{\pi(1-\tau)\sqrt{1-\tau^2}} \sum_{k=0}^{N-1} \frac{(\tau/2)^{k+1/2}}{(2k+1)!!} H_{2k+1}\left(\frac{z}{\sqrt{\tau}} \right) \sum_{l=0}^{k} \frac{(\tau/2)^l}{(2l)!!} H_{2l}\left(\frac{w}{\sqrt{\tau}} \right)$$
$$- \frac{\sqrt{2}}{\pi(1-\tau)\sqrt{1-\tau^2}} \sum_{k=0}^{N-1} \frac{(\tau/2)^{k+1/2}}{(2k+1)!!} H_{2k+1}\left(\frac{w}{\sqrt{\tau}} \right) \sum_{l=0}^{k} \frac{(\tau/2)^l}{(2l)!!} H_{2l}\left(\frac{z}{\sqrt{\tau}} \right). \tag{10.46}$$

It reduces to the pre-kernel κ^g_N in (10.28) in the limit $\tau \to 0^+$. In the weakly non-Hermitian regime when $\tau = 1 - \alpha^2/N$, the scaling limit of the correlation functions at the origin was derived in [353]. The analogous result at the edge of the spectrum was later obtained in [35]. (We also remark that a mapping between the elliptic GinSE with a fermion field theory was suggested by Hastings [318].) Fairly recently, it was shown in [26] that for a fixed τ (also called the regime of strong non-Hermiticity), the universal scaling limit (10.31) appears at the origin. Let us mention that the analysis in [35, 353] was based on proper Riemann sum approximations, whereas a double contour integral representation was used in [26]. These methods provide a short way to find an explicit formula of the limiting pre-kernel, but it is not easy to perform the asymptotic analysis in a more general setup or to precisely control the error term. For these purposes, extending Proposition 10.4, the idea of using a proper differential equation was established in [118, 209], which reads as follows.

Proposition 10.6 *We have*

$$
\partial_z \kappa_N^g(z, w) = \frac{2z}{1+\tau} \kappa_N^g(z, w) + \frac{2}{\pi(1-\tau^2)^{3/2}} \sum_{k=0}^{2N-1} \frac{(\tau/2)^k}{k!} H_k\left(\frac{z}{\sqrt{\tau}}\right) H_k\left(\frac{w}{\sqrt{\tau}}\right)
$$

$$
- \frac{2}{\pi(1-\tau^2)^{3/2}} \frac{(\tau/2)^N}{(2N-1)!!} H_{2N}\left(\frac{z}{\sqrt{\tau}}\right) \sum_{l=0}^{N-1} \frac{(\tau/2)^l}{(2l)!!} H_{2l}\left(\frac{w}{\sqrt{\tau}}\right).
$$

$$
(10.47)
$$

Proof (Sketch) The general idea to derive such an identity is the same as that used in the proof of Proposition 10.4; differentiate the pre-kernel and properly rearrange the indices in the summations to extract the pre-kernel itself multiplied by z up to proportionality, and then collect all the remaining additive terms. Contrary to the proof of Proposition 10.4, the well-known functional relations

$$
H_j'(z) = 2j H_{j-1}(z), \qquad H_{j+1}(z) = 2z H_j(z) - H_j'(z) \tag{10.48}
$$

of the Hermite polynomials are crucially used in the computations and we refer to [118] for more details. □

We also refer to [127] for the Laguerre version of such an identity.

We now bring to attention the fact that the first inhomogeneous term in (10.47) corresponds to the kernel of the elliptic GinUE with $N \mapsto 2N$; see Proposition 2.5. (A similar feature for the elliptic GinOE is highlighted above Proposition 7.12.) As will be discussed below, such a relation can be observed in further extensions to GinSE. We also refer the reader to [6] for a similar relation for the Hermitian random matrix models.

Using Proposition 10.6, one can derive the scaling limits of the elliptic GinSE correlation functions in various regimes. For τ fixed, it was shown in [118] that for the real axis centred bulk and edge of the spectrum, the universal scaling limits (10.31) and (10.34) arise. Furthermore, in the edge scaling limits, as a counterpart of Proposition 2.6, the subleading correction term was derived. In the weakly non-Hermitian regime when $\tau = 1 - \alpha^2/N$, the bulk and edge scaling limits were obtained in [119], extending previous results [35, 353]; see also [209]. The first paper on this topic [369] used supersymmetry techniques to deduce the density profile perpendicular to the

real axis in the bulk, and was motivated by numerical findings in quantum chromodynamics (QCD) [310]. A structural finding in [119] shows that the limiting pre-kernel
in the bulk scaling limit is of the unified Wronskian form (10.37) with

$$f_z(u) := \frac{1}{2\pi} \int_{-C(\alpha)}^{C(\alpha)} e^{-t^2/2} \sin(2t(z-u)) \frac{dt}{t}, \qquad E := \mathbb{R}, \tag{10.49}$$

where $C(\alpha)$ is an explicit constant depending on α and on the position where we
zoom the point process. In the same spirit, the edge scaling limit is again of the form
(10.37) with

$$f_z(u) := 2\alpha \int_0^u e^{\alpha^3(z-t)+\frac{\alpha^6}{12}} \operatorname{Ai}\left(2\alpha(z-t) + \frac{\alpha^4}{4}\right) dt, \qquad E := (-\infty, 0). \tag{10.50}$$

Remark 10.2 Note that the bulk scaling limit (10.49) has again the translation invariance. Conversely, it was shown in [22] that if a scaling limit of (10.4) satisfies the
translation invariance, then it is of the form (10.49). The main idea for this characterisation was a use of Ward's identity for the ensemble (10.4), which says that
$\mathbb{E}_N W_N^+[\psi] = 0$, where ψ is a test function and

$$W_N^+[\psi] := \sum_{j \neq k} \psi(z_j)\left(\frac{1}{z_j - z_k} + \frac{1}{z_j - \bar{z}_k}\right) + 2\sum_{j=1}^N \frac{\psi(z_j)}{z_j - \bar{z}_j}$$

$$- 2\sum_{j=1}^N [\partial_z W \cdot \psi](z_j) + \sum_{j=1}^N \partial \psi(z_j); \tag{10.51}$$

cf. (5.21). Here, $[\partial_z W \cdot \psi](z_j) := \partial_z W(z)|_{z=z_j} \psi(z_j)$ and $\partial \psi(z_j) = \partial_z \psi(z)|_{z=z_j}$.

10.6   Partition Functions and Gap Probabilities

Recall that the normal matrix ensemble (10.5) forms a determinantal point process.
This integrable structure allows an explicit expression of the partition function; see
(5.15). Similarly, using the Pfaffian structure (10.26) and de Bruijn type formulas
[110], one can express the partition function $Z_N^{\mathbb{H}}(W)$ in terms of the skew norms
(10.11) as

$$Z_N^{\mathbb{H}}(W) = \prod_{j=0}^{N-1} r_j; \tag{10.52}$$

see e.g. [26, Remark 2.5]. For instance, for the elliptic GinSE, it follows from (10.23)
that the associated partition function $Z_N^{\mathbb{H}}(W^{\mathrm{e}})$ is given by

$$Z_N^{\mathbb{H}}(W^{\mathrm{e}}) = \frac{((1-\tau)\sqrt{1-\tau^2}\pi)^N}{2^{N^2}} \prod_{k=0}^{N-1} (2k+1)!. \tag{10.53}$$

This explicit expression leads to the following asymptotic expansion; cf. Proposition 4.1 for its counterpart for $Z_N^{\mathbb{C}}(W^g)$.

Proposition 10.7 *We have*

$$
\log Z_N^{\mathbb{H}}(W^e) = N^2 \log N - \frac{3}{2}N^2 + \frac{1}{2}N \log N + \left(\frac{\log(4\pi^3(1-\tau)^3(1+\tau))}{2} - \frac{1}{2}\right)N
$$
$$
- \frac{1}{24}\log N + \frac{5\log 2}{24} + \frac{1}{2}\zeta'(-1) - \frac{1}{48N} - \frac{1}{1920N^2} + O(\frac{1}{N^3}).
$$
(10.54)

In particular, we have

$$
\log Z_N^{\mathbb{H}}(NW^g) = -\frac{3}{2}N^2 - \frac{1}{2}N \log N + \left(\frac{\log(4\pi^3)}{2} - \frac{1}{2}\right)N
$$
$$
- \frac{1}{24}\log N + \frac{5\log 2}{24} + \frac{1}{2}\zeta'(-1) - \frac{1}{48N} - \frac{1}{1920N^2} + O(\frac{1}{N^3}).
$$
(10.55)

Proof One can rewrite (10.53) in terms of the Barnes G-function as

$$
Z_N^{\mathbb{H}}(W^e) = \frac{((1-\tau)\sqrt{1-\tau^2}\pi)^N}{2^{3N^2/2-N/2}}\frac{G(2N+1)}{G(N+1)}
$$
$$
= ((1-\tau)^3(1+\tau)\pi)^{N/2}G(N+1)\frac{G(N+\frac{3}{2})}{G(\frac{3}{2})}.
$$

Then the asymptotic behaviour (10.54) follows from the knowledge of the known asymptotic expansion of the G-function (see e.g. [226, Th. 1]). The second expansion (10.55) immediately follows from (10.54) with $\tau = 0$, where the additional difference $(N^2 + N)\log N$ is due to the simple scaling $z_j \mapsto \sqrt{N}z_j$. □

We now discuss the asymptotics of the partition functions in a more general setup. For a fixed Q, the asymptotic expansion of the partition function $Z_N^{\mathbb{C}}(NQ)$ in (10.5) was discussed in Sect. 5.3 in detail. For radially symmetric potentials, the use of (10.52) and (10.13) was made in [125] to show that

$$
\log Z_N^{\mathbb{H}}(NQ) = -2N^2 I_Q[\mu_Q] - \frac{1}{2}N \log N
$$
$$
+ \left(\frac{\log(4\pi^2)}{2} - \frac{1}{2}E_Q[\mu_Q] - U_{\mu_Q}(0)\right)N \qquad (10.56)
$$
$$
- \frac{\chi}{24}\log N + \left(\frac{\chi}{2}\zeta'(-1) + \frac{5}{24}\log 2\right) + O(1), \qquad (10.57)
$$

where

$$
E_Q[\mu_Q] := \int_{\mathbb{C}} \mu_Q(z) \log \mu_Q(z)\, d^2z \qquad (10.58)
$$

is the entropy of the equilibrium measure μ_Q. Here χ is the Euler index of the droplet; for instance $\chi = 1$ for the disk and $\chi = 0$ for the annulus. Compared to the expansion (5.17) of $Z_N^{\mathbb{C}}(NQ)$, a notable difference is the additional $U_{\mu_Q}(0)$ in the $O(N)$ term, which is the logarithmic potential

$$U_\mu(z) = \int \log \frac{1}{|z - w|} \, d\mu(w) \tag{10.59}$$

evaluated at the origin. This term is closely related to the notion of renormalised energy of the Hamiltonian (10.3) (cf. [394]) since it can be checked that

$$U_{\mu_Q}(0) = - \int \log |w - \bar{w}| \, d\mu_Q(w). \tag{10.60}$$

For a radially symmetric potential $q(r) = Q(|z| = r)$ with the droplet specified by the radii (10.8), we have

$$I_Q[\mu_Q] = q(R_1) - \log R_1 - \frac{1}{4} \int_{R_0}^{R_1} r q'(r)^2 \, dr,$$

$$U_{\mu_Q}(0) = - \log R_1 + \frac{q(R_1) - q(R_0)}{2}. \tag{10.61}$$

This gives that for $Q = W^{\mathrm{g}}$,

$$I_Q[\mu_Q] = \frac{3}{4}, \qquad U_{\mu_Q}(0) = \frac{1}{2}, \qquad E_Q[\mu_Q] = - \log \pi. \tag{10.62}$$

Substituting these in the formula (10.57) reclaims (10.55).

We now discuss a relation between $Z_N^{\mathbb{C}}$ and $Z_N^{\mathbb{H}}$.

Proposition 10.8 *For a radially symmetric potential* $W_{\mathbb{C}}(z, \bar{z}) \equiv \omega_{\mathbb{C}}(|z|)$, *let*

$$W_{\mathbb{H}}(z, \bar{z}) \equiv \omega_{\mathbb{H}}(|z|) := \frac{1}{2} \omega_{\mathbb{C}}(|z|^2). \tag{10.63}$$

Then we have

$$Z_N^{\mathbb{H}}(W_{\mathbb{H}}) = Z_N^{\mathbb{C}}(W_{\mathbb{C}}). \tag{10.64}$$

Proof By the change of variables,

$$4 \int_0^\infty r^{4k+3} e^{-2\omega_{\mathbb{H}}(r)} \, dr = 2 \int_0^\infty r^{2k+1} e^{-2\omega_{\mathbb{H}}(\sqrt{r})} \, dr = 2 \int_0^\infty r^{2k+1} e^{-\omega_{\mathbb{C}}(r)} \, dr.$$

Then result now follows from Proposition 10.1, (5.15) and (10.52).   □

As an example, note that by (5.15), we have

$$Z_N^{\mathbb{C}}(2|z|) = \prod_{k=0}^{N-1} 2\pi \int_0^\infty r^{2k+1} e^{-2r} \, dr = \prod_{k=0}^{N-1} \frac{(2k+1)!}{2^{2k+1}} \pi. \tag{10.65}$$

Then one can observe that this coincides with (10.53) with $\tau = 0$. We also mention that if the droplet $S_{\mathbb{H}}$ associated with $W_{\mathbb{H}}$ is

$$S_{\mathbb{H}} = \{z \in \mathbb{C} : R_0 \le |z| \le R_1\}, \tag{10.66}$$

then its counterpart $S_{\mathbb{C}}$ for $W_{\mathbb{C}}$ is given by

$$S_{\mathbb{C}} = \{z \in \mathbb{C} : R_0^2 \le |z| \le R_1^2\}. \tag{10.67}$$

This can be directly checked using (10.8). Such a relation holds in general beyond the radially symmetric potentials; see [73, Lem. 1].

As a consequence of Proposition 10.8, one can obtain various statistics of the ensemble (10.4) from the analogous results for (10.5). To be more concrete, let us focus on the gap probabilities.

Proposition 10.9 *For a radially symmetric domain D, let $E_N^{W_{\mathbb{H}}}(0; D)$ be the probability that the ensemble (10.4) with a potential $W_{\mathbb{H}}$ has no particle inside D. We define $E_N^{W_{\mathbb{C}}}$ in a same way for (10.5). Then we have*

$$E_N^{W_{\mathbb{H}}}(0; D) = E_N^{W_{\mathbb{C}}}(0; \tilde{D}), \tag{10.68}$$

where $\tilde{D}$ is the image of D under the map $z \mapsto z^2$.

Proof Let us write

$$W_{\mathbb{H},D}(z, \bar{z}) := \begin{cases} W_{\mathbb{H}}(z, \bar{z}) & \text{if } z \in D^c, \\ +\infty & \text{otherwise,} \end{cases}, \qquad W_{\mathbb{C},\tilde{D}}(z, \bar{z}) := \begin{cases} W_{\mathbb{C}}(z, \bar{z}) & \text{if } z \in \tilde{D}^c, \\ +\infty & \text{otherwise.} \end{cases}$$

Then by Proposition 10.8,

$$E_N^{\mathbb{H}}(0; D) = \frac{Z_N^{\mathbb{H}}(W_{\mathbb{H},D})}{Z_N^{\mathbb{H}}(W_{\mathbb{H}})} = \frac{Z_N^{\mathbb{C}}(W_{\mathbb{C},\tilde{D}})}{Z_N^{\mathbb{C}}(W_{\mathbb{C}})} = E_N^{W_{\mathbb{C}}}(0; \tilde{D}), \tag{10.69}$$

which completes the proof. $\qquad\square$

This proposition is particularly helpful in the context of the Mittag–Leffler ensembles. They are two-parameter generalisations of the GinU/SE for which the associated potential is of the form

$$\omega^{\mathrm{ML}}(|z|) = |z|^{2b} - 2\alpha \log |z|, \qquad b > 0, \quad \alpha > -1. \tag{10.70}$$

For the complex Mittag–Leffler ensemble, the precise asymptotic behaviours of the gap probabilities were obtained in [149]; cf. see Sect. 3.1.1 for a summary and further references for the GinUE case when $\alpha = 0, b = 1$. Then as a consequence of Proposition 10.9, the analogous results for the symplectic Mittag–Leffler ensemble immediately follow. In particular, the gap probabilities of the GinSE can be obtained from the result of a complex Mittag–Leffler ensemble with $\alpha = 0, b = 1/2$. (See also [36] for an earlier work.) Beyond the gap probabilities, Proposition 10.8 can be used to investigate various counting statistics [20, 116, 150] as well as fluctuations of the maximal modulus [199, 469].

Remark 10.3 1. The Neumann boundary conditions disk Coulomb gas with Boltzmann factor (10.9) is exactly solvable for $\beta = 2$ [504]. In distinction to the expansion (10.55), it is found that the large N form of the logarithm of the partition function is at order N and order $\log N$ the same for the GinUE (4.3), although now there is also an $O(\sqrt{N})$ surface tension term [518, Eq. (3.42)]. Let us also mention that a generalisation to a two-component Coulomb gas and its Pfaffian structure have been studied in [346].

2. The simple explicit formula involving (10.12) of Proposition 10.1 for the skew-orthogonal polynomials normalisations in the radial case facilitates the derivation of a central limit theorem for a radially symmetric linear statistic in GinSE [120, Appendix B]. The covariance is again given by (3.21) except that the boundary term therein is not present (as a consequence of the restriction to a radial potential), and with the bulk term now multiplied by a factor of $1/2$.

10.7 GinSE Singular Values

Consider a $p \times N$ $(p \geq N)$ rectangular GinSE matrix X. As a complex matrix, X is of size $2n \times 2N$. Forming $X^\dagger X$ gives the well known construction of quaternion Wishart matrices [237, Sect. 3.2.1]. The $2N$ eigenvalues of $X^\dagger X$—which are the square singular values of X—are doubly degenerate. Let the N independent eigenvalues be denoted by $\{s_j\}$. Their PDF is proportional to (see [237, Proposition 3.2.2])

$$\prod_{l=1}^{N} s_l^{2(p-N)+1} e^{-2s_l} \prod_{1 \leq j < k \leq N} (s_k - s_j)^4. \tag{10.71}$$

Upon the global scaling $X^\dagger X \mapsto \frac{1}{N} X^\dagger X$, and with $p = \alpha N$ $(\alpha > 1)$, as for (8.18) the smallest and largest tend almost surely to $(\sqrt{\alpha \mp 1} + 1)^2$, as for the real case. This coincidence can be understood as being a consequence of $\{s_j\}$ in both (9.43) and (10.71) as being well approximated by the zeros of the Laguerre polynomials $L_N^{p-N}(px)$ [192, 328]. As in the real case, we thus have that the condition number tends to a constant in the circumstance that $\alpha > 1$. On the other hand, this breaks down for $\alpha = 1$. Specifically, we will consider the square case $p = N$. It seems that the limiting condition number has not previously been reported in the literature. The starting point is to calculate the PDF for the smallest eigenvalue. This is equal to the differentiation operation $-\frac{d}{ds}$ applied to the gap probability $E_N^{\mathrm{qW}}(0, (0, s))$ of there being no eigenvalues from the origin to a point s. The latter is defined by integrating the PDF (10.71) over $s_l \in (s, \infty)$, $(l = 1, \ldots, N)$. Changing variables $s_l \mapsto s_l + s$ then shows

$$E_N^{\mathrm{qW}}(0, (0, s)) = \frac{e^{-2Ns}}{C_N} \int_0^\infty ds_1 \cdots \int_0^\infty ds_N \prod_{l=1}^N (s + s_l) e^{-2s_l} \prod_{1 \le j < k \le N} (s_k - s_j)^4,$$

where C_N is such that the LHS equals unity for $s = 0$. According to [237, Eq. (13.44) with $a = 0, m = 1, \beta = 4, t_1 = -s$], the normalised multiple integral is equal to the hypergeometric polynomial $_1F_1(-N; 1/2; -s)$, which in turn is proportional to the Laguerre polynomial $L_N^{(-1/2)}(-s)$. From the confluent limit of the hypergeometric polynomial, we conclude the simple result

$$\lim_{N \to \infty} E_N^{\mathrm{qW}}(0, (0, x/N)) = e^{-2x} {}_0F_1(1/2; x) = e^{-2x} \cosh(2\sqrt{x}); \qquad (10.72)$$

this is equivalent to [233, Eq. (2.15a)]. Consequently, the scaled condition number $\kappa_N/(2N)$ has the limiting PDF

$$-\frac{2}{y^3} \frac{d}{dx} \left(e^{-2x} \cosh(2\sqrt{x}) \right) \Big|_{x=1/y^2}.$$

In keeping with our previous discussion relating to singular values, we record here too the explicit form of the distribution of $|\det X|^2$ for X a square GinSE matrix [269, Proposition 2 with $\beta = 4, \sigma_l^2 = 1/4$],

$$|\det X|^2 \overset{d}{=} \prod_{j=1}^N \frac{1}{4} \chi_{4j}^2. \qquad (10.73)$$

Here we have defined $|\det X|^2 = \prod_{l=1}^N s_l$, even though the eigenvalues of X are doubly degenerate. One approach to the derivation of (10.73) is to make use of knowledge of the joint PDF (10.71) and the Laguerre weight version of Selberg's integral to compute first the Mellin transform of the distribution; recall (6.12).

10.8 GinSE Eigenvectors

Consider an eigenvalue z_j of a GinSE matrix, and let ℓ_j, r_j denote the corresponding left and right eigenvectors. As for GinUE and GinOE eigenvectors, of interest is the matrix of overlaps (6.19). In particular, the considerations of the paragraph above

Proposition 6.5 are again valid, to allow for the quaternion structure. Thus while r_j can be chosen as the first standard basis vector (now in $\mathbb{C}^{2N}$), we now require $\ell_1 = (b_1, c_1, \ldots, b_N, c_N)$ where $b_1 := 1$, $c_1 := 0$ so that

$$O_{11} = \langle \ell_1, \ell_1 \rangle = \sum_{k=1}^{N} (|b_k|^2 + |c_k|^2).$$

Moreover, with $Z_{pq} = \begin{bmatrix} z_{pq} & w_{pq} \\ -\bar{w}_{pq} & \bar{z}_{pq} \end{bmatrix}$ now a 2×2 block element of the strictly upper triangular portion of the triangular matrix Z in the block Schur decomposition $\{b_k, c_k\}$, and regarding now ℓ_1 as a left eigenvector of Z with eigenvalue z_1 (this does not effect O_{11}), it follows that $\{b_k, c_k\}$ satisfy the coupled recurrence

$$b_p = \frac{1}{z_1 - z_p} \sum_{k=1}^{p-1} (b_k z_{k,p} - c_k \bar{w}_{k,p}), \quad c_p = \frac{1}{z_1 - \bar{z}_p} \sum_{k=1}^{p-1} (b_k w_{k,p} + c_k \bar{z}_{p,q}),$$

subject to the initial conditions $b_1 = 1$ and $c_1 = 0$.

Writing now $\ell_1 = \ell_1^{(N)}$ and denoting by $\ell_1^{(n)}$ the truncation of this vector ending with b_n, c_n, it follows that

$$||\ell_1^{(n+1)}||^2 = ||\ell_1^{(n)}||^2 \left(1 + \frac{1}{|z_1 - z_{n+1}|^2} \left| \sum_{q=1}^{N} (\tilde{b}_k z_{k,p} - \tilde{c}_k w_{k,p}) \right|^2 \right.$$

$$\left. + \frac{1}{|z_1 - \bar{z}_{n+1}|^2} \left| \sum_{q=1}^{N} (\tilde{b}_k w_{k,p} + \tilde{c}_k \bar{z}_{k,p}) \right|^2 \right),$$

where

$$\tilde{b}_k = b_k \Big/ \left(\sum_{q=1}^{n} (|b_q|^2 + |c_q|^2) \right)^{1/2}, \quad \tilde{c}_k = c_k \Big/ \left(\sum_{q=1}^{n} (|b_q|^2 + |c_q|^2) \right)^{1/2}.$$

Taking into consideration that $(\tilde{b}_1, \tilde{c}_1, \ldots, \tilde{b}_N, \tilde{c}_N)$ is a unit vector and each $z_{k,p}$, $w_{k,p}$ is an independent standard complex Gaussian, the analogue of (6.22) now follows [27, 199]:

$$O_{11} \overset{\text{d}}{=} \prod_{n=2}^{N} \left(1 + \frac{|X_n|^2}{|z_1 - z_n|^2} + \frac{|Y_n|^2}{|z_1 - \bar{z}_n|^2} \right). \tag{10.74}$$

Here X_n, Y_n are independent standard complex Gaussians.

Remark 10.4 1. A Pfaffian formula for the average of (10.74) over $\{z_j\}_{j=2}^{N}$ is given in [27]. To obtain from (10.74) the analogue of (6.22) one requires knowledge of the distribution of $\{|z_n|^2\}_{n=2}^{N}$ conditioned on $z_1 = 0$. The first result of this type, which is an easy consequence of (10.52) and the normalisation formula in Proposition 10.1 gives that [469] $\{|z_n|^2\}_{n=1}^N \overset{\mathrm{d}}{=} \{\Gamma[2n;1]\}_{n=1}^N$. As shown in [199], this same result holds true conditioned on $z_1 = 0$ with the sets now beginning at $n = 2$. From this, conditioning on $z_1 = 0$ and averaging over $\{|z_n|^2\}_{n=2}^N$ in (10.74) gives [199] $\langle O_{11}\rangle|_{z_1=0} \overset{\mathrm{d}}{=} 1/\mathrm{B}[4,2N]$ and hence $\frac{1}{2N}\langle O_{11}\rangle|_{z_1=0} \overset{\mathrm{d}}{\to} 1/\Gamma[4;1]$.

2. In [27], evidence is given for the validity of the analogue of the large N forms (6.28) (with $M = 1$) and (6.22) (with respect to the leading-order term at least).

Chapter 11
Further Extensions to GinSE

The elliptic extension of GinSE considered in the previous chapter gives rise to a potential which is not spherically symmetric. In contrast, the extensions of GinSE to be considered in this chapter—specifically the induced GinSE relating to a polar decomposition involving a rectangular generalisation of GinSE and a random Haar symplectic unitary matrix, the quaternion spherical ensemble of matrices $G_1^{-1}G_2$ with G_1, G_2 both GinSE matrices, a sub-block of a Haar symplectic unitary matrix, and products of GinSE matrices—all have the common feature that the corresponding potential is spherically symmetric. This allows for a unified treatment in relation to the corresponding skew-orthogonal polynomials. Furthermore, in each case the pre-kernel can be shown to satisfy an inhomogeneous partial differential equation, from which scaled limits can be computed.

11.1 Common Structures

The eigenvalue PDF of several extensions to GinSE discussed in this section is of the form (10.4) with a radially symmetric potential $W(z, \bar{z}) = \omega(|z|)$. Before moving on to each example, let us draw together some common structures.

We first note the following consequence of Proposition 10.1 and (10.25).

Proposition 11.1 *For a radially symmetric potential $W(z, \bar{z}) = \omega(|z|)$ the associated pre-kernel is given by*

$$\kappa_N^s(z, w) = \frac{1}{\pi} \sum_{0 \le l \le k \le N-1} a_{2k+1} a_{2l} (z^{2k+1} w^{2l} - z^{2l} w^{2k+1}), \tag{11.1}$$

© The Author(s) 2025
S.-S. Byun and P. J. Forrester, *Progress on the Study of the Ginibre Ensembles*, KIAS Springer Series in Mathematics 3, https://doi.org/10.1007/978-981-97-5173-0_11

where

$$a_{2k+1} := \frac{1}{\sqrt{2}} \frac{h_{2k}}{h_{2k+1}} \frac{h_{2k-2}}{h_{2k-1}} \cdots \frac{h_0}{h_1}, \qquad a_{2l} := \frac{1}{\sqrt{2}} \frac{h_{2l-1}}{h_{2l}} \frac{h_{2l-3}}{h_{2l-2}} \cdots \frac{h_1}{h_2} \frac{1}{h_0}. \qquad (11.2)$$

Due to Proposition 11.1, we have

$$\rho^{s}_{(1),N}(z) = \frac{e^{-2\omega(|z|)}}{\pi}$$

$$\times \sum_{0 \le l \le k \le N-1} a_{2k+1} a_{2l} (z^{2k+1} \bar{z}^{2l+1} - z^{2l} \bar{z}^{2k+2} - z^{2k+2} \bar{z}^{2l} + z^{2l+1} \bar{z}^{2k+1}). \qquad (11.3)$$

The asymptotic behaviour of $\rho^{s}_{(1),N}(z)$ in the global scaling $W = NQ$ can be deduced from the Coulomb gas approach previously discussed in Sect. 10.2, and so is independent of knowledge of (11.3). It gives

$$\lim_{N \to \infty} \frac{\rho^{s}_{(1),N}(z)}{N} \bigg|_{W=NQ} = \frac{\partial_z \partial_{\bar{z}} Q(z)}{\pi} \chi_{z \in S_Q}, \qquad (11.4)$$

where we recall that S_Q is the droplet.

As a feature dependent on (11.3), we now consider the radial density

$$\widehat{\rho}^{s}_{(1),N}(r) := \frac{1}{2\pi} \int_0^{2\pi} \rho^{s}_{(1),N}(re^{i\theta}) \, d\theta = \frac{e^{-2\omega(r)}}{\pi} \sum_{k=0}^{N-1} \frac{r^{4k+2}}{h_{2k+1}}. \qquad (11.5)$$

Here the second identity readily follows from (11.3) and $2a_k a_{k+1} = 1/h_{k+1}$. This same formula, with $2k + 1$ in the summation replaced by k, holds for the radial density of the normal matrix ensemble (10.5), due to the determinantal structure (5.18). A consequence of (11.5) is that the expected number of eigenvalues $E^{s}_N(R)$ in the disk of radius $R > 0$ can be expressed as

$$E^{s}_N(R) = 2\pi \sum_{k=0}^{N-1} \frac{\int_0^R r^{2k+1} e^{-2\omega(r)} \, dr}{h_{2k+1}}; \qquad (11.6)$$

see [20, Proposition 1.1] for a similar expression for the number variance. Furthermore, (11.6) is consistent with the fact that in the case of a general radially symmetric potential, the joint density of moduli of the eigenvalues forms a permanental process; see e.g. [29].

In the asymptotic analysis of the pre-kernel, the differential equations (10.29) and (10.47) have been key. This technique will be shown to be of further utility in the study of the extensions of GinSE considered below.

11.2 Induced GinSE

The induced GinSE can be constructed in a similar way described in Sect. 9.2. For this, we begin with an $n \times N$ ($n \geq N$) Gaussian random matrix G with independent real quaternion elements. Then we define $\tilde{G} = (G^\dagger G)^{1/2} U$, where U is a Haar unitary matrix with elements from the real quaternion field (see e.g. [194]). Then letting $L := n - N$, the element distribution on $N \times N$ matrices is proportional to

$$(\det \tilde{G}^\dagger \tilde{G})^{2L} e^{-2\mathrm{Tr}\, \tilde{G}^\dagger \tilde{G}}; \tag{11.7}$$

see [245, Sect. 2.4]. Its eigenvalue PDF follows (10.4) with

$$W^i(z, \bar{z}) = |z|^2 - 2L \log |z|. \tag{11.8}$$

This can be regarded as a special case of (10.70). We remark here that when we consider the model as a Coulomb particle system governed by the law (10.4), we can allow L to be an arbitrary real number as long as $L > -1$. Let us also mention that the system (10.4) with (11.8) can be interpreted as a distribution of N random eigenvalues of $(N + L) \times (N + L)$ GinSE, conditioned to have zero eigenvalues with multiplicity L.

We first discuss the global scaling $z \mapsto \sqrt{N}z$ in the regime $L = N\alpha$. As discussed in Sect. 10.2, the empirical measure of the eigenvalues converges to the equilibrium measure μ_{Q^i}, where

$$Q^i(z) = |z|^2 - 2\alpha \log |z|. \tag{11.9}$$

Then it follows from (10.7) and (10.8) that

$$\lim_{N \to \infty} \rho^i_{(1),N}(\sqrt{N}z) = \frac{1}{\pi} \left(\chi_{|z| < \sqrt{\alpha+1}} - \chi_{|z| < \sqrt{\alpha}} \right), \tag{11.10}$$

which agrees with the limiting density (9.7) for the induced GinOE. Notice that for $\alpha = 0$, we recover the circular law.

Turning to the higher correlation functions, we first note that due to Proposition 10.1, the associated skew-orthogonal polynomials are given by

$$q^i_{2k+1}(z) = z^{2k+1}, \qquad q^i_{2k}(z) = \sum_{l=0}^{k} \frac{\Gamma(k+1+L)}{\Gamma(l+1+L)} z^{2l},$$

$$r^i_k = \frac{\Gamma(2k+2+2L)}{2^{2k+1+2L}} \pi; \tag{11.11}$$

cf. (10.15). Combining this with (10.25) gives rise to the pre-kernel

$$\kappa^i_N(z, w) = \frac{2^{2L+1/2}}{\pi} \sum_{k=0}^{N-1} \frac{(\sqrt{2}z)^{2k+1}}{(2k+1+2L)!!} \sum_{l=0}^{k} \frac{(\sqrt{2}w)^{2l}}{(2l+2L)!!}$$

$$-\frac{2^{2L+1/2}}{\pi}\sum_{k=0}^{N-1}\frac{(\sqrt{2}w)^{2k+1}}{(2k+1+2L)!!}\sum_{l=0}^{k}\frac{(\sqrt{2}z)^{2l}}{(2l+2L)!!}.\qquad(11.12)$$

Taking into consideration the integrable structure, we further define the transformation

$$\widehat{\kappa}_N^{\mathrm{i}}(z,w)=(zw)^{2L}\kappa_N^{\mathrm{i}}(z,w),$$

$$\widehat{\mathcal{K}}_N^{\mathrm{i}}(z,w)=e^{-|z|^2-|w|^2}\begin{bmatrix}\widehat{\kappa}_N^{\mathrm{i}}(z,w)&\widehat{\kappa}_N^{\mathrm{i}}(z,\bar{w})\\\widehat{\kappa}_N^{\mathrm{i}}(\bar{z},w)&\widehat{\kappa}_N^{\mathrm{i}}(\bar{z},\bar{w})\end{bmatrix}.\qquad(11.13)$$

Then by Proposition 10.3 together with basic properties of the Pfaffian, we have

$$\rho_{(k),N}^{\mathrm{i}}(z_1,\dots,z_k)=\prod_{j=1}^{k}(\bar{z}_j-z_j)\mathrm{Pf}\,[\widehat{\mathcal{K}}_N^{\mathrm{i}}(z_j,z_l)]_{j,l=1,\dots,k}.\qquad(11.14)$$

The following was found in [117].

Proposition 11.2 *We have*

$$\partial_z\widehat{\kappa}_N^{\mathrm{i}}(z,w)=2z\,\kappa_N^{\mathrm{i}}(z,w)+\frac{2}{\pi}e^{2zw}\left(\frac{\Gamma(2L+2N;2zw)}{\Gamma(2L+2N)}-\frac{\Gamma(2L;2zw)}{\Gamma(2L)}\right)$$
$$-\frac{2^{3/2}}{\pi}\frac{z^{2N+2L}}{\Gamma(N+L+1/2)}e^{w^2}\left(\frac{\Gamma(N+L;w^2)}{\Gamma(N+L)}-\frac{\Gamma(L;w^2)}{\Gamma(L)}\right)$$
$$-\frac{2}{\sqrt{\pi}}\frac{z^{2L-1}}{\Gamma(L)}e^{w^2}\left(\frac{\Gamma(N+L+1/2;w^2)}{\Gamma(N+L+1/2)}-\frac{\Gamma(L+1/2;w^2)}{\Gamma(L+1/2)}\right).$$
$$(11.15)$$

Note that Proposition 11.2 with $L=0$ reduces to Proposition 10.4. For $L=0$, the last term in the RHS of (11.15) vanishes since $\lim_{L\to0}1/\Gamma(L)=0$. As previously remarked below Proposition 10.6, one can notice again that the first inhomogeneous term in (11.15) coincides with the kernel of the induced GinUE with $N\mapsto 2N$ up to a weight function; see Sect. 2.4.

As in Sect. 10.4, the uniform asymptotic expansion (3.31) can be used to derive various scaling limits. It then follows that for $L=\alpha N$ with $\alpha>0$, the universal scaling limits (10.31) and (10.34) appear in the bulk and edge of the spectrum. Another interesting regime is the case when L is fixed while $N\to\infty$. In this case, the scaling limit at the origin should be treated separately since the potential (11.8) reveals a conical singularity. This was investigated in [22].

Proposition 11.3 *For a fixed $L>-1$, we have*

$$\lim_{N\to\infty}\rho_{(k),N}^{\mathrm{i}}(z_1,\dots,z_k)=\prod_{j=1}^{k}(\bar{z}_j-z_j)\mathrm{Pf}\,[\mathcal{K}_\infty^{\mathrm{s},L}(z_j,z_l)]_{j,l=1,\dots,k},\qquad(11.16)$$

where

$$\mathcal{K}_\infty^{s,L}(z, w) := e^{-|z|^2 - |w|^2} \begin{bmatrix} \kappa_\infty^{s,L}(z, w) & \kappa_\infty^{s,L}(z, \bar{w}) \\ \kappa_\infty^{s,L}(\bar{z}, w) & \kappa_\infty^{s,L}(\bar{z}, \bar{w}) \end{bmatrix}. \tag{11.17}$$

Here,

$$\kappa_\infty^{s,L}(z, w) = \frac{2}{\pi} (2zw)^{2L} \int_0^1 s^{2L} \left(z e^{(1-s^2)z^2} - w e^{(1-s^2)w^2} \right) E_{2,1+2L}((2szw)^2) \, ds, \tag{11.18}$$

where $E_{a,b}(z) := \sum_{k=0}^\infty z^k / \Gamma(ak + b)$ is the two-parametric Mittag–Leffler function. This can be rewritten in terms of the incomplete gamma function as

$$\kappa_\infty^{s,L}(z, w) = \frac{1}{\pi} \int_0^1 (z e^{(1-s^2)z^2} - w e^{(1-s^2)w^2})$$
$$\times \left(e^{2szw} \frac{\gamma(2L; 2szw)}{\Gamma(2L)} + (-1)^{-2L} e^{-2szw} \frac{\gamma(2L; -2szw)}{\Gamma(2L)} \right) ds. \tag{11.19}$$

We mention that if $2L$ is a non-negative integer, the pre-kernel $\kappa_\infty^{s,L}$ can also be expressed in terms of error functions. (Cf. See also [14] for a different way to express the correlation functions as a ratio of certain Pfaffians.) Namely, if L is a non-negative integer, we have

$$\kappa_\infty^{s,L}(z, w) = \frac{e^{z^2 + w^2}}{\sqrt{\pi}} \operatorname{erf}(z - w) + \frac{1}{\pi} \sum_{0 \le l < k \le L-1} \frac{w^{2k} z^{2l+1} - z^{2k} w^{2l+1}}{k!(1/2)_{l+1}}$$
$$+ \frac{e^{w^2}}{\sqrt{\pi}} \operatorname{erf}(w) \sum_{k=0}^{L-1} \frac{z^{2k}}{k!} - \frac{e^{z^2}}{\sqrt{\pi}} \operatorname{erf}(z) \sum_{k=0}^{L-1} \frac{w^{2k}}{k!}. \tag{11.20}$$

Note that if $L = 0$, this recovers (10.31), where we have used the convention that the summation with an empty index equals zero. Here $(a)_n = a(a + 1) \cdots (a + n - 1)$ is Pochhammer's symbol. On the other hand, if $L - 1/2$ is a non-negative integer, we have

$$\kappa_\infty^{s,L}(z, w) = \frac{e^{z^2 + w^2}}{\sqrt{\pi}} (\operatorname{erf}(z - w) - \operatorname{erf}(z) + \operatorname{erf}(w))$$
$$+ \frac{1}{\pi} \sum_{1 \le l < k \le L-1/2} \frac{w^{2k-1} z^{2l} - z^{2k-1} w^{2l}}{l!(1/2)_k}$$
$$+ \frac{e^{w^2} - 1}{\pi} \sum_{k=1}^{L-1/2} \frac{z^{2k-1}}{(1/2)_k} - \frac{e^{z^2} - 1}{\pi} \sum_{k=1}^{L-1/2} \frac{w^{2k-1}}{(1/2)_k}. \tag{11.21}$$

Setting $w = \bar{z}$ gives the density, in particular reclaiming (10.32) for $L = 0$. We mention that other than $L = 0$, the limiting 1-point function is no longer translation invariant along the real axis.

Another interesting double scaling limit arises in the so-called almost-circular regime, where the spectrum tends to form a thin annulus of width $O(1/N)$. In this case, the scaling limits both at the real axis as well as away from the real axis were obtained in [117]. For the former case, the limiting correlation functions are Pfaffians, which interpolate the bulk scaling limits of the GinSE and of the antisymmetric Gaussian Hermitian ensemble. For the latter case, the limiting correlation functions are determinants, and the kernel is the same as the one appearing in the bulk scaling limit of the weakly non-Hermitian elliptic GinUE [24, 276].

11.3 Spherical Induced GinSE

Following [228, 434], we first consider an $(N + L) \times N$ rectangular GinSE X and an $N \times N$ Wishart matrix with quaternion entries A (constructed as $A = B^\dagger B$ with B an $n \times N$ ($n \geq N$) rectangular GinSE matrix) to introduce a particular $(N + L) \times N$ random matrix $Y = XA^{-1/2}$ with each entry itself a 2×2 matrix representation of a quaternion (1.1). In terms of such Y together with a Haar distributed unitary random matrix U with quaternion entries, we define $G = U(Y^\dagger Y)^{1/2}$. Then the matrix distribution of G is given by

$$\frac{\det(GG^\dagger)^{2L}}{\det(\mathbb{I}_N + GG^\dagger)^{2(n+N+L)}}. \tag{11.22}$$

It was shown in [227, 433, 434] that its eigenvalue PDF follows (10.4) with

$$W^{\text{si}}(z, \bar{z}) = (n + L + 1)\log(1 + |z|^2) - 2L\log|z|. \tag{11.23}$$

We refer to [120, Appendix A] for a Coulomb gas picture of the spherical induced ensemble.

With L, n scaled with N in a way that $L/N \to \alpha_1 - 1 \geq 0$ and $n/N \to \alpha_2 \geq 1$,

$$\frac{W^{\text{si}}(z, \bar{z})}{N} \sim (\alpha_1 + \alpha_2 - 1)\log(1 + |z|^2) - 2(\alpha_1 - 1)\log|z|. \tag{11.24}$$

Then by (10.8) one can specify inner and outer radii of the limiting spectrum, which leads to

$$\lim_{N \to \infty} \frac{1}{N}\rho^{\text{si}}_{(1),N}(z) = \frac{1}{\pi}\frac{\alpha_1 + \alpha_2 - 1}{(1 + |z|^2)^2}\left(\chi_{|z| < \sqrt{\alpha_1/(\alpha_2 - 1)}} - \chi_{|z| < \sqrt{(\alpha_1 - 1)/\alpha_2}}\right), \tag{11.25}$$

where the limiting eigenvalue density is obtained by taking the Laplacian of the RHS of (11.24); cf. (10.7). This limiting distribution is in consistent with the spherical induced GinUE; see Sect. 2.5. Due to the limiting form (11.25), one again notices that the stereographically projected eigenvalues tend to uniformly occupy a spherical annulus, whence the name spherical.

Beyond the leading-order asymptotic of the eigenvalue density, a fluctuation formula can be found [120, Appendix B].

Proposition 11.4 *Let* $\Phi = \sum_{j=1}^{N} \phi(|z_j|)$ *be a radially symmetric linear statistic, where* z_j*'s are drawn from the spherical induced ensemble as specified by (10.4) and (11.23). It is required that* ϕ *be smooth and subject to a growth condition as* $|z| \to \infty$. *Then the corresponding characteristic function* $\hat{P}_{N,B}(k)$ *satisfies*

$$\log \hat{P}_{N,\Phi}(k) = ik\tilde{\mu}_{N,\Phi} - k^2\tilde{\sigma}_B^2/2 + o(1), \tag{11.26}$$

where

$$\tilde{\mu}_{N,\Phi} = \frac{n+L}{\pi} \int_{S^{si}} \frac{\phi(z)}{(1+|z|^2)^2} d^2z, \qquad \tilde{\sigma}_\Phi^2 = \frac{1}{8\pi} \int_{S^{si}} ||\nabla\phi(|z|)||^2 d^2z. \tag{11.27}$$

Here S^{si} *is the droplet specified in (11.25).*

As a consequence of this proposition, the asymptotic normality of the centred linear statistic $\Phi - \tilde{\mu}_{N,\Phi}$ follows. The variance is equal to one half times that for the GinUE in the case of a radially symmetric test statistic; see (3.21).

We now discuss the higher-order correlation functions and their scaling limits. Due to the radial symmetry of the potential (11.23), the associated skew-orthogonal polynomials can be again constructed by using Proposition 10.1. This in turn gives that [240, Proposition 4]

$$q_{2k+1}^{si}(z) = z^{2k+1}, \qquad q_{2k}^{si}(z) = \frac{\Gamma(k+L+1)}{\Gamma(k-n+\frac{1}{2})} \sum_{l=0}^{k}(-1)^{k-l}\frac{\Gamma(l-n+\frac{1}{2})}{\Gamma(l+L+1)}z^{2l}, \tag{11.28}$$

with the skew norm

$$r_k^{si} = \frac{2\Gamma(2k+2L+2)\Gamma(2n-2k)}{\Gamma(2n+2L+2)}. \tag{11.29}$$

Now it follows from (10.25) and basic functional identities of the gamma function that

$$\kappa_N^{si}(z,w)$$
$$= \frac{\Gamma(2n+2L+2)}{2^{2L+2n+1}} \sum_{k=0}^{N-1}\sum_{l=0}^{k} \frac{z^{2k+1}w^{2l}}{\Gamma(k+L+\frac{3}{2})\Gamma(n-k)\Gamma(n-l+\frac{1}{2})\Gamma(l+L+1)}$$

$$-\frac{\Gamma(2n+2L+2)}{2^{2L+2n+1}}\sum_{k=0}^{N-1}\sum_{l=0}^{k}\frac{w^{2k+1}z^{2l}}{\Gamma(k+L+\frac{3}{2})\Gamma(n-k)\Gamma(n-l+\frac{1}{2})\Gamma(l+L+1)};$$

$$\tag{11.30}$$

see [120, Lem 3.1]. Next, we define

$$\widetilde{\kappa}_N^{\mathrm{si}}(z,w)=\frac{(zw)^{2L}}{((1+z^2)(1+w^2))^{n+L-\frac{1}{2}}}\kappa_N^{\mathrm{si}}(z,w).\tag{11.31}$$

As in Sect. 9.3, the correlation kernel can be effectively analysed using the incomplete beta function I_x, which admits the expression

$$I_x(m,n-m+1)=\sum_{j=m}^{n}\binom{n}{j}x^j(1-x)^{n-j}.\tag{11.32}$$

The following identity was found in [120, Proposition 1.1].

Proposition 11.5 *Let*

$$\zeta:=\frac{zw}{1+zw},\qquad \eta:=\frac{w^2}{1+w^2}.\tag{11.33}$$

Then we have

$$\partial_z\widetilde{\kappa}_N^{\mathrm{si}}(z,w)=I_N(z,w)-II_N(z,w)-III_N(z,w),\tag{11.34}$$

where

$$I_N(z,w)=\frac{(1+zw)^{2n+2L-1}}{(1+z^2)^{n+L+\frac{1}{2}}(1+\eta^2)^{n+L-\frac{1}{2}}}(2n+2L+1)(n+L)$$

$$\times\Big(I_\zeta(2L,2n)-I_\zeta(2N+2L,2n-2N)\Big),\tag{11.35}$$

$$II_N(z,w)=\frac{z^{2N+2L}}{2^{2L+2n}(1+z^2)^{n+L+\frac{1}{2}}}\frac{\pi\,\Gamma(2n+2L+2)}{\Gamma(N+L+\frac{1}{2})\Gamma(n-N)\Gamma(n+L+\frac{1}{2})}$$

$$\times\Big(I_\eta(L,n+\tfrac{1}{2})-I_\eta(N+L,n-N+\tfrac{1}{2})\Big),\tag{11.36}$$

and

$$III_N(z,w)=\frac{z^{2L-1}}{2^{2L+2n}(1+z^2)^{n+L+\frac{1}{2}}}\frac{\pi\,\Gamma(2n+2L+2)}{\Gamma(n+\frac{1}{2})\Gamma(L)\Gamma(n+L+\frac{1}{2})}$$

$$\times\Big(I_\eta(L+\tfrac{1}{2},n)-I_\eta(N+L+\tfrac{1}{2},n-N)\Big).\tag{11.37}$$

From this result, we can again observe a common feature relating the correlation kernels of the symplectic and complex ensembles; namely, up to a weight function, the term I_N agrees with the correlation kernel of the complex spherical induced ensemble; see Sect. 2.5. Also, one can use the known asymptotic behaviours of the incomplete beta functions (that can be found for instance in [522, Sect. 11.3.3]) to derive various scaling limits. As a consequence, the bulk and edge universality with the limiting pre-kernels (10.31) and (10.34) were shown in [120]. Furthermore, in the regime where $L \geq -1$ is fixed, the scaling limit at the origin turns out to be again universal, with the limiting form (11.18).

Remark 11.1 The recent work [308] on winding number statistics for chiral random matrices encounters the average over GinSE matrices K_1, K_2 of the ratio of determinants

$$\det(\alpha_1 \mathbb{I}_N + K_1^{-1} K_2)/\det(\alpha_2 \mathbb{I}_N + K_1^{-1} K_2)$$

(as well as higher-order products). As noted in this reference, $K_1^{-1} K_2$ is a member of spherical GinSE, and so the eigenvalues have the PDF (11.22) with $n = N, L = 0$, facilitating the computation of the average.

Using (10.52) and (11.29), one can express the partition function $Z_N^{\mathbb{H}}(W^{\text{si}})$ in terms of the Barnes G-function as

$$Z_N^{\mathbb{H}}(W^{\text{si}}) = \left(\frac{2^{2n+2L+1}}{\Gamma(2n+2L+2)}\right)^N$$
$$\times \frac{G(N+L+1)}{G(L+1)} \frac{G(N+L+3/2)}{G(L+3/2)} \frac{G(n+1)}{G(n-N+1)} \frac{G(n+3/2)}{G(n-N+3/2)}.$$
$$(11.38)$$

Then the well-known asymptotic behaviour of the Barnes G-function (see [226, Th. 1]) allows the large N expansion of $Z_N^{\mathbb{H}}(W^{\text{si}})$ to be calculated explicitly. In the scaling $L = (\alpha_1 - 1)N$ and $n = \alpha_2 N - 1$, where the associated droplet is an annulus, one can directly check that the general asymptotic result (10.57) with the Euler characteristic $\chi = 0$ holds. On the other hand, for the spherical case when $L = 0, n = N$, the limiting eigenvalue support is the whole complex plane with density $1/(\pi(1+|z|^2)^2)$. In this case, it follows from straightforward computations that

$$\log Z_N^{\mathbb{H}}(W^{\text{si}})\Big|_{L=0,n=N} = -N^2 - \frac{1}{2}N\log N + \left(\frac{\log(4\pi^3)}{2} - 1\right)N - \frac{1}{12}\log N$$
$$- \frac{13}{24} + \frac{5\log 2}{12} + \zeta'(-1) + \frac{1}{12N} - \frac{61}{1440N^2} + O(\frac{1}{N^3}).$$
$$(11.39)$$

This exhibits the universal coefficient $-\chi/24$ of the $\log N$ in (10.57), with the Euler characteristic $\chi = 2$ for the sphere. Also, we check that the $O(N)$ term is consistent with the general form in (10.57); see [229, Sect. 4] for the complex counterparts.

11.4   Truncations of Haar Unitary Symplectic Matrices

We now consider the truncations of unitary symplectic matrices, or equivalently unitary matrices with quaternion elements each realised as a 2×2 complex block (1.1). These unitary matrices are to be chosen with Haar measure—see [194] for an extended discussion on this. Their eigenvalues are all on the unit circle in the complex plane and are doubly degenerate. Let A_N be the $N \times N$ sub-block of an $(n + N) \times (n + N)$ quaternion unitary matrix drawn with Haar measure. It was derived in [245, Sect. 2.3, conditional of $n \geq N$], and later in [361] unconditionally, that the eigenvalue PDF of A_N follows the law (10.4) with

$$
W^{\mathrm{t}}(z, \bar{z}) = \begin{cases} -(n - 1/2) \log(1 - |z|^2) & \text{if } |z| < 1, \\ \infty & \text{otherwise.} \end{cases} \tag{11.40}
$$

This form of potential makes all the eigenvalues completely contained inside the unit disk $|z| < 1$, as is consistent with the effect of the truncation. We remark too that after the truncation the eigenvalues are no longer doubly degenerate, but rather occur in complex conjugate pairs.

From the Coulomb gas viewpoint, the potential (11.40) can be interpreted as a situation under the presence of a hard edge. In this situation, the equilibrium problem should be treated depending on the position of the hard edge. To be more precise, if the hard edge is built inside the droplet, the mass outside the hard edge then lies on the boundary of the droplet, which makes the equilibrium measure no longer absolutely continuous with respect to the area measure. Namely, in this case, the equilibrium measure is a combination of two and one-dimensional measure, where the latter is given in terms of the balayage measure. We refer to [55, Appendix A] for an explicit formula. Here, the balayage measure is a unique measure supported on the boundary of the droplet whose logarithmic energy is the same as the equilibrium measure; see [479, Ch. II] for further details. On the other hand, if the hard edge is built outside the droplet, the resulting equilibrium measure is not affected by the hard edge.

In the scaling $n = (1 - \alpha)/\alpha N$, the empirical measure converges to the equilibrium measure associated with the potential

$$
Q^{\mathrm{t}}(z) = \frac{\alpha - 1}{\alpha} \log(1 - |z|^2). \tag{11.41}
$$

Then it follows from (10.7) and (10.8) that

$$
\lim_{N \to \infty} \frac{\rho^{\mathrm{t}}_{(1),N}(z)}{N} \bigg|_{\alpha = N/(N+n)} = \frac{1 - \alpha}{\pi \alpha} \frac{1}{(1 - |z|^2)^2} \chi_{|z| < \sqrt{\alpha}}; \tag{11.42}
$$

cf. (9.33). We mention that

$$
\lim_{\alpha \to 0} Q^{\mathrm{t}}(\sqrt{\alpha} z) = |z|^2. \tag{11.43}
$$

This means that after the scaling $z_j \mapsto z_j/\sqrt{\alpha}$, the eigenvalue statistics of truncated ensembles tend to the GinSE as $\alpha \to 0$, equivalently, $n \gg N$. On the other hand, in the regime when n is fixed while $N \to \infty$, the limiting global scaled potential has its value 0 inside the unit disk, and ∞ outside the disk. The associated equilibrium measure is a uniform distribution on the unit circle. This is consistent with the fact that when $n \to 0$, the truncated ensemble corresponds to the circular symplectic ensemble [237, Sect. 2.6].

Turning to the correlation functions, Proposition 11.1 gives that [361, Eq. (14)]

$$\kappa_N^t(z, w) = \frac{B(1/2, n)}{\pi} \sum_{0 \le l \le k \le N-1} \frac{z^{2k+1} w^{2l} - z^{2l} w^{2k+1}}{B(k + 3/2, n) B(l + 1, n)}. \tag{11.44}$$

As an analogue of Propositions 10.6, 11.2 and 11.5, we have the following. Let us stress that this proposition has not been reported in previous literature.

Proposition 11.6 *We have*

$$\frac{1 - z^2}{2n} \partial_z \kappa_N^t(z, w) = \frac{n+1}{n} z \kappa_N^t(z, w) + \frac{2n + 1}{2\pi} \frac{1 - I_{zw}(2N, 2n + 2)}{(1 - zw)^{2n+2}}$$

$$- \frac{z^{2N}}{\sqrt{\pi}} \frac{\Gamma(n + N + 3/2)}{\Gamma(n + 1/2)\Gamma(N + 1/2)} \frac{1 - I_{w^2}(N, n + 1)}{(1 - w^2)^{n+1}}. \tag{11.45}$$

Proof The proof is similar in spirit to that of Proposition 11.5, which can be found in [120]. Let us write

$$G_N^t(z, w) := \sum_{k=0}^{N-1} \sum_{l=0}^{k} \frac{\Gamma(n + k + 3/2)\Gamma(n + l + 1)}{\Gamma(k + 3/2)\Gamma(l + 1)} z^{2k+1} w^{2l}. \tag{11.46}$$

By differentiating this expression, we have

$$\partial_z G_N^t(z, w) = 2z \sum_{k=0}^{N-1} \sum_{l=0}^{k} \frac{\Gamma(n + k + 3/2)\Gamma(n + l + 1)}{\Gamma(k + 1/2)\Gamma(l + 1)} z^{2k-1} w^{2l}$$

$$= 2\frac{\Gamma(n + 3/2)\Gamma(n + 1)}{\Gamma(1/2)} + 2z \sum_{k=0}^{N-2} \sum_{l=0}^{k+1} \frac{\Gamma(n + k + 5/2)\Gamma(n + l + 1)}{\Gamma(k + 3/2)\Gamma(l + 1)} z^{2k+1} w^{2l}.$$

Rearranging the summations, the last term can be rewritten as

$$2z \sum_{k=0}^{N-1} \sum_{l=0}^{k} \frac{\Gamma(n + k + 5/2)\Gamma(n + l + 1)}{\Gamma(k + 3/2)\Gamma(l + 1)} z^{2k+1} w^{2l}$$

$$- 2 \sum_{l=0}^{N-1} \frac{\Gamma(n + N + 3/2)\Gamma(n + l + 1)}{\Gamma(N + 1/2)\Gamma(l + 1)} z^{2N} w^{2l}$$

$$+ 2 \sum_{k=1}^{N-1} \frac{\Gamma(n+k+3/2)\Gamma(n+k+1)}{\Gamma(k+1/2)\Gamma(k+1)} (zw)^{2k}.$$

Here, we have

$$\sum_{k=0}^{N-1} \sum_{l=0}^{k} \frac{\Gamma(n+k+5/2)\Gamma(n+l+1)}{\Gamma(k+3/2)\Gamma(l+1)} z^{2k+1} w^{2l}$$

$$= \sum_{k=0}^{N-1} \sum_{l=0}^{k} ((n+1)+(k+1/2)) \frac{\Gamma(n+k+3/2)\Gamma(n+l+1)}{\Gamma(k+3/2)\Gamma(l+1)} z^{2k+1} w^{2l}$$

$$= (n+1)G_N^t(z,w) + \frac{z}{2} \partial_z G_N^t(z,w).$$

Combining all of the above, we obtain

$$(1-z^2)\partial_z G_N^t(z,w) = 2(n+1)zG_N^t(z,w)$$

$$- 2 \sum_{l=0}^{N-1} \frac{\Gamma(n+N+3/2)\Gamma(n+l+1)}{\Gamma(N+1/2)\Gamma(l+1)} z^{2N} w^{2l}$$

$$+ 2 \sum_{k=0}^{N-1} \frac{\Gamma(n+k+3/2)\Gamma(n+k+1)}{\Gamma(k+1/2)\Gamma(k+1)} (zw)^{2k}. \qquad (11.47)$$

We now compute $\partial_z G_N^t(w,z)$. We begin by writing

$$\partial_z G_N^t(w,z) = 2z \sum_{k=0}^{N-1} \sum_{l=0}^{k} \frac{\Gamma(n+k+3/2)\Gamma(n+l+2)}{\Gamma(k+3/2)\Gamma(l+1)} w^{2k+1} z^{2l}$$

$$- 2 \sum_{k=0}^{N-1} \frac{\Gamma(n+k+3/2)\Gamma(n+k+2)}{\Gamma(k+3/2)\Gamma(k+1)} (zw)^{2k+1}.$$

Note here that

$$\sum_{k=0}^{N-1} \sum_{l=0}^{k} \frac{\Gamma(n+k+3/2)\Gamma(n+l+2)}{\Gamma(k+3/2)\Gamma(l+1)} w^{2k+1} z^{2l}$$

$$= \sum_{k=0}^{N-1} \sum_{l=0}^{k} ((n+1)+l) \frac{\Gamma(n+k+3/2)\Gamma(n+l+1)}{\Gamma(k+3/2)\Gamma(l+1)} w^{2k+1} z^{2l}$$

$$= (n+1)G_N^t(w,z) + \frac{z}{2} \partial_z G_N^t(w,z).$$

We have shown that

$$(1 - z^2)\partial_z G_N^t(w, z) = 2(n + 1)z G_N^t(w, z)$$
$$- 2\sum_{k=0}^{N-1} \frac{\Gamma(n + k + 3/2)\Gamma(n + k + 2)}{\Gamma(k + 3/2)\Gamma(k + 1)}(zw)^{2k+1}. \quad (11.48)$$

Then by combining (11.44), (11.47) and (11.48), we conclude the desired identity. $\qquad\square$

As before, an important point to note here is the form of the second term in (11.45), which agrees with the kernel of the complex counterpart up to a weight function; see [120, Eq. (2.59)]. Furthermore, as in Sect. 11.3, Proposition 11.6 and the knowledge of the uniform asymptotic expansion of the incomplete beta functions can be used to derive scaling limits. In the regime of strong non-unitarity when $n = O(N)$, the universal bulk and edge scaling limits (10.31) and (10.34) again appear. On the other hand, in the regime of weak non-unitarity when n is fixed while $N \to \infty$, a qualitatively distinct scaling limit arises at the edge of the spectrum. In particular, it was shown in [361, Th. 6.13] that for $x > 0$ and $y \in \mathbb{R}$,

$$\lim_{N\to\infty} \frac{1}{N^2}\rho_{(1),N}^t\left(1 - \frac{x + iy}{N}\right)$$
$$= \frac{2^{4n}yx^{2n+1}}{\pi\Gamma(2n)} \int_{(0,1)^2} s^{2n+1}t^n e^{-2s(1+t)x} \sin(2s(1 - t)y)\, ds\, dt. \quad (11.49)$$

In [361], instead of Proposition 11.6, a contour integral representation of κ_N^t was used to derive various scaling limits including those away from the real axis. We mention too that the scaling limit (11.49) also appears in the context of the GinSE with hard edge; see [119, Th. 2.4].

Remark 11.2 Distinct from Haar distributed symplectic unitary matrices are matrices from Dyson's circular symplectic ensemble (see e.g. [237, Sect. 2.2.2] or [206] for the original work). However, both have quaternion elements represented as 2×2 complex blocks (1.1), and each eigenvalue is doubly degenerate. It has been shown in [368] that the deletion of a single (quaternion) row and column gives for the PDF of the independent eigenvalues, taken to be in the upper half plane as now the eigenvalues appear in complex conjugate pairs, is proportional to

$$\prod_{1\le j<k\le N} |z_k - z_j|^2 |1 - z_j\bar{z}_k|^2 \prod_{j=1}^N (1 - |z_j|^2)\chi_{|z_j|<1, z_j\in\mathbb{C}_+}; \quad (11.50)$$

cf. (10.9) with $\beta = 2$. As noted in [245, Sect. 5] this corresponds to a Pfaffian point process, with correlation kernel given by that computed in [237, Sect. 15.9 with $\eta \to 0^+$].

11.5 Products of GinSE

As in Sect. 9.5, to describe the products of GinSE in the most general setup, we begin with the $N \times N$ square matrices $\tilde{G}_i$ whose element distribution is proportional to

$$|\det \tilde{G}_i \tilde{G}_i^\dagger|^{\nu_i} e^{-\mathrm{Tr}\, \tilde{G}_i \tilde{G}_i^\dagger},$$

where $\nu_i > 0$ are the differences between matrix dimensions. Then the eigenvalue PDF can be computed for a general M and ν_j (see [30, 329]). It is of the form (10.4) with

$$W^\nu(z, \bar{z}) = -\frac{1}{2} \log G_{0,M}^{M,0}\left(2\nu_1, \ldots, 2\nu_{M-1}, 2\nu_M \,\middle|\, 2^M |z|^2\right), \tag{11.51}$$

where $G_{0,M}^{M,0}$ is the Meijer G-function as previously discussed in Remark 9.2. For the special case when $M = 2$ and $\nu_1 = \nu_2 = 0$, this can be written in terms of the Bessel function K_0. Furthermore, as mentioned in Remark 2.6.3, the case $M = 2$ permits a generalisation using a non-Hermiticity parameter; see [11].

Turning to the Coulomb gas perspective, we first note that the well-known asymptotic behaviour (2.80) of the Meijer G-function implies

$$-\frac{1}{2N} \log G_{0,M}^{M,0}\left(2\nu_1, \ldots, 2\nu_{M-1}, 2\nu_M \,\middle|\, (2N)^M |z|^2\right)$$
$$\sim M|z|^{2/M} - \frac{2(\nu_1 + \cdots + \nu_M)}{MN} \log |z|.$$

Therefore, in the scaling $(\nu_1 + \cdots + \nu_M)/N \sim M\alpha$ $(\alpha \geq 0)$, the general formula (11.4) can again be applied to the product ensembles, which gives rise to

$$\lim_{N\to\infty} N^{M-1} \rho_{(1),N}(N^{M/2} z) = \frac{|z|^{-2+2/M}}{\pi M}\left(\chi_{|z|<(\alpha+1)^{M/2}} - \chi_{|z|<\alpha^{M/2}}\right). \tag{11.52}$$

The associated skew-orthogonal polynomials can be constructed by Proposition 10.1 with the evaluation

$$\int_{\mathbb{C}} |z|^{2k} e^{-2W^\nu(z,\bar{z})} \, d^2 z = \frac{\pi}{2^{M(k+1)}} \prod_{l=1}^{M} \Gamma(2\nu_l + k + 1). \tag{11.53}$$

Furthermore, by Proposition 11.1, we have

$$\kappa_N^{M,\nu}(z, w) = \frac{2^{M-1-2\sum_j \nu_j}}{\pi^{2-M/2}} \sum_{0 \le l \le k \le N-1} \frac{z^{2k+1} w^{2l} - z^{2l} w^{2k+1}}{\prod_{j=1}^{M} \Gamma(\nu_j + k + 3/2)\Gamma(\nu_j + l + 1)}. \tag{11.54}$$

As we have illustrated, the main idea to analyse various pre-kernels presented above is to write down proper differential equations and then derive their large N limit. This technique can be extended to the present case. Contrary to the previous cases, the resulting differential equation is of order M. In the case of $M = 2$, the associated second-order differential equation was found and used in [11] to derive the limiting correlation kernel at the origin; see also [26]. A similar situation arises in the study of the Mittag–Leffler ensemble with the potential (10.70) and it was found in [22] that the associated pre-kernel satisfies a fractional differential equation of order $1/b$. On the other hand, as expected from the structure (11.5), the analysis for the radial density is considerably simplified; see [29, 329, 333]. We also mention that in the same spirit as Remark 9.3, the Lyapunov and stability exponents of the products of GinSE are available in the literature [269, 330, 356].

11.6   A Summary of GinSE Analogues with GinUE and GinOE

Comparing the development of GinSE theory as presented above with that of GinUE and GinOE, the following relative features—both quantitative and qualitative—are evident.

- Like both GinUE and GinOE, the joint eigenvalue PDF of GinSE has a Coulomb gas interpretation (Sect. 10.2). This gives relevance to the large N expansion of the associated partition function (Sect. 10.6).
- Like GinOE, GinSE eigenvalues form a Pfaffian point process (Sect. 10.2). The pre-kernel then involves particular skew-orthogonal polynomials. The latter can be explicitly determined for radial potentials (Sect. 10.3).
- The generalisation of GinSE considered in earlier sections of this chapter all relate to radial potentials, and scaling limits can be analysed in detail. These generalisations, already listed in the introductory paragraph of this chapter, each have analogues for GinOE and GinUE. The radial symmetry facilitates the analysis of gap probabilities and a central limit theorem for smooth radially symmetric linear statistics (Sect. 10.6).
- The elliptic generalisation of GinSE provides for a continuous deformation to the Hermitian ensemble of GSE matrices. An analogous generalisation also holds for GinOE and GinUE. Now the associated potential is not radially symmetric, and the theory relating to the skew-orthogonal polynomials is more involved (Sect. 10.3). The analysis of a weakly non-Hermitian scaling limit is possible; see the paragraph including (10.49) for references.
- Singular values and eigenvectors of GinSE matrices can be studied along the same lines as for GinUE and GinOE.

A distinction between the GinSE analysis, and that of GinUE and GinOE, is the extra challenge of the asymptotic analysis imposed by the particular structure of the pre-kernel. Here, an approach based on an inhomogenous partial differential equation satisfied by the pre-kernel is shown to provide a unified treatment. We remark too that to date there is no theory relating to Wigner generalisations of GinSE.

References

1. Abreu, L.D.: Entanglement entropy and hyperuniformity of Ginibre and Weyl-Heisenberg ensembles. Lett. Math. Phys. **113**, 54 (2023)
2. Abreu, L.D., Pereira, J.M., Romero, J.L., Torquato, S.: The Weyl-Heisenberg ensemble: hyperuniformity and higher Landau Levels. J. Stat. Mech. Theory Exp. **2017**, 043103 (2017)
3. Adhikari, K.: Hole probabilities for β-ensembles and determinantal point processes in the complex plane. Electron. J. Probab. **23**, 21pp (2018)
4. Adhikari, K., Reddy, N.K.: Hole probabilities for finite and infinite Ginibre ensembles. Int. Math. Res. Not. **2017**, 6694–6730 (2017)
5. Adhikari, K., Reddy, N.K., Reddy, T.R., Saha, K.: Determinantal point processes in the plane from products of random matrices. Ann. Inst. H. Poincaré Probab. Statist. **52**, 16–46 (2016)
6. Adler, M., Forrester, P., Nagao, T., van Moerbeke, P.: Classical skew orthogonal polynomials and random matrices. J. Stat. Phys. **99**, 141–170 (2000)
7. Agam, O., Bettelheim, E., Wiegmann, P., Zabrodin, A.: Viscous fingering and a shape of an electronic droplet in the quantum Hall regime. Phys. Rev. Lett. **88**, 236801 (2002)
8. Aharonov, Y., Casher, A.: Ground state of a spin-1/2 charged particle in a two-dimensional magnetic field. Phys. Rev. A **19**, 2461–2462 (1979)
9. Ahn, A., Van Peski, R.: Lyapunov exponents for truncated unitary and Ginibre matrices. Ann. Inst. H. Poincaré Probab. Statist. **59**, 1029–1039 (2023)
10. Akemann, G.: Microscopic correlations for non-Hermitian Dirac operators in three-dimensional QCD. Phys. Rev. D **64**, 114021 (2001)
11. Akemann, G.: The complex Laguerre symplectic ensemble of non-Hermitian matrices. Nuclear Phys. B **730**, 253–299 (2005)
12. Akemann, G., Baake, M., Chakarov, N., Krüger, O., Mielke, A., Ottensmann, M., Werdehausen, R.: Territorial behaviour of buzzards versus random matrix spacing distributions. J. Theor. Biol. **509**, 110475 (2021)
13. Akemann, G., Baik, J., Di Francesco, P. (eds.): The Oxford Handbook of Random Matrix Theory. Oxford University Press, Oxford (2011)
14. Akemann, G., Basile, F.: Massive partition functions and complex eigenvalue correlations in matrix models with symplectic symmetry. Nuclear Phys. B **766**, 150–177 (2007)
15. Akemann, G., Burda, Z.: Universal microscopic correlations for products of independent Ginibre matrices. J. Phys. A **45**, 465210 (2012)
16. Akemann, G., Burda, Z., Kieburg, M.: Universal distribution of Lyapunov exponents for products of Ginibre matrices. J. Phys. A **47**, 395202 (2014)
17. Akemann, G., Burda, Z., Kieburg, M., Nagao, T.: Universal microscopic correlation functions for products of truncated unitary matrices. J. Phys. A **47**, 255202 (2013)

© The Editor(s) (if applicable) and The Author(s) 2025
S.-S. Byun and P. J. Forrester, *Progress on the Study of the Ginibre Ensembles*, KIAS
Springer Series in Mathematics 3, https://doi.org/10.1007/978-981-97-5173-0

18. Akemann, G., Byun, S.-S.: The high temperature crossover for general 2D Coulomb gases. J. Stat. Phys. **175**, 1043–1065 (2019)
19. Akemann, G., Byun, S.-S.: The Product of m real $N \times N$ Ginibre matrices: real eigenvalues in the critical regime $m = O(N)$. Constr. Approx. **59**, 31–59 (2024)
20. Akemann, G., Byun, S.-S., Ebke, M.: Universality of the number variance in rotational invariant two-dimensional Coulomb gases. J. Stat. Phys. **190**, 9 (2023)
21. Akemann, G., Byun, S.-S., Kang, N.-G.: A non-Hermitian generalisation of the Marchenko-Pastur distribution: from the circular law to multi-criticality. Ann. Henri Poincaré **22**, 1035–1068 (2021)
22. Akemann, G., Byun, S.-S., Kang, N.-G.: Scaling limits of planar symplectic ensembles. SIGMA Symmetry Integrability Geom. Methods Appl. **22**, 007, 40pp (2021)
23. Akemann, G., Cikovic, M.: Products of random matrices from fixed trace and induced Ginibre ensembles. J. Phys. A **51**, 184002 (2018)
24. Akemann, G., Cikovic, M., Venker, M.: Universality at weak and strong non-Hermiticity beyond the elliptic Ginibre ensemble. Comm. Math. Phys. **362**, 1111–1141 (2018)
25. Akemann, G., Duits, M., Molag, L.D.: The elliptic Ginibre ensemble: a unifying approach to local and global statistics for higher dimensions. J. Math. Phys. **64**, 023503 (2023)
26. Akemann, G., Ebke, M., Parra, I.: Skew-orthogonal polynomials in the complex plane and their Bergman-like kernels. Comm. Math. Phys. **389**, 621–659 (2022)
27. Akemann, G., Förster, Y., Kieburg, M.: Universal eigenvector correlations in quaternionic Ginibre ensembles. J. Phys. A **53**, 145201 (2020)
28. Akemann, G., Götze, F., Neuschel, T.: Characteristic polynomials of products of non-Hermitian Wigner matrices: finite-N results and Lyapunov universality. Electron. Commun. Probab. **26**, 1–13 (2021)
29. Akemann, G., Ipsen, J.R., Strahov, E.: Permanental processes from products of complex and quaternionic induced Ginibre ensembles. Random Matrices Theory Appl. **54**, 1450014 (2014)
30. Akemann, G., Ipsen, J.R.: Recent exact and asymptotic results for products of independent random matrices. Acta Phys. Polon. B **46**, 1747–1784 (2015)
31. Akemann, G., Kanzieper, E.: Integrable structure of Ginibre's ensemble of real random matrices and a Pfaffian integration theorem. J. Stat. Phys. **129**, 1159–1231 (2007)
32. Akemann, G., Kieburg, M., Mielke, A., Prosen, T.: Universal signature from integrability to chaos in dissipative open quantum systems. Phys. Rev. Lett. **123**, 254101 (2019)
33. Akemann, G., Kieburg, M., Phillips, M.J.: Skew-orthogonal Laguerre polynomials for chiral real asymmetric random matrices. J. Phys. A **43**, 375207 (2010)
34. Akemann, G., Nagao, T., Parra, I., Vernizzi, I.: Gegenbauer and other planar orthogonal polynomials on an ellipse in the complex plane. Constr. Approx. **53**, 441–478 (2021)
35. Akemann, G., Phillips, M.J.: The interpolating Airy kernels for the $\beta = 1$ and $\beta = 4$ elliptic Ginibre ensembles. J. Stat. Phys. **155**, 421–465 (2014)
36. Akemann, G., Phillips, M.J., Shifrin, L.: Gap probabilities in non-Hermitian random matrix theory. J. Math. Phys. **50**, 063504 (2009)
37. Akemann, G., Phillips, M.J., Sommers, H.-J.: Characteristic polynomials in real Ginibre ensembles. J. Phys. A **42**, 012001 (2009)
38. Akemann, G., Phillips, M.J., Sommers, H.-J.: The chiral Gaussian two-matrix ensemble of real asymmetric matrices. J. Phys. A **43**, 085211 (2010)
39. Akemann, G., Tribe, R., Tsareas, A., Zaboronski, O.: On the determinantal structure of conditional overlaps for the complex Ginibre ensemble. Random Matrices Theory Appl. **09**, 2050015 (2020)
40. Akemann, G., Vernizzi, G.: Characteristic polynomials of complex random matrix models. Nucl. Phys. B **660**, 532–556 (2003)
41. Alastuey, A., Jancovici, B.: On the two-dimensional one-component Coulomb plasma. J. Physique **42**, 1–12 (1981)
42. Alastuey, A., Jancovici, B.: On potential and field fluctuations in two-dimensional classical charged systems. J. Stat. Phys. **34**, 557–569 (1984)

43. Alishahi, K., Zamani, M.: The spherical ensemble and uniform distribution of points on the sphere. Electron. J. Probab. **20**, 1–27 (2015)
44. Allesina, S., Tang, S.: The stability complexity relationship at age 40: a random matrix perspective. Popul. Ecol. **57**, 63–75 (2015)
45. Allez, R., Touboul, J., Wainrib, G.: Index distribution of the Ginibre ensemble. J. Phys. A **47**, 042001 (2014)
46. Alt, J., Erdős, L., Krüger, T.: Spectral radius of random matrices with independent entries. Probab. Math. Phys. **2**, 221–280 (2021)
47. Ameur, Y.: Near-boundary asymptotics of correlation kernels. J. Geom. Anal. **23**, 73–95 (2013)
48. Ameur, Y.: Repulsion in low temperature β-ensembles. Comm. Math. Phys. **359**, 1079–1089 (2018)
49. Ameur, Y.: A localization theorem for the planar Coulomb gas in an external field. Electron. J. Probab. **26**, 1–21 (2021)
50. Ameur, Y., Byun, S.-S.: Almost-Hermitian random matrices and bandlimited point processes. Anal. Math. Phys. **13**, 52 (2023)
51. Ameur, Y., Charlier, C., Cronvall, J.: The two-dimensional Coulomb gas: fluctuations through a spectral gap. arXiv:2210.13959
52. Ameur, Y., Charlier, C., Cronvall, J.: Random normal matrices: eigenvalue correlations near a hard wall. arXiv:2306.14166
53. Ameur, Y., Charlier, C., Cronvall, J.: Free energy and fluctuations in the random normal matrix model with spectral gaps. arXiv:2312.13904
54. Ameur, Y., Charlier, C., Cronvall, J., Lenells, J.: Exponential moments for disk counting statistics at the hard edge of random normal matrices. Nonlinearity **36**, 1593–1616 (2023)
55. Ameur, Y., Charlier, C., Cronvall, J., Lenells, J.: Disk counting statistics near hard edges of random normal matrices: the multi-component regime. Adv. Math. **441**, 109549 (2024)
56. Ameur, Y., Charlier, C., Moreillon, P.: Eigenvalues of truncated unitary matrices: disk counting statistics. Monatsh Math (2023). https://doi.org/10.1007/s00605-023-01920-4, arXiv:2305.08976
57. Ameur, Y., Cronvall, J.: Szegő type asymptotics for the reproducing kernel in spaces of full-plane weighted polynomials. Comm. Math. Phys. **398**, 1291–1348 (2023)
58. Ameur, Y., Hedenmalm, H., Makarov, N.: Fluctuations of eigenvalues of random normal matrices. Duke Math. J. **159**, 31–81 (2011)
59. Ameur, Y., Hedenmalm, H., Makarov, N.: Random normal matrices and Ward identities. Ann. Probab. **43**, 1157–1201 (2015)
60. Ameur, Y., Kang, N.-G.: On a problem for Ward's equation with a Mittag-Leffler potential. Bull. Sci. Math. **137**, 968–975 (2013)
61. Ameur, Y., Kang, N.-G., Makarov, N.: Rescaling Ward identities in the random normal matrix model. Constr. Approx. **50**, 63–127 (2019)
62. Ameur, Y., Kang, N.-G., Makarov, N., Wennman, A.: Scaling limits of random normal matrix processes at singular boundary points. J. Funct. Anal. **278**, 108340 (2020)
63. Ameur, Y., Kang, N.-G., Seo, S.-M.: On boundary confinements for the Coulomb gas. Anal. Math. Phys. **10**, 68 (2020)
64. Ameur, Y., Kang, N.-G., Seo, S.-M.: The random normal matrix model: insertion of a point charge. Potential Anal. **58**, 331–372 (2023)
65. Ameur, Y., Romero, J.L.: The planar low temperature coulomb gas: separation and equidistribution. Rev. Mat. Iberoam. **39**, 611–648 (2023)
66. Armstrong, S., Serfaty, S.: Local laws and rigidity for Coulomb gases at any temperature. Ann. Probab. **49**, 46–121 (2021)
67. Ashida, Y., Gong, Z., Ueda, M.: Non-Hermitian physics. Adv. Phys. **69**, 249–435 (2020)
68. Bai, Z.D.: Circular law. Ann. Probab. **25**, 494–529 (1997)
69. Baik, J., Bothner, T.: The largest real eigenvalue in the real Ginibre ensemble and its relation to the Zakharov-Shabat system. Ann. Appl. Probab. **30**, 460–501 (2020)

70. Baik, J., Bothner, T.: Edge distribution of thinned real eigenvalues in the real Ginibre ensemble. Ann. Henri Poincaré **23**, 4003–4056 (2022)
71. Balogh, F., Bertola, M., Lee, S.-Y., McLaughlin, K.D.T.-R.: Strong asymptotics of the orthogonal polynomials with respect to a measure supported on the plane. Comm. Pure Appl. Math. **68**, 112–172 (2015)
72. Balogh, F., Grava, T., Merzi, D.: Orthogonal polynomials for a class of measures with discrete rotational symmetries in the complex plane. Constr. Approx. **46**, 109–169 (2017)
73. Balogh, F., Merzi, D.: Equilibrium measures for a class of potentials with discrete rotational symmetries. Constr. Approx. **42**, 399–424 (2015)
74. Bartlett, M.S.: The vector representation of a sample. Math. Proc. Cambridge Philos. Soc. **30**, 327–340 (1934)
75. Bauerschmidt, R., Bourgade, P., Nikula, M., Yau, H.-T.: The two-dimensional Coulomb plasma: quasi-free approximation and central limit theorem. Adv. Theor. Math. Phys. **23**, 841–1002 (2019)
76. Beck, J.: Irregularities of distribution: I. Acta Math. **159**, 1–49 (1987)
77. Beenakker, C.W.J., Edge, J.M., Dahlhaus, J.P., Pikulin, D.I., Mi, S., Wimmer, M.: Wigner-Poisson statistics of topological transitions in a Josephson junction. Phys. Rev. Lett. **111**, 037001 (2013)
78. Belinschi, S., Nowak, M.A., Speicher, R., Tarnowski, W.: Squared eigenvalue condition numbers and eigenvector correlations from the single ring theorem. J. Phys. A **50**, 105204 (2017)
79. Beltrán, C., Hardy, A.: Energy of the Coulomb gas on the sphere at low temperature. Arch. Ration. Mech. Anal. **231**, 2007–2017 (2019)
80. ben Avraham, D.: Complete exact solution of diffusion-limited coalescence $A + A \to A$, Phys. Rev. Lett. **81**, 4756–4759 (1998)
81. Ben Arous, G., Péché, S.: Universality of local eigenvalue statistics for some sample covariance matrices. Comm. Pure Appl. Math. **58**, 1316–1357 (2005)
82. Benaych-Georges, F., Chapon, F.: Random right eigenvalues of Gaussian quaternionic matrices. Random Matrices Theory Appl. **1**, 1150009 (2012)
83. Bender, C.M., Boettcher, S.: Real spectra in non-Hermitian Hamiltonians having P T symmetry. Phys. Rev. Lett. **80**, 5243 (1998)
84. Bender, E.A.: Central and local limit theorems applied to asymptotic enumeration. J. Combin. Theory Ser. A **15**, 91–111 (1973)
85. Bender, M.: Edge scaling limits for a family of non-Hermitian random matrix ensembles. Probab. Theory Related Fields **147**, 241–271 (2010)
86. Berezin, S., Kuijlaars, A.B.J., Parra, I.: Planar orthogonal polynomials as type I multiple orthogonal polynomials. SIGMA Symmetry Integrability Geom. Methods Appl. **19**, 020, 18pp (2023)
87. Berman, R.J.: Bergman kernels for weighted polynomials and weighted equilibrium measures of $\mathbb{C}^n$. Indiana Univ. Math. J. **58**, 1921–1946 (2009)
88. Berman, R.J.: Sharp asymptotics for Toeplitz determinants and convergence towards the Gaussian free field on Riemann surfaces. Int. Math. Res. Not. **2012**, 5031–5062 (2012)
89. Berman, R.J.: Determinantal point processes and Fermions on polarized complex manifolds: bulk universality. In: Algebraic and Analytic Microlocal Analysis, pp. 341–393. Springer (2013)
90. Berman, R.J.: Determinantal point processes and fermions on complex manifolds: large deviations and bosonization. Comm. Math. Phys. **327**, 1–47 (2014)
91. Bernard, D., LeClair, A.: A classification of non-Hermitian random matrices. In: Cappelli, A., Mussardo, G. (eds.) Statistical Field Theories. NATO Science Series II, vol. 73, pp. 207–214. Springer, Berlin (2002)
92. Bertola, M., Elias Rebelo, J.G., Grava, T.: Painlevé IV critical asymptotics for orthogonal polynomials in the complex plane. SIGMA Symmetry Integrability Geom. Methods Appl. **14**, 091, 34pp (2018)
93. Bétermin, L., Sandier, E.: Renormalized energy and symptotic expansion of optimal logarithmic energy on the sphere. Const. Approx. **47**, 39–74 (2018)

94. Bleher, P.M., Di, X.: Correlations between zeros of non-Gaussian random polynomials. Int. Math. Res. Not. **2004**, 2443–2484 (2004)
95. Bleher, P.M., Kuijlaars, A.B.J.: Orthogonal polynomials in the normal matrix model with a cubic potential. Adv. Math. **230**, 1272–1321 (2012)
96. Bleher, P.M., Silva, G.L.F.: The mother body phase transition in the normal matrix model. Mem. Amer. Math. Soc. **265**, no. 1289 (2020)
97. Blum, L., Henderson, D., Lebowitz, J.L., Gruber, Ch., Martin, Ph.A.: A sum rule for an inhomogeneous electrolyte. J. Chem. Phys. **75**, 5974–5975 (1981)
98. Bohigas, O., Giannoni, M.J., Schmit, C.: Characterization of chaotic quantum spectra and universality of level fluctuation laws. Phys. Rev. Lett. **52**, 1–4 (1984)
99. Bojanczyk, A., Golub, G., Van Dooren, P.: The periodic Schur form. Algorithms and applications. In: Proceedings SPIE Conference, pp. 31–42, San Diego (1992)
100. Bordenave, C., Chafaï, D.: Around the circular law. Probab. Surv. **9**, 1–89 (2012)
101. Bornemann, F.: On the numerical evaluation of Fredholm determinants. Math. Comp. **79**, 871–915 (2010)
102. Bornemann, F.: On the numerical evaluation of distributions in random matrix theory: a review. Markov Processes Relat. Fields **16**, 803–866 (2010)
103. Borodin, A.: Determinantal point processes. In: Akemann, G., Baik, J., di Francesco, P. (eds.) The Oxford Handbook of Random Matrix Theory, pp. 231–249. Oxford University Press, Oxford (2011)
104. Borodin, A., Sinclair, C.D.: The Ginibre ensemble of real random matrices and its scaling limit. Comm. Math. Phys. **291**, 177–224 (2009)
105. Bothner, T., Little, A.: Complex elliptic Ginibre ensemble at weak non-Hermiticity: edge spacing distributions. arXiv:2208.04684
106. Bothner, T., Little, A.: Complex elliptic Ginibre ensemble at weak non-Hermiticity: bulk spacing distributions. arXiv:2212.00525
107. Bourgade, P., Dubach, G.: The distribution of overlaps between eigenvectors of Ginibre matrices. Probab. Theory Relat. Fields **177**, 397–464 (2020)
108. Brauchart, J.S., Hardin, D.P., Saff, E.B.: The next-order term for optimal Riesz and logarithmic energy asymptotics on the sphere. Contemp. Math. **578**, 31–61 (2012)
109. Bray, A.J., Majumdar, S.N., Schehr, G.: Persistence and first-passage properties in nonequilibrium systems. Adv. Phys. **62**, 225–361 (2013)
110. de Bruijn, N.G.: On some multiple integrals involving determinants. J. Indian Math. Soc. **19**, 133–151 (1955)
111. Burda, Z., Grela, J., Nowak, M.A., Tarnowski, W., Warchol, P.: Unveiling the significance of eigenvectors in diffusing non-Hermitian matrices by identifying the underlying Burgers dynamics. Nucl. Phys. B **897**, 421–447 (2015)
112. Burda, Z., Nowak, M.A., Swiech, A.: Spectral relations between products and powers of isotropic random matrices. Phys. Rev. E **86**, 061137 (2012)
113. Burda, Z., Spisak, B.J., Vivo, P.: Eigenvector statistics of the product of Ginibre matrices. Phys. Rev. E **95**, 022134 (2017)
114. Byun, S.-S.: Planar equilibrium measure problem in the quadratic fields with a point charge. Comput. Methods Funct. Theory (2023). https://doi.org/10.1007/s40315-023-00494-4, arXiv:2301.00324
115. Byun, S.-S.: Harer-Zagier type recursion formula for the elliptic GinOE. arXiv:2309.11185
116. Byun, S.-S., Charlier, C.: On the characteristic polynomial of the eigenvalue moduli of random normal matrices. arXiv:2205.04298
117. Byun, S.-S., Charlier, C.: On the almost-circular symplectic induced Ginibre ensemble. Stud. Appl. Math. **150**, 184–217 (2023)
118. Byun, S.-S., Ebke, M.: Universal scaling limits of the symplectic elliptic Ginibre ensembles. Random Matrices Theory Appl. **12**, 2250047 (2023)
119. Byun, S.-S., Ebke, M., Seo, S.-M.: Wronskian structures of planar symplectic ensembles. Nonlinearity **36**, 809–844 (2023)

120. Byun, S.-S., Forrester, P.J.: Spherical induced ensembles with symplectic symmetry. SIGMA Symmetry Integrability Geom. Methods Appl. **19**, 033, 28pp (2023)
121. Byun, S.-S., Forrester, P.J.: Progress on the study of the Ginibre ensembles I: GinUE. arXiv:2211.16223
122. Byun, S.-S., Forrester, P.J.: Progress on the study of the Ginibre ensembles II: GinOE and GinSE. arXiv:2301.05022
123. Byun, S.-S., Forrester, P.J.: Spectral moments of the real Ginibre ensemble. Ramanujan J. (to appear). arXiv:2312.08896
124. Byun, S.-S., Kang, N.-G., Lee, J.O., Lee, J.: Real eigenvalues of elliptic random matrices. Int. Math. Res. Not. **2023**, 2243–2280 (2023)
125. Byun, S.-S., Kang, N.-G., Seo, S.-M.: Partition functions of determinantal and Pfaffian Coulomb gases with radially symmetric potentials. Comm. Math. Phys. **401**, 1627–1663 (2023)
126. Byun, S.-S., Molag, L.D., Simm, N.: Large deviations and fluctuations of real eigenvalues of elliptic random matrices. arXiv:2305.02753
127. Byun, S.-S., Noda, K.: Scaling limits of complex and symplectic non-Hermitian Wishart ensembles. arXiv:2402.18257
128. Byun, S.-S., Seo, S.-M.: Random normal matrices in the almost-circular regime. Bernoulli **29**, 1615–1637 (2023)
129. Byun, S.-S., Seo, S.-M., Yang, M.: Free energy expansions of a conditional GinUE and large deviations of the smallest eigenvalue of the LUE. arXiv:2402.18983
130. Byun, S.-S., Yang, M.: Determinantal Coulomb gas ensembles with a class of discrete rotational symmetric potentials. SIAM J. Math. Anal. **55**, 6867–6897 (2023)
131. Caillol, J.M.: Exact results for a two-dimensional one-component plasma on a sphere. J. Phys. Lett. (Paris) **42**, L245–L247 (1981)
132. Caillol, J.M., Levesque, D., Weis, J.J., Hansen, J.P.: A Monte Carlo study of the classical two-dimensional one-component plasma. J. Stat. Phys. **28**, 325–349 (1982)
133. Can, T.: Random Lindblad dynamics. J. Phys. A **52**, 485302 (2019)
134. Can, T., Forrester, P.J., Tellez, G., Wiegmann, P.: Singular behavior at the edge of Laughlin states. Phys. Rev. B **89**, 235137 (2015)
135. Can, T., Forrester, P.J., Tellez, G., Wiegmann, P.: Exact and asymptotic features of the edge density profile for the one component plasma in two dimensions. J. Stat. Phys. **158**, 1147–1180 (2015)
136. Can, T., Laskin, M., Wiegmann, P.: Geometry of quantum Hall states: gravitational anomaly and transport coefficients. Ann. Phys. **362**, 752–794 (2015)
137. Can, V.H., Pham, V.H.: Persistence probability of random Weyl polynomial. J. Stat. Phys. **176**, 262–277 (2019)
138. Cardoso, G., Stéphan, J.-M., Abanov, A.G.: The boundary density profile of a Coulomb droplet. Freezing at the edge. J. Phys. A **54**, 015002 (2021)
139. Carnie, S.L., Chan, D.Y.C.: The Stillinger-Lovett condition for non-uniform electrolytes. Chem. Phys. Lett. **77**, 437–440 (1981)
140. Carnie, S.L.: On sum rules and Stillinger-Lovett conditions for inhomogeneous Coulomb systems. J. Chem. Phys. **78**, 2742–2745 (1983)
141. Chafaï, D.: Aspects of Coulomb gases. arXiv:2108.10653
142. Chafaï, D., García-Zelada, D., Jung, P.: Macroscopic and edge behavior of a planar jellium. J. Math. Phys. **61**, 033304 (2020)
143. Chafaï, D., Gozlan, N., Zitt, P.-A.: First-order global asymptotics for confined particles with singular pair repulsion. Ann. Appl. Probab. **24**, 2371–2413 (2014)
144. Chafaï, D., Hardy, A., Maïda, M.: Concentration for Coulomb gases and Coulomb transport inequalities. J. Funct. Anal. **275**, 1447–1483 (2018)
145. Chafaï, D., Péché, S.: A note on the second order universality at the edge of Coulomb gases on the plane. J. Stat. Phys. **156**, 368–383 (2014)
146. Chalker, J.T., Mehlig, B.: Eigenvector statistics in non-Hermitian random matrix ensembles. Phys. Rev. Lett. **81**, 3367–3370 (1998)

147. Chalker, J.T., Mehlig, B.: Statistical properties of eigenvectors in non-Hermitian Gaussian random matrix ensembles. J. Math. Phys. **41**, 3233–3256 (2000)

148. Charles, L., Estienne, B.: Entanglement entropy and Berezin-Toeplitz operators. Comm. Math. Phys. **376**, 521–554 (2020)

149. Charlier, C.: Large gap asymptotics on annuli in the random normal matrix model. Math. Ann. (2023). https://doi.org/10.1007/s00208-023-02603-z, arXiv:2110.06908

150. Charlier, C.: Asymptotics of determinants with a rotation-invariant weight and discontinuities along circles. Adv. Math. **408**, 108600 (2022)

151. Charlier, C., Lenells, J.: Exponential moments for disk counting statistics of random normal matrices in the critical regime. Nonlinearity **3**, 1593–1616 (2023)

152. Chau, L.-L., Zaboronsky, O.: On the structure of correlation functions in the normal matrix model. Comm. Math. Phys. **196**, 202–247 (1998)

153. Chen, Z., Dongarra, J.J.: Condition numbers of Gaussian random matrices. SIAM J. Matrix Anal. Appl. **27**, 603–620 (2005)

154. Choquard, Ph., Favre, P., Gruber, Ch.: On the equation of state of classical one-component systems with long-range forces. J. Stat. Phys. **23**, 405–442 (1980)

155. Choquard, Ph., Piller, B., Rentsch, R.: On the dielectric susceptibility of classical Coulomb systems II. J. Stat. Phys. **46**, 599–633 (1987)

156. Cipolloni, G.: Fluctuations in the spectrum of non-Hermitian i.i.d. matrices. J. Math. Phys. **63**, 053503 (2022)

157. Cipolloni, G., Erdős, L., Schröder, D.: Optimal lower bound on the least singular value of the shifted Ginibre ensemble. Prob. Math. Phys. **1**, 101–146 (2020)

158. Cipolloni, G., Erdős, L., Schröder, D.: Fluctuation around the circular law for random matrices with real entries. Electron. J. Probab. **26**, 1–61 (2021)

159. Cipolloni, G., Erdős, L., Schröder, D.: Edge universality for non-Hermitian random matrices. Probab. Theory Related Fields **179**, 1–28 (2021)

160. Cipolloni, G., Erdős, L., Schröder, D.: On the condition number of the shifted real Ginibre ensemble. SIAM J. Matrix Anal. Appl. **43**, 1469–1487 (2022)

161. Cipolloni, G., Erdős, L., Schröder, D., Xu, Y.: Directional extremal statistics for Ginibre eigenvalues. J. Math. Phys. **63**, 103303 (2022)

162. Cipolloni, G., Erdős, L., Xu, Y.: Precise asymptotics for the spectral radius of a large random matrix. arXiv:2210.15643

163. Cipolloni, G., Grometto, N.: The dissipative spectral form factor for i.i.d. matrices. J. Stat. Phys. **191**, 21 (2024)

164. Cipolloni, G., Kudler-Flam, J.: Entanglement entropy of non-Hermitian eigenstates and Ginibre ensemble. Phys. Rev. Lett. **130**, 010401 (2023)

165. Cipolloni, G., Kudler-Flam, J.: Non-Hermitian Hamiltonians violate the eigenstate thermalization hypothesis. Phys. Rev. B **109**, L020201 (2024)

166. Cohen-Tannoudji, C., Diu, B., Laloë, F.: Quantum Mechanics. Wiley, Paris (1977)

167. Collins, B.: Product of random projections, Jacobi ensembles and universality problems arising from free probability. Probab. Theory Related Fields **133**, 315–344 (2005)

168. Collins, B., Matsumoto, S., Saad, N.: Integration of invariant matrices and moments of inverses of Ginibre and Wishart matrices. J. Multivariate Anal. **126**, 1–13 (2014)

169. Constantine, A.G.: Some noncentral distribution problems in multivariate analysis. Ann. Math. Statist. **34**, 1270–1285 (1963)

170. Cornish, E.A.: The multivariate t-distribution associated with a set of normal sample deviates. Austr. J. Phys. **7**, 531–542 (1954)

171. Coston, N., O'Rourke, S.: Gaussian fluctuations for linear eigenvalue statistics of products of independent iid random matrices. J. Theoret. Probab. **33**, 1541–1612 (2020)

172. Costin, O., Lebowitz, J.L.: Gaussian fluctuations in random matrices. Phys. Rev. Lett. **75**, 69–72 (1995)

173. Crawford, N., Rosenthal, R.: Eigenvector correlations in the complex Ginibre ensemble. Ann. Appl. Probab. **32**, 2706–2754 (2022)

174. Criado del Rey, G., Kuijlaars, A.: A vector equilibrium problem for symmetrically located point charges on a sphere. Constr. Approx. **55**, 775–827 (2022)
175. Cunden, F.D., Mezzadri, F., Vivo, P.: Large deviations of radial statistics in the two-dimensional one-component plasma. J. Stat. Phys. **164**, 1062 (2016)
176. Dartois, S., Forrester, P.J.: Schwinger-Dyson and loop equations for a product of square Ginibre random matrices. J. Phys. A **53**, 175201 (2020)
177. De Tomasi, G., Khaymovich, I.M.: Non-Hermitian Rosenzweig-Porter random-matrix ensemble: obstruction to the fractal phase. Phys. Rev. B **106**, 094204 (2022)
178. Dean, D.S., Le Doussal, P., Majumdar, S.N., Schehr, G.: Non-interacting fermions at finite temperature in a d -dimensional trap: universal correlations. Phys. Rev. A **94**, 063622 (2016)
179. Dean, D.S., Le Doussal, P., Majumdar, S.N., Schehr, G.: Non-interacting fermions in a trap and random matrix theory. J. Phys. A **52**, 144006 (2019)
180. Deaño, A., Simm, N.: Characteristic polynomials of complex random matrices and Painlevé transcendents. Int. Math. Res. Not. **2022**, 210–264 (2022)
181. Decreusefond, L., Flint, I., Vergne, A.: Efficient simulation of the Ginibre point process. Adv. Appl. Probab. **52**, 1003–1012 (2015)
182. Decreusefond, L., Flint, I., Privault, N., Torrisi, G.L.: Determinantal point processes. In: Peccati, G., Reitzner, M. (eds.), Stochastic Analysis for Poisson Point Processes: Malliavin Calculus, Wiener-Itô Chaos Expansions and Stochastic Geometry, Bocconi & Springer Series, vol. 7, pp. 311–342 (2016)
183. Deift, P.A.: Application of a commutation formula. Duke Math. J. **45**, 267–310 (1978)
184. Deleporte, A., Lambert, G.: Universality for free Fermions and the local Weyl law for semi-classical Schrödinger operators. J. Eur. Math. Soc. (to appear). arXiv:2109.02121
185. del Molino, L.C.G., Pakdaman, K., Touboul, J., Wainrib, G.: The real Ginibre ensemble with $k = O(n)$ real eigenvalues. J. Stat. Phys. **163**, 303–323 (2016)
186. Dembo, A., Poonen, B., Shao, Q.-M., Zeitouni, O.: Random polynomials having few or no real zeros. J. Am. Math. Soc. **15**, 857–892 (2002)
187. Dembo, A., Mukherjee, S.: No zero-crossings for random polynomials and the heat equation. Ann. Probab. **43**, 85–118 (2015)
188. Deng, N., Zhou, W., Haenggi, M.: The Ginibre point process as a model for wireless networks with repulsion. IEEE Trans. Wirel. Commun. **14**, 107–121 (2015)
189. Denisov, S., Laptyeva, T., Tarnowski, W., Chruściński, D., Życzkowski, K.: Universal spectra of random Lindblad operators. Phys. Rev. Lett. **123**, 140403 (2019)
190. Derrida, B., Zeitak, R.: Distribution of domain sizes in the zero temperature Glauber dynamics of the one-dimensional Potts model. Phys. Rev. E **54**, 2513–2525 (1996)
191. Dessertaine, T.: Some mixed-moments of Gaussian elliptic matrices and Ginibre matrices. arXiv:2212.05793
192. Dette, H., Imhof, L.A.: Uniform approximation of eigenvalues in Laguerre and Hermite β -ensembles by roots of orthogonal polynomials. Trans. Am. Math. Soc. **359**, 4999–5018 (2007)
193. Di Francesco, P., Gaudin, M., Itzykson, C., Lesage, F.: Laughlin's wave functions, Coulomb gases and expansions of the discriminant. Int. J. Mod. Phys. A **9**, 4257–4351 (1994)
194. Diaconis, P., Forrester, P.J.: Hurwitz and the origin of random matrix theory in mathematics. Random Matrices Theory Appl. **6**, 1730001 (2017)
195. Dieng, M.: Distribution functions for edge eigenvalues in orthogonal and symplectic ensembles: Painlevé representations. Int. Math. Res. Not. **2005**, 2263–2287 (2005)
196. Ding, X., Jiang, T.: Spectral distributions of adjacency and Laplacian matrices of random graphs. Ann. Appl. Probab. **20**, 2086–2117 (2010)
197. Dornic, I.: Universal Painlevé VI probability distribution in Pfaffian persistence and Gaussian first-passage problems with a sech-kernel. arXiv:1810.06957
198. Dubach, G.: Powers of Ginibre eigenvalues. Electron. J. Probab. **23**, 1–31 (2018)
199. Dubach, G.: Symmetries of the quaternionic Ginibre ensemble. Random Matrices Theory Appl. **10**, 2150013 (2021)

200. Dubach, G.: On eigenvector statistics in the spherical and truncated unitary ensembles. Electron. J. Probab. **26**, 1–29 (2021)
201. Dubach, G., Peled, Y.: On words of non-Hermitian random matrices. Ann. Probab. **49**, 1886–1916 (2021)
202. Dubach, G., Reker, J.: Dynamics of a rank-one multiplicative perturbation of a unitary matrix. arXiv:2212.14638
203. Dunne, G.V.: Hilbert space for charged particles in perpendicular magnetic fields. Ann. Phys. **215**, 233–263 (1992)
204. Dusa, I.G., Wettig, T.: Approximation formula for complex spacing ratios in the Ginibre ensemble. Phys. Rev. E **105**, 044144 (2022)
205. Dyson, F.J.: The three fold way. Algebraic structure of symmetry groups and ensembles in quantum mechanics. J. Math. Phys. **3**, 1199–1215 (1962)
206. Dyson, F.J.: Statistical theory of energy levels of complex systems I. J. Math. Phys. **3**, 140–156 (1962)
207. Dyson, F.J.: Statistical theory of energy levels of complex systems II. J. Math. Phys. **3**, 157–165 (1962)
208. Dyson, F.J.: Correlations between eigenvalues of a random matrix. Comm. Math. Phys. **29**, 235–250 (1970)
209. Ebke, M.: Universal scaling limits of the symplectic elliptic Ginibre ensemble, Ph.D. thesis, Bielefeld University (2021)
210. Edelman, A.: Eigenvalues and condition numbers of random matrices. SIAM J. Matrix Anal. Appl. **9**, 543–560 (1988)
211. Edelman, A.: The probability that a random real Gaussian matrix has k real eigenvalues, related distributions, and the circular law. J. Multivariate Anal. **60**, 203–232 (1997)
212. Edelman, A., Kostlan, E., Shub, M.: How many eigenvalues of a random matrix are real? J. Am. Math. Soc. **7**, 247–267 (1994)
213. Edelman, A., Raj Rao, N.: Random matrix theory. In: Iserles, A. (ed.) Acta Numerica, vol. 14. Cambridge University Press, Cambridge (2005)
214. Efetov, K.B.: Directed quantum chaos. Phys. Rev. Lett. **79**, 491 (1997)
215. van Eijndhoven, S.J.L., Meyers, J.L.H.: New orthogonality relations for the Hermite polynomials and related Hilbert spaces. J. Math. Anal. Appl. **146**, 89–98 (1990)
216. Elbau, P.: Random Normal Matrices and Polynomial Curves, Ph.D. thesis, ETH Zürich (2006). arXiv:0707.0425
217. Elbau, P., Felder, G.: Density of eigenvalues of random normal matrices. Comm. Math. Phys. **259**, 433–450 (2005)
218. Erdős, L.: The matrix Dyson equation and its applications for random matrices. arXiv:1903.10060
219. Erdős, L., Yau, H.-T.: A dynamical approach to random matrix theory. Courant Lecture Notes in Mathematics, vol. 28. American Mathematical Society, Providence (2017)
220. Esaki, S., Katori, M., Yabuoku, S.: Eigenvalues, eigenvector-overlaps, and regularized Fuglede-Kadison determinant of the non-Hermitian matrix-valued Brownian motion. arXiv:2306.00300
221. Feinberg, J., Riser, R.: Pseudo-hermitian random matrix theory: a review. J. Phys. Conf. Ser. **2038**, 012009 (2021)
222. Feinberg, J., Zee, A.: Non-Gaussian non-Hermitian random matrix theory: phase transition and addition formalism. Nucl. Phys. B **501**, 643–669 (1997)
223. Feller, W.: An introduction to probability theory and its applications, 2nd edn. Wiley, New York (1971)
224. Fenzl, M., Lambert, G.: Precise deviations for disk counting statistics of invariant determinantal processes. Int. Math. Res. Not. **2022**, 7420–7494 (2022)
225. Ferrari, P.L., Spohn, H.: A determinantal formula for the GOE Tracy-Widom distribution. J. Phys. A **38**, 557–561 (2005)
226. Ferreira, C., López, J.L.: An asymptotic expansion of the double gamma function. J. Approx. Theory **111**, 298–314 (2001)

227. Fischmann, J.A.: Eigenvalue distributions on a single ring, Ph.D. thesis, Queen Mary University of London (2012)
228. Fischmann, J., Bruzda, W., Khoruzhenko, B.A., Sommers, H.-J., Zyczkowski, K.: Induced Ginibre ensemble of random matrices and quantum operations. J. Phys. A **45**, 075203 (2012)
229. Fischmann, J., Forrester, P.J.: One-component plasma on a spherical annulus and a random matrix ensemble. J. Stat. Mech. Theory Exp. **2011**, 10003 (2011)
230. FitzGerald, W., Tribe, R., Zaboronski, O.: Asymptotic expansions for a class of Fredholm Pfaffians and interacting particle systems. Ann. Probab. **50**, 2409–2474 (2022)
231. FitzGerald, W., Simm, N.: Fluctuations and correlations for products of real asymmetric random matrices. Ann. Inst. Henri Poincaré Probab. Stat. **59**, 2308–2342 (2023)
232. Forrester, P.J.: Some statistical properties of the eigenvalues of complex random matrices. Phys. Lett. A **169**, 21–24 (1992)
233. Forrester, P.J.: Exact results and universal asymptotics in the Laguerre random matrix ensemble. J. Math. Phys. **35**, 2539–2551 (1994)
234. Forrester, P.J.: Exact results for two-dimensional Coulomb systems. Phys. Reports **301**, 235–270 (1998)
235. Forrester, P.J.: Fluctuation formula for complex random matrices. J. Phys. A **32**, 159–163 (1999)
236. Forrester, P.J.: Hard and soft edge spacing distributions for random matrix ensembles with orthogonal and symplectic symmetry. Nonlinearity **19**, 2989–3002 (2006)
237. Forrester, P.J.: Log-Gases and Random Matrices. Princeton University Press, Princeton, NJ (2010)
238. Forrester, P.J.: The limiting Kac random polynomial and truncated random orthogonal polynomials. J. Stat. Mech. Theory Exp. **2010**, P12018 (2010)
239. Forrester, P.J.: Lyapunov exponents of products of complex Gaussian random matrices. J. Stat. Phys. **151**, 796–808 (2013)
240. Forrester, P.J.: Skew orthogonal polynomials for the real and quaternion real Ginibre ensembles and generalizations. J. Phys. A **46**, 245203 (2013)
241. Forrester, P.J.: Probability of all eigenvalues real for products of standard Gaussian matrices. J. Phys. A **47**, 065202 (2014)
242. Forrester, P.J.: Asymptotics of spacing distributions 50 years later. In: Deift, P., Forrester, P. (eds.) Random Matrix Theory, Interacting Particle Systems and Integrable Systems, vol. 65, pp. 199–222. MSRI Publications, Berkeley (2014)
243. Forrester, P.J.: Asymptotics of finite system Lyapunov exponents for some random matrix ensembles. J. Phys. A **48**, 215205 (2015)
244. Forrester, P.J.: Diffusion processes and the asymptotic bulk gap probability for the real Ginibre ensemble. J. Phys. A **48**, 324001 (2015)
245. Forrester, P.J.: Analogies between random matrix ensembles and the one-component plasma in two-dimensions. Nucl. Phys. B **904**, 253–281 (2016)
246. Forrester, P.J.: Meet Andréief, Bordeaux 1886, and Andreev, Kharkov 1882–83. Random Matrices Theory Appl. **8**, 1930001 (2019)
247. Forrester, P.J.: Differential identities for the structure function of some random matrix ensembles. J. Stat. Phys. **183**, 33 (2021)
248. Forrester, P.J.: Circulant L -ensembles in the thermodynamic limit. J. Phys. A **54**, 444003 (2021)
249. Forrester, P.J.: Rank 1 perturbations in random matrix theory — a review of exact results. Random Matrices Theory Appl. 12, no. 4, Paper No. 2330001, 48 pp (2023)
250. Forrester, P.J.: A review of exact results for fluctuation formulas in random matrix theory. Probab. Surv. **20**, 170–225 (2023)
251. Forrester, P.J.: Local central limit theorem for real eigenvalue fluctuations of elliptic GinOE matrices. arXiv:2305.09124
252. Forrester, P.J., Honner, G.: Exact statistical properties of the zeros of complex random polynomials. J. Phys. A **32**, 2961–2981 (1999)

253. Forrester, P.J., Gamburd, A.: Counting formula associated with some random matrix averages. J. Comb. Th. A **113**, 934–951 (2006)
254. Forrester, P.J., Ipsen, J.R.: Real eigenvalue statistics for products of asymmetric real Gaussian matrices. Lin. Algebra Appl. **510**, 259–290 (2016)
255. Forrester, P.J., Ipsen, J.R.: A generalisation of the relation between zeros of the complex Kac polynomial and eigenvalues of truncated unitary matrices. Prob. Theory Related Fields **175**, 833–847 (2019)
256. Forrester, P.J., Ipsen, J.R., Kumar, S.: How many eigenvalues of a product of truncated orthogonal matrices are real? Exp. Math. **29**, 276–290 (2020)
257. Forrester, P.J., Jancovici, B.: Two-dimensional one-component plasma in a quadrupolar field. Int. J. Mod. Phys. A **11**, 941–949 (1996)
258. Forrester, P.J., Jancovici, B., McAnally, D.S.: Analytic properties of the structure function for the one-dimensional one-component log-gas. J. Stat. Phys. **102**, 737–780 (2000)
259. Forrester, P.J., Kieburg, M.: Relating the Bures measure to the Cauchy two-matrix model. Comm. Math. Phys. **342**, 151–187 (2016)
260. Forrester, P.J., Krishnapur, M.: Derivation of an eigenvalue probability density function relating to the Poincaré disk. J. Phys. A **42**, 385204 (2009)
261. Forrester, P.J., Kumar, S.: The probability that all eigenvalues are real for products of truncated real orthogonal random matrices. J. Theoret. Probab. **31**, 2056–2071 (2018)
262. Forrester, P.J., Lebowitz, J.: Local central limit theorem for determinantal point processes. J. Stat. Phys. **157**, 60–69 (2014)
263. Forrester, P.J., Liu, D.-Z.: Singular values for products of complex Ginibre matrices with a source: hard edge limit and phase transition. Comm. Math. Phys. **344**, 333–368 (2016)
264. Forrester, P.J., Mays, A.: A method to calculate correlation functions for $\beta = 1$ random matrices of odd size. J. Stat. Phys. **134**, 443–462 (2009)
265. Forrester, P.J., Mays, A.: Pfaffian point processes for the Gaussian real generalised eigenvalue problem. Prob. Theory and Rel. Fields **154** (2012)
266. Forrester, P.J., Nagao, T.: Eigenvalue statistics of the real Ginibre ensemble. Phys. Rev. Lett. **99**, 050603 (2007)
267. Forrester, P.J., Nagao, T.: Skew orthogonal polynomials and the partly symmetric real Ginibre ensemble. J. Phys. A **41**, 375003 (2008)
268. Forrester, P.J., Rains, E.M.: Matrix averages relating to the Ginibre ensemble. J. Phys. A **42**, 385205 (2009)
269. Forrester, P.J., Zhang, J.: Lyapunov exponents for some isotropic random matrix ensembles. J. Stat. Phys. **180**, 558–575 (2020)
270. Förster, Y.P., Kieburg, M., Kösters, H.: Polynomial ensembles and Pólya frequency functions. J. Theor. Probab. **34**, 1917–1950 (2021)
271. Fyodorov, Y.V.: Spectra of random matrices close to unitary and scattering theory for discrete-time systems. In: Disordered and Complex Systems. AIP Conference Proceedings 553, pp. 191–196. American Institute of Physics, Melville, NY (2001)
272. Fyodorov, Y.V.: On statistics of bi-orthogonal eigenvectors in real and complex Ginibre ensembles: combining partial Schur decomposition with supersymmetry. Comm. Math. Phys. **363**, 579–603 (2018)
273. Fyodorov, Y.V., Khoruzhenko, B.A.: Systematic analytical approach to correlation functions of resonances in quantum chaotic scattering. Phys. Rev. Lett. **83**, 65–68 (1999)
274. Fyodorov, Y.V., Khoruzhenko, B.A.: Nonlinear analogue of the May-Wigner instability transition. Proc. Nat. Acad. Sci. **113**, 6827–6832 (2016)
275. Fyodorov, Y.V., Khoruzhenko, B.A., Sommers, H.-J.: Almost-Hermitian random matrices: crossover from Wigner-Dyson to Ginibre eigenvalue statistics. Phys. Rev. Lett. **79**, 557–560 (1997)
276. Fyodorov, Y.V., Khoruzhenko, B.A., Sommers, H.-J.: Universality in the random matrix spectra in the regime of weak non-Hermiticity. Ann. Inst. H. Poincaré Phys. Théor. **68**, 449–489 (1998)

277. Fyodorov, Y.V., Sommers, H.-J.: Random matrices close to Hermitian or unitary: overview of methods and results. J. Phys. A **36**, 3303–3347 (2003)
278. Fyodorov, Y.V., Tarnowski, W.: Condition numbers for real eigenvalues in the real elliptic Gaussian ensemble. Ann. Henri Poincaré **22**, 309–330 (2021)
279. Fyodorov, Y.V., Titov, M., Sommers, H.-J.: Statistics of S-matrix poles for chaotic systems with broken time reversal invariance: a conjecture. Phys. Rev. E **58**, R1195 (1998)
280. García-García, A.M., Nishigaki, S.M., Verbaarschot, J.J.M.: Critical statistics for non-Hermitian matrices. Phys. Rev. E **66**, 016132 (2002)
281. García-García, A.M., Sá, L., Verbaarschot, J.J.M.: Symmetry classification and universality in non-Hermitian many-body quantum chaos by the Sachdev-Ye-Kitaev model. Phys. Rev. X **12**, 021040 (2022)
282. García-García, A.M., Sá, L., Verbaarschot, J.J.M.: Universality and its limits in non-Hermitian many-body quantum chaos using the Sachdev-Ye-Kitaev model. Phys. Rev. D **107**, 066007 (2023)
283. García-Zelada, D.: Edge fluctuations for random normal matrix ensembles. Random Matrices: Theor. Appl. **11**, 2250040 (2022)
284. Garcia del Molino, L.C., Pakdaman, K., Touboul, J., Wainrib, G.: The real Ginibre ensemble with $k = o(n)$ real eigenvalues. J. Stat. Phys. **162**, 303–323 (2016)
285. Garrod, B., Poplavskyi, M., Tribe, R., Zaboronski, O.: Examples of interacting particle systems on Z as Pfaffian point processes: annihilating and coalescing random walks. Ann. Henri Poincaré **19**, 3635–3662 (2018)
286. Gaudin, M.: Sur la loi limite de l'espacement des valeurs propres d'une matrice aléatoire. Nucl. Phys. **25**, 447–458 (1961)
287. Gebert, M., Poplavskyi, M.: On pure complex spectrum for truncations of random orthogonal matrices and Kac polynomials. arXiv:1905.03154
288. Geman, S.: A limit theorem for the norm of random matrices. Ann. Probab. **8**, 252–261 (1980)
289. Ghosh, S., Lebowitz, J.L.: Fluctuations, large deviations and rigidity in hyperuniform systems: a brief survey. Indian J. Pure Appl. Math. **48**, 609–631 (2017)
290. Ghosh, S., Lebowitz, J.L.: Number rigidity in superhomogeneous random point fields. J. Stat. Phys. **166**, 1016–1027 (2017)
291. Ghosh, S., Miyoshi, N., Shirai, T.: Disordered complex networks: energy optimal lattices and persistent homology. IEEE Trans. Inf. Theory **68**, 5513–5534 (2022)
292. Ghosh, S., Peres, Y.: Rigidity and tolerance in point processes: Gaussian zeros and Ginibre eigenvalues. Duke Math. J. **166**, 1789–1858 (2017)
293. Ginibre, J.: Statistical ensembles of complex, quaternion, and real matrices. J. Math. Phys. **6**, 440 (1965)
294. Girko, V.L.: The circular law. Teor. Veroyatn. Primen. **29**, 669–679 (1984)
295. Girko, V.L.: Elliptic law. Theory Probab. Appl. **30**, 677–690 (1986)
296. Girvin, S.M., MacDonald, A.H., Platzman, P.M.: Magneto-roton theory of collective excitations in the fractional quantum Hall effect. Phys. Rev. B **33**, 2481–2494 (1986)
297. Goel, A., Lopatto, P., Xie, X.: Central limit theorem for the complex eigenvalues of Gaussian random matrices. Electron. Commun. Probab. **29**, 1–13 (2024)
298. Gorin, V., Sun, Y.: Gaussian fluctuations for products of random matrices. Am. J. Math. **144**(2), 287–393 (2022)
299. Gorini, V., Kossakowski, A., Sudarshan, E.C.G.: Completely positive dynamical semigroups of N-level systems. J. Math. Phys. **17**, 821–825 (1976)
300. Götze, F., Jalowy, J.: Rate of convergence to the Circular Law via smoothing inequalities for log-potentials. Random Matrices Theory Appl. **10**, 25 pp (2021)
301. Götze, F., Kösters, H., Tikhomirov, A.: Asymptotic spectra of matrix-valued functions of independent random matrices and free probability. Random Matrices Theory Appl. **4**, 1550005 (2015)
302. Götze, F., Tikhomirov, A.: The circular law for random matrices. Ann. Probab. **38**, 1444–1491 (2010)

303. Grela, J., Warchol, P.: Full Dysonian dynamics of the complex Ginibre ensemble. J. Phys. A **51**, 42 (2018)
304. Grobe, R., Haake, F., Sommers, H.-J.: Quantum distinction of regular and chaotic dissipative motion. Phys. Rev. Lett. **61**, 1899 (1988)
305. Guionnet, A., Krishnapur, M., Zeitouni, O.: The single ring theorem. Ann. Math. **174**, 1189–1217 (2017)
306. Haagerup, U., Larsen, F.: Brown's spectral distribution measure for R-diagonal elements in finite von Neumann algebras. J. Funct. Anal. **176**, 331–367 (2000)
307. Haake, F.: Quantum Signatures of Chaos, 2nd edn. Springer, Berlin (2000)
308. Hahn, N., Kieburg, M., Gat, O., Guhr, T.: Winding number statistics for chiral random matrices: averaging ratios of determinants with parametric dependence. J. Math. Phys. **64**, 021901 (2023)
309. Haimi, A., Hedenmalm, H.: The polyanalytic Ginibre ensembles. J. Stat. Phys. **153**, 10–47 (2013)
310. Halasz, M.A., Osborn, J.C., Verbaarschot, J.J.M.: Random matrix triality at nonzero chemical potential. Phys. Rev. D **56**, 7059–7062 (1997)
311. Haldane, F.D.M.: Fractional quantization of the Hall effect: a hierarchy of incompressible quantum fluid states. Phys. Rev. Lett. **55**, 2095–2098 (1983)
312. Halmagyi, H., Lai, S.: Mixed moments for the product of Ginibre matrices. arXiv:2007.10181
313. Hamazaki, R., Kawabata, K., Kura, N., Ueda, M.: Universality classes of non-Hermitian random matrices. Phys. Rev. Res. **2**, 023286 (2020)
314. Hameed, S., Jain, K., Lakshminarayan, A.: Real eigenvalues of non-Gaussian random matrices and their products. J. Phys. A **48**, 385204 (2015)
315. Hammersley, J.M.: The zeros of random polynomials. In: Neyman, J. (ed.) Proceedings of the Third Berkeley Symposium on Probability and Statistics, vol. 2, pp. 89–111. University of California Press, Berekeley, CA (1956)
316. Hannay, J.H.: Chaotic analytic zero points: exact statistics for those of a random spin state. J. Phys. A **29**, L101–L105 (1996)
317. Harper, L.H.: Stirling behaviour is asymptotically normal. Ann. Math. Statist. **38**, 410–414 (1967)
318. Hastings, M.B.: Fermionic mapping for eigenvalue correlation functions of weakly non-Hermitian symplectic ensemble. Nuclear Phys. B **572**, 535–546 (2000)
319. Hastings, M.B.: Eigenvalue distribution in the self-dual non-Hermitian ensemble. J. Stat. Phys. **103**, 903–913 (2001)
320. Hatano, N., Nelson, D.R.: Localization transitions in Non-Hermitian quantum mechanics. Phys. Rev. Lett. **77**, 570–573 (1996)
321. Hedenmalm, H.: Soft Riemann-Hilbert problems and planar orthogonal polynomials. Comm. Pure Appl. Math. (to appear). arXiv:2108.05270
322. Hedenmalm, H., Makarov, N.: Coulomb gas ensembles and Laplacian growth. Proc. London Math Soc. **106**, 859–907 (2013)
323. Hedenmalm, H., Wennman, A.: Riemann-Hilbert hierarchies for hard edge planar orthogonal polynomials. Am. J. Math. (to appear). arXiv:2008.02682
324. Hedenmalm, H., Wennman, A.: Planar orthogonal polynomials and boundary universality in the random normal matrix model. Acta Math. **227**, 309–406 (2021)
325. Ho, T.-L., Ciobanu, C.: Rapidly rotating Fermi gases. Phys. Rev. Lett. **85**, 4648 (2000)
326. Hough, J.B., Krishnapur, M., Peres, Y., Virág, B.: Zeros of Gaussian Analytic Functions and Determinantal Point Processes. American Mathematical Society, Providence, RI (2009)
327. Hsu, P.L.: On the distribution of the roots of certain determinantal equations. Ann. Eugen. **9**, 250–258 (1939)
328. Huang, Y., Harrow, A.W.: Improved concentration of Laguerre and Jacobi ensembles. SIAM J. Math. Anal. **56**, 554–567 (2024)
329. Ipsen, J.R.: Products of independent quaternion Ginibre matrices and their correlation functions. J. Phys. A **46**, 265201 (2013)

330. Ipsen, J.R.: Products of independent Gaussian random matrices, Ph.D. thesis, Bielefeld University (2015)
331. Ipsen, J.R.: Lyapunov exponents for products of rectangular real, complex and quaternionic Ginibre matrices. J. Phys. A **48**, 155204 (2015)
332. Ipsen, J.R., Forrester, P.J.: Kac-Rice fixed point analysis for single-and multi-layered complex systems. J. Phys. A **51**, 474003 (2018)
333. Ipsen, J.R., Kieburg, M.: Weak commutation relations and eigenvalue statistics for products of rectangular random matrices. Phys. Rev. E **89**, 032106 (2014)
334. Jalowy, J.: Rate of convergence for non-Hermitian random matrices and their products, Ph.D. thesis, Bielefeld University (2020)
335. Jalowy, J.: Rate of convergence for products of independent non-Hermitian random matrices. Electron. J. Probab. **26**, 24 pp (2021)
336. Jalowy, J.: The Wasserstein distance to the circular law. Ann. Inst. H. Poincaré Probab. Stat. **59**, 2285–2307 (2023)
337. Jancovici, B.: Exact results for the two-dimensional one-component plasma. Phys. Rev. Lett. **46**, 386–388 (1981)
338. Jancovici, B.: Classical Coulomb systems near a plane wall. I. J. Stat. Phys. **28**, 43–65 (1982)
339. Jancovici, B.: Classical Coulomb systems near a plane wall. II. J. Stat. Phys. **29**, 263–280 (1982)
340. Jancovici, B.: Charge correlations and sum rules in Coulomb systems I. In: Rogers, F.J., Dewitt, H.E. (ed.) Strongly Coupled Plasma Physics, pp. 349–356. Plenum Publishing Corporation (1987)
341. Jancovici, B.: Classical Coulomb systems: screening and correlations revisited. J. Stat. Phys. **80**, 445–459 (1995)
342. Jancovici, B., Lebowitz, J.L., Manificat, G.: Large charge fluctuations in classical Coulomb systems. J. Stat. Phys. **72**, 773–787 (1993)
343. Jancovici, B., Lebowitz, J.L., Martin, Ph.A.: Time-dependent correlations in an inhomogeneous one-component plasma. J. Stat. Phys. **41**, 941–974 (1985)
344. Jancovici, B., Manificat, G., Pisani, C.: Coulomb systems seen as critical systems: finite-size effects in two dimensions. J. Stat. Phys. **76**, 307–330 (1994)
345. Jancovici, B., Samaj, L.: Charge correlations in a Coulomb system along a plane wall: a relation between asymptotic behavior and dipole moment. J. Stat. Phys. **105**, 193–209 (2001)
346. Jancovici, B., Šamaj, L.: Coulomb systems with ideal dielectric boundaries: free fermion point and universality. J. Stat. Phys. **104**, 753–775 (2001)
347. Jimbo, M., Miwa, T., Môri, Y., Sato, M.: Density matrix of an impenetrable Bose gas and the fifth Painlevé transcendent. Phys. D **1**, 80–158 (1980)
348. Johansson, K.: On fluctuations of eigenvalues of random Hermitian matrices. Duke Math. J. **91**, 151–204 (1998)
349. Kac, M.: On the average number of real roots of a random algebraic equation. Bull. Am. Math. Soc. **49**, 314–320 (1943)
350. Kalinay, P., Markoš, P., Šamaj, L., Travěnec, I.: The sixth-moment sum rule for the pair correlations of the two-dimensional one-component plasma: exact result. J. Stat. Phys. **98**, 639–666 (2000)
351. Kanazawa, T., Kieburg, M.: GUE-chGUE transition preserving chirality at finite matrix size. J. Phys. A **51**, 345202 (2018)
352. Kang, N.-G., Makarov, N.: Gaussian free field and conformal field theory. Astérisque **353**, viii+136 (2013)
353. Kanzieper, E.: Eigenvalue correlations in non-Hermitean symplectic random matrices. J. Phys. A **35**, 6631–6644 (2002)
354. Kanzieper, E.: Exact replica treatment of non-Hermitean complex random matrices. In: Kovras, O. (ed.) Frontiers in Field Theory, pp. 23–51. Nova Science Publishers, New York (2005)
355. Kanzieper, E., Poplavskyi, M., Timm, C., Tribe, R., Zaboronski, O.: What is the probability that a large random matrix has no real eigenvalues? Ann. Appl. Probab. **26**, 2733–2753 (2016)

356. Kargin, V.: On the largest Lyapunov exponent for products of Gaussian matrices. J. Stat. Phys. **157**, 70–83 (2014)
357. Katori, M.: Two-dimensional elliptic determinantal point processes and related systems. Comm. Math. Phys. **371**, 1283–1321 (2019)
358. Katori, M., Shirai, T.: Partial isometries, duality, and determinantal point processes. Random Matrices Theory Appl. **11**, 2250025 (2022)
359. Kawabata, K., Shiozaki, K., Ueda, M., Sato, M.: Symmetry and topology in non-Hermitian physics. Phys. Rev. X **9**, 041015 (2019)
360. Kawabata, K., Xiao, Z., Ohtsuki, T., Shindou, R.: Singular-value statistics of non-Hermitian random matrices and open quantum systems. PRX Quantum **4**, 040312 (2023)
361. Khoruzhenko, B.A., Lysychkin, S.: Truncations of random symplectic unitary matrices. arXiv:2111.02381
362. Khoruzhenko, B.A., Sommers, H.-J.: Non-Hermitian ensembles. In: Akemann, G., Baik, J., Di Francesco, P. (eds.) Chapter 18 The Oxford Handbook of Random Matrix Theory. Oxford University Press (2011)
363. Khoruzhenko, B.A., Sommers, H.-J., Zyczkowski, K.: Truncations of random orthogonal matrices. Phys. Rev. E **82**, 040106 (2010)
364. Kieburg, M., Kösters, H.: Exact relation between singular value and eigenvalue statistics. Random Matrices Theory Appl. **5**, 1650015 (2016)
365. Kieburg, M., Kösters, H.: Products of random matrices from polynomial ensembles. Ann. Inst. H. Poincaré Probab. Stat. **55**, 98–126 (2019)
366. Kieburg, M., Kuijlaars, A.B.J., Stivigny, D.: Singular value statistics of matrix products with truncated unitary matrices. Int. Math. Res. Not. **2016**, 3392–3424 (2016)
367. Kiessling, M.K.-H., Spohn, H.: A note on the eigenvalue density of random matrices. Comm. Math. Phys. **199**, 683–695 (1999)
368. Killip, R., Kozhan, R.: Matrix models and eigenvalue statistics for truncations of classical ensembles of random unitary matrices. Comm. Math. Phys. **349**, 991–1027 (2017)
369. Kolesnikov, A.V., Efetov, K.B.: Distribution of complex eigenvalues for symplectic ensembles of non-Hermitian matrices. Waves Random. Media **9**, 71–82 (1999)
370. Kopel, P.: Linear statistics of non-Hermitian matrices matching the real or complex Ginibre ensemble to four moments. arXiv:1510.02987
371. Kopel, P., O'Rourke, S., Vu, V.: Random matrix products: universality and least singular values. Ann. Probab. **48**, 1372–1410 (2020)
372. Kostlan, E.: On the spectra of Gaussian matrices. Linear Algebra Appl. **162**, 385–388 (1992)
373. Krishnapur, M.: From random matrices to random analytic functions. Ann. Prob. **37**, 314–346 (2009)
374. Kristjansen, C., Plefka, J., Semenoff, G.W., Staudacher, M.: A new double-scaling limit of $\mathcal{N} = 4$ super Yang-Mills theory and PP-wave strings. Nucl. Phys. B **643**, 3–30 (2002)
375. Kuijlaars, A.B.J.: Universality. In: Akemann, G., Baik, J., Di Francesco, P. (eds.) Chapter 6 in The Oxford Handbook of Random Matrix Theory. Oxford University Press (2011)
376. Kuijlaars, A.B.J., López-García, A.: The normal matrix model with a monomial potential, a vector equilibrium problem, and multiple orthogonal polynomials on a star. Nonlinearity **28**, 347–406 (2015)
377. Kuijlaars, A.B.J., Martinez-Finkelshtein, A., Wielonsky, F.: Non-intersecting squared Bessel paths and multiple orthogonal polynomials for modified Bessel weights. Comm. Math. Phys. **286**, 217–275 (2009)
378. Kuijlaars, A.B.J., Stivigny, D.: Singular values of products of random matrices and polynomial ensembles. Random Matrices Theory Appl. **3**, 1450011 (2014)
379. Kuijlaars, A.B.J., Tovbis, A.: The supercritical regime in the normal matrix model with cubic potential. Adv. Math. **283**, 530–587 (2015)
380. Kulkarni, M., Majumdar, S.N., Schehr, G.: Multilayered density profile for noninteracting fermions in a rotating two-dimensional trap. Phys. Rev. A **103**, 033321 (2021)
381. Kulkarni, M., Le Doussal, P., Majumdar, S.N., Schehr, G.: Density profile of noninteracting fermions in a rotating 2d trap at finite temperature. Phys. Rev. A **107**, 023302 (2023)

382. Lacroix-A-Chez-Toine, B., Grabsch, A., Majumdar, S.N., Schehr, G.: Extremes of 2d Coulomb gas: universal intermediate deviation regime. J. Stat. Mech. Theory Exp. **2018**, 013203 (2018)

383. Lacroix-A-Chez-Toine, B., Garzón, J.A.M., Calva, C.S.H., Castillo, I.P., Kundu, A., Majumdar, S.N., Schehr, G.: Intermediate deviation regime for the full eigenvalue statistics in the complex Ginibre ensemble. Phys. Rev. E **100**, 012137 (2019)

384. Lacroix-A-Chez-Toine, B., Majumdar, S.N., Schehr, G.: Rotating trapped fermions in two dimensions and the complex Ginibre ensemble: exact results for the entanglement entropy and number variance. Phys. Rev. A **99**, 021602(R) (2019)

385. Lakshminarayan, A.: On the number of real eigenvalues of products of random matrices and an application to quantum entanglement. J. Phys. A **46**, 152003 (2013)

386. Lambert, G.: Incomplete determinantal processes: from random matrix to Poisson statistics. J. Stat. Phys. **176**, 1343–1374 (2019)

387. Lambert, G.: Maximum of the characteristic polynomial of the Ginibre ensemble. Comm. Math. Phys. **378**, 943–985 (2020)

388. Lambert, G.: Poisson statistics for Gibbs measures at high temperature. Ann. Inst. H. Poincaré Probab. Statist. **57**, 326–350 (2021)

389. Laughlin, R.B.: Anomalous quantum Hall effect: an incompressible quantum fluid with fractionally charged excitations. Phys. Rev. Lett. **50**, 3383 (1983)

390. Le Caër, G.: Do Swedish pines diagonalise complex random matrices? Internal Report, LSG2M (Nancy, 1990), unpublished

391. Le Caër, G., Delannay, R.: The administrative divisions of mainland France as 2d random cellular structures. J. Phys. I (France) **3**, 1777–1800 (1993)

392. Leblé, T.: The two-dimensional one-component plasma is hyperuniform. arXiv:2104.05109

393. Leblé, T., Serfaty, S.: Large deviation principle for empirical fields of log and Riesz gases. Invent. Math. **210**, 645–757 (2017)

394. Leblé, T., Serfaty, S.: Fluctuations of two dimensional Coulomb gases. Geom. Funct. Anal. **28**, 443–508 (2018)

395. Lebowitz, J.L.: Charge fluctuations in Coulomb systems. Phys. Rev. A **27**, 1491–1494 (1983)

396. Lee, S.-Y., Makarov, N.G.: Topology of quadrature domains. J. Am. Math. Soc. **29**, 333–369 (2016)

397. Lee, S.-Y., Riser, R.: Fine asymptotic behavior for eigenvalues of random normal matrices: ellipse case. J. Math. Phys. **57**, 023302 (2016)

398. Lee, S.-Y., Yang, M.: Discontinuity in the asymptotic behavior of planar orthogonal polynomials under a perturbation of the Gaussian weight. Comm. Math. Phys. **355**, 303–338 (2017)

399. Lee, S.-Y., Yang, M.: Planar orthogonal polynomials as Type II multiple orthogonal polynomials. J. Phys. A **52**, 275202 (2019)

400. Lee, S.-Y., Yang, M.: Strong asymptotics of planar orthogonal polynomials: Gaussian weight perturbed by finite number of point charges. Comm. Pure Appl. Math. **76**, 2888–2956 (2023)

401. Lehmann, N., Sommers, H.-J.: Eigenvalue statistics of random real matrices. Phys. Rev. Lett. **67**, 941 (1991)

402. Lenard, A.: Correlation functions and the uniqueness of the state in classical statistical mechanics. Comm. Math. Phys. **30**, 35–44 (1973)

403. Lewin, M.: Coulomb and Riesz gases: the known and the unknown. J. Math. Phys. **63**, 061101 (2022)

404. Li, J., Prosen, T., Chan, A.: Spectral statistics of non-Hermitian matrices and dissipative quantum chaos. Phys. Rev. Lett. **127**, 170602 (2021)

405. Lieb, E.H., Narnhofer, H.: The thermodynamic limit for jellium. J. Stat. Phys. **12**, 291–310 (1975)

406. Lindblad, G.: On the generators of quantum dynamical semigroups. Comm. Math. Phys. **48**, 119–130 (1976)

407. Little, A., Mezzadri, F., Simm, N.: On the number of real eigenvalues of a product of truncated orthogonal random matrices. Electron. J. Probab. **27**, 1–32 (2021)

408. Litvak, A., Collins, B., Gawron, P., Zyczkowski, K.: Numerical range of random matrices. J. Math. Anal. Appl. **418**, 516–533 (2014)
409. Liu, D.-Z., Wang, Y.: Universality for products of random matrices I: Ginibre and truncated unitary cases. Int. Math. Res. Not. 3473–3524 (2016)
410. Liu, D.-Z., Wang, Y.: Phase transitions for infinite products of large non-Hermitian random matrices. arXiv:1912.11910
411. Liu, D.-Z., Wang, D., Wang, Y.: Lyapunov exponent, universality and phase transition for products of random matrices. Comm. Math. Phys. **399**, 1811–1855 (2023)
412. Liu, D.-Z., Zhang, L.: Phase transition of eigenvalues in deformed Ginibre ensembles. arXiv:2204.13171
413. Luh, K., O'Rourke, S.: Eigenvector delocalization for non-Hermitian random matrices and applications. Random Struct. Algorithms **57**, 169–210 (2020)
414. Luke, Y.L.: The special functions and their approximations, vol. I. Academic Press, New York-London (1969)
415. Lust, K.: Improved numerical Floquet multipliers. Internat. J. Bifur. Chaos Appl. Sci. Engrg. **11**, 2389–2410 (2001)
416. Lysychkin, S.: Complex eigenvalues of high dimensional quaternion random matrices, Ph.D. thesis, Queen Mary University of London (2021)
417. Lytova, A., Tikhomirov, K.: On delocalization of eigenvectors of random non-Hermitian matrices. Probab. Theory Related Fields **177**, 465–524 (2020)
418. Magnea, U.: Random matrices beyond the Cartan classification. J. Phys. A **71**, 045203 (2008)
419. Mahoux, G., Mehta, M.L.: A method of integration over matrix variables IV. J. Phys. I (France) **1**, 1093–1108 (1991)
420. Majumdar, S.N., Schehr, G.: Top eigenvalue of a random matrix: large deviations and third order phase transition. J. Stat. Mech. Theory Exp. **2014**, 01012 (2014)
421. Markum, H., Pullirsch, R., Wettig, T.: Non-Hermitian random matrix theory and lattice QCD with chemical potential. Phys. Rev. Lett. **83**, 484 (1999)
422. Martin, Ph.A.: Sum rules in charged fluids. Rev. Mod. Phys. **60**, 1075–1127 (1988)
423. Martin, Ph.A., Gruber, Ch.: A new proof of the Stillinger-Lovett complete shielding condition. J. Stat. Phys. **31**, 691–710 (1983)
424. Martin, Ph., Yalcin, T.: The charge fluctuations in classical Coulomb systems. J. Stat. Phys. **22**, 435–463 (1980)
425. Martínez-Finkelshtein, A., Silva, G.L.F.: Critical measures for vector energy: asymptotics of non-diagonal multiple orthogonal polynomials for a cubic weight. Adv. Math. **349**, 246–315 (2019)
426. Matsui, T., Katori, M., Shirai, T.: Local number variances and hyperuniformity of the Heisenberg family of determinantal point processes. J. Phys. A **54**, 165201 (2021)
427. Masser, T.O., ben Avraham, D.: Correlation functions for diffusion-limited annihilation, $A + A \rightarrow 0$. Phys. Rev. E **64**, 062101 (2001)
428. Mathai, A.M.: Random p-content of a p-parallelotope in Euclidean n-space. Adv. Appl. Probab. **31**, 343–354 (1999)
429. Matsumoto, S.: General moments of the inverse real Wishart distribution and orthogonal Weingarten functions. J. Theor. Probab. **25**, 798–822 (2012)
430. Matsumoto, S., Shirai, T.: Correlation functions for zeros of a Gaussian power series and Pfaffians. Electron. J. Probab. **18**, 1–18 (2013)
431. May, R.M.: Will a large complex system be stable? Nature **238**, 413–414 (1972)
432. Mays, A.: A geometrical triumvirate of real random matrices, Ph.D. thesis, University of Melbourne (2012). arXiv:1202.1218
433. Mays, A.: A real quaternion spherical ensemble of random matrices. J. Stat. Phys. **153**, 48–69 (2013)
434. Mays, A., Ponsaing, A.: An induced real quaternion spherical ensemble of random matrices. Random Matrices Theory Appl. **6**, 1750001 (2017)
435. Mehta, M.L.: Random Matrices and the Statistical Theory of Energy Levels. Academic Press, New York (1967)

436. Mehta, M.L.: Random Matrices, 2nd edn. Academic Press, New York (1991)
437. Mehta, M.L., Srivastava, P.K.: Correlation functions for eigenvalues of real quaternian matrices. J. Math. Phys. **7**, 341–344 (1966)
438. Mezzadri, F., Taylor, H.: A matrix model of a non-Hermitian β-ensemble. arXiv:2305.13184
439. Miles, R.E.: Isotropic random simplifies. Adv. Appl. Probab. **3**, 353–382 (1971)
440. Miyoshi, N., Shirai, T.: A cellular network model with Ginibre configured base stations. Adv. Appl. Probab. **46**, 832–845 (2014)
441. Molag, L.D.: Edge behavior of higher complex-dimensional determinantal point processes. Ann. Henri Poincaré **24**, 4405–4437 (2023)
442. Muirhead, R.J.: Aspects of Multivariate Statistical Theory. Wiley, New York (1982)
443. Nagao, T., Akemann, G., Kieburg, M., Parra, I.: Families of two-dimensional Coulomb gases on an ellipse: correlation functions and universality. J. Phys. A **53**, 075201 (2020)
444. Nemes, G., Olde Daalhuis, A.B.: Asymptotic expansions for the incomplete gamma function in the transition regions. Math. Comp. **88**, 1805–1827 (2019)
445. Newman, C.M.: The distribution of Lyapunov exponents: exact results for random matrices. Comm. Math. Phys. **103**, 121–126 (1986)
446. Nguyen, H.H., O'Rourke, S.: The elliptic law. Int. Math. Res. Not. **2015**, 7620–7689 (2015)
447. Nicolaescu, L.I.: Counting zeros of random functions, online resource (2014)
448. Niculescu, C.P.: A new look at Newton's inequalities. J. Inequal. Pure Appl. Math. **1**, Article 17, 14pp (2000)
449. Nishigaki, S.M., Kamenev, A.: Replica treatment of non-Hermitian disordered Hamiltonians. J. Phys. A **35**, 4571–4590 (2002)
450. NIST Digital Library of Mathematical Functions (https://dlmf.nist.gov/)
451. Nowak, M.A., Tarnowski, W.: Probing non-orthogonality of eigenvectors in non-Hermitian matrix models: diagrammatic approach. J. High Energy Phys. **2018**, 152 (2018)
452. Oblak, B., Lapierre, B., Moosavi, P., Stéphan, J.M., Estienne, B.: Anisotropic quantum Hall droplets. Phys. Rev. X **14**, 011030 (2024)
453. O'Rourke, S., Renfrew, D.: Central limit theorem for linear eigenvalue statistics of elliptic random matrices. J. Theoret. Probab. **29**, 1121–1191 (2016)
454. O'Rourke, S., Renfrew, D., Sohnikov, A., Vu, V.: Products of independent elliptic random matrices. J. Stat. Phys. **160**, 89–119 (2015)
455. O'Rourke, S., Williams, N.: Partial linear eigenvalue statistics for non-Hermitian random matrices. Theory Probab. Appl. **67**, 613–632 (2023)
456. Osborn, J.C.: Universal results from an alternative random matrix model for QCD with a baryon chemical potential. Phys. Rev. Lett. **93**, 222001 (2004)
457. Ouellette, D.V.: Schur complements and statistics. Lin. Algebra Appl. **36**, 187–295 (1981)
458. Outhwaite, C.W.: Comment on the second moment condition of Stillinger and Lovett. Chem. Phys. Lett. **24**, 73–74 (1974)
459. Pastur, L., Shcherbina, M.: Eigenvalue Distribution of Large Random Matrices. American Mathematical Society, Providence, RI (2011)
460. Peron, T., de Resende, B.M.F., Rodrigues, F.A., Costa, L.d.F., Méndez-Bermúdez, J.A.: Spacing ratio characterization of the spectra of directed random networks. Phys. Rev. E **102**, 062305 (2020)
461. Poplavskyi, M., Tribe, R., Zaboronski, O.: On the distribution of the largest real eigenvalue for the real Ginibre ensemble. Ann. Appl. Probab. **27**, 1395–1413 (2017)
462. Poplavskyi, M., Schehr, G.: Exact persistence exponent for the 2d-diffusion equation and related Kac polynomials. Phys. Rev. Lett. **121**, 150601 (2018)
463. Porter, C.E.: Statistical Theories of Spectra: Fluctuations. Academic Press, New York (1965)
464. Potters, M., Bouchaud, J.-P.: A First Course in Random Matrix Theory. Cambridge University Press (2020)
465. Prosen, T.: Exact statistics of complex zeros for Gaussian random polynomials with real coefficients. J. Phys. A **29**, 4417–4423 (1996)
466. Rains, E.M.: Correlations for symmetrized increasing subsequences. math.CO/0006097 (2000)

467. Reddy, T.R.: Probability that product of real random matrices have all eigenvalues real tend to 1. Statist. Probab. Lett. **124**, 30–32 (2017)
468. Reddy, N.K.: Equality of Lyapunov and stability exponents for products of isotropic random matrices. Int. Math. Res. Not. **2019**, 606–624 (2019)
469. Rider, B.: A limit theorem at the edge of a non-Hermitian random matrix ensemble. J. Phys. A **36**, 3401–3410 (2003)
470. Rider, B., Sinclair, C.D.: Extremal laws for the real Ginibre ensemble. Ann. Appl. Probab. **24**, 1621–1651 (2014)
471. Rider, B., Virág, B.: Complex determinantal processes and H1 noise. Electron. J. Probab. **12**, 1238–1257 (2007)
472. Rider, B., Virág, B.: The noise in the circular law and the Gaussian free field. Int. Math. Res. Not. **2007**, rnm006, 33 pp (2007)
473. von Rosen, D.: Moments for the inverted Wishart distribution. Scand J. Statist. **15**, 97–109 (1988)
474. Rota Nodari, S., Serfaty, S.: Renormalized energy equidistribution and local charge balance in 2D Coulomb systems. Int. Math. Res. Not. **2015**, 3035–3093 (2015)
475. Rouault, A.: Asymptotic behavior of random determinants in the Laguerre, Gram and Jacobi ensembles. ALEA **3**, 181–230 (2007)
476. Rudelson, M., Vershynin, R.: No-gaps delocalization for general random matrices. Geom. Funct. Anal. **26**, 1716–1776 (2016)
477. Sá, L., Ribeiro, P., Prosen, T.: Complex spacing ratios: a signature of dissipative quantum chaos. Phys. Rev. X **10**, 021019 (2020)
478. Sá, L., Ribeiro, P., Prosen, T.: Spectral and steady-state properties of random Liouvillians. J. Phys. A **53**, 305303 (2020)
479. Saff, E.B., Totik, V.: Logarithmic Potentials with External Fields. Springer, Berlin (1997)
480. Salazar, R., Téllez, G.: Exact energy computation of the one component plasma on a sphere for even values of the coupling parameter. J. Stat. Phys. **164**, 969–999 (2016)
481. Samaj, L.: Is the two-dimensional one-component plasma exactly solvable? J. Stat. Phys. **117**, 131–158 (2004)
482. Samaj, L.: Short-Distance symmetry of pair correlations in two-dimensional jellium. J. Stat. Phys. **178**, 247–264 (2019)
483. Samaj, L., Percus, J.K.: A functional relation among the pair correlations of the two-dimensional one-component plasma. J. Stat. Phys. **80**, 495–512 (1995)
484. Sandier, E., Serfaty, S.: From the Ginzburg-Landau model to vortex lattice problems. Comm. Math. Phys. **313**, 635–743 (2012)
485. Sandier, E., Serfaty, S.: 2D Coulomb gases and the renormalized energy. Ann. Probab. **43**, 2026–2083 (2015)
486. Sari, R.R., Merlini, D.: On the ν-dimensional one-component classical plasma: the thermodynamic limit problem revisited. J. Stat. Phys. **14**, 91–100 (1976)
487. Schehr, G., Majumdar, S.N.: Statistics of the number of zero crossings: from random polynomials to the diffusion equation. Phys. Rev. Lett. **99**, 060603 (2007)
488. Schehr, G., Majumdar, S.N.: Real roots of random polynomials and zero crossing properties of diffusion equation. J. Stat. Phys. **132**, 235–273 (2008)
489. Seo, S.-M.: Edge scaling limit of the spectral radius for random normal matrix ensembles at hard edge. J. Stat. Phys. **181**, 1473–1489 (2020)
490. Seo, S.-M.: Edge behavior of two-dimensional Coulomb gases near a hard wall. Ann. Henri Poincaré **23**, 2247–2275 (2022)
491. Serfaty, S.: Microscopic description of Log and Coulomb gases. In: Random Matrices, IAS/Park City Mathematics Series, vol. 26, pp. 341–387. American Mathematical Society, Providence, RI (2019)
492. Serfaty, S.: Gaussian fluctuations and free energy expansion for Coulomb gases at any temperature. Ann. Inst. H. Poincaré Probab. Statist. **59**, 1074–1142 (2023)
493. Shail, R.: Some logarithmic lattice sums. J. Phys. A **28**, 6999–7009 (1995)

494. Shakirov, Sh.: Exact solution for mean energy of 2d Dyson gas at $\beta = 1$. Phys. Lett. A **375**, 984–989 (2011)
495. Shi, D., Jiang, Y.: Smallest gaps between eigenvalues of random matrices with complex Ginibre, Wishart and universal unitary ensembles. arXiv:1207.4240
496. Shirai, T.: Large deviations for the Fermion point process associated with the exponential kernel. J. Stat. Phys. **123**, 615–629 (2006)
497. Shirai, T.: Ginibre-type point processes and their asymptotic behavior. J. Math. Soc. Japan **67**, 763–787 (2015)
498. Shirai, T., Takahashi, Y.: Random point fields associated with certain Fredholm determinants. I. fermion, Poisson and boson point processes. J. Funct. Anal. **205**, 414–463 (2003)
499. Shivam, S., De Luca, A., Huse, D.A., Chan, A.: Many-body quantum chaos and emergence of Ginibre ensemble. Phys. Rev. Lett. **130**, 140403 (2023)
500. Simm, N.: Central limit theorems for the real eigenvalues of large Gaussian random matrices. Random Matrices Theory Appl. **6**, 1750002 (2017)
501. Simm, N.: On the real spectrum of a product of Gaussian matrices. Electron. Commun. Probab. **22**, 11 (2017)
502. Sinclair, C.D.: Averages over Ginibre's ensemble of random real matrices. Int. Math. Res. Not. **2007**, rnm015, 15 pp (2007)
503. Sinclair, C.D., Yattselev, M.L.: The reciprocal Mahler ensembles of random polynomials. Random Matrices Theory Appl. **8**, 1950012 (2019)
504. Smith, E.R.: Effects of surface charge on the two-dimensional one-component plasma: I. Single double layer. J. Phys. A **15**, 3861–3868 (1982)
505. Smith, N.R., Le Doussal, P., Majumdar, S.N., Schehr, G.: Counting statistics for noninteracting fermions in a rotating trap. Phys. Rev. A **105**, 043315 (2022)
506. Sommers, H.-J.: Symplectic structure of the real Ginibre ensemble. J. Phys. A **40**, F671 (2007)
507. Sommers, H.-J., Crisanti, A., Sompolinsky, H., Stein, Y.: Spectrum of large random asymmetric matrices. Phys. Rev. Lett. **60**, 1895–1898 (1988)
508. Sommers, H.-J., Khoruzhenko, B.A.: Schur function averages for the real Gininbre ensemble. J. Phys. A **42**, 222002 (2009)
509. Sommers, H.-J., Wieczorek, W.: General eigenvalue correlations for the real Ginibre ensemble. J. Phys. A **41**, 405003 (2008)
510. Soshnikov, A.: Determinantal random point fields. Russ. Math. Surv. **55**, 923–975 (2000)
511. Steinerberger, S.: On the logarithmic energy of points on $\mathbb{S}^2$. JAMA **148**, 187–211 (2022)
512. Stillinger, F.H., Lovett, R.: General restriction on the distribution of ions in electrolytes. J. Chem. Phys. **49**, 1991–1994 (1968)
513. Szegö, G.: Orthogonal Polynomials, 4th edn. American Mathematical Society, Providence R.I. (1975)
514. Tao, T., Vu, V.: Random matrices: Universality of ESD and the Circular Law. Ann. Probab. **38**, 2023–2065 (2010). With an appendix by Manjunath Krishnapur
515. Tao, T., Vu, V.: Random matrices: universality of local spectral statistics of non-Hermitian matrices. Ann. Probab. **43**, 782–874 (2015)
516. Tarnowski, W.: Real spectra of large real asymmetric random matrices. Phys. Rev. E **105**, L012104 (2022)
517. Téllez, G.: Debye-Hückel theory for two-dimensional Coulomb systems living on a finite surface without boundaries. Physica A **349**, 155 (2005)
518. Téllez, G.: Two-dimensional Coulomb systems in a disk with ideal dielectric boundaries. J. Stat. Phys. **104**, 945–970 (2001)
519. Téllez, G.: Debye-Hückel theory for two-dimensional Coulomb systems living on a finite surface without boundaries. Physica A **349**, 155 (2005)
520. Téllez, G., Forrester, P.J.: Finite size study of the 2dOCP at $\Gamma = 4$ and $\Gamma = 6$. J. Stat. Phys. **97**, 489–521 (1999)
521. Téllez, G., Forrester, P.J.: Expanded Vandermonde powers and sum rules for the two-dimensional one-component plasma. J. Stat. Phys. **148**, 824–855 (2012)

522. Temme, N.M.: Special Functions, An Introduction to the Classical Functions of Mathematical Physics. A Wiley Interscience Publication. Wiley, New York (1996)
523. Torquato, S.: Hyperuniformity and its generalizations. Phys. Rev. E **94**, 022122 (2016)
524. Torquato, S., Stillinger, F.: Local density fluctuations, hyperuniformity, and order metrics. Phys. Rev. E **68**, 041113 (2003)
525. Torres, A., Téllez, G.: Finite size corrections for Coulomb systems in the Debye-Hückel regime. J. Phys. A **37**, 2121 (2004)
526. Tracy, C.A., Widom, H.: Correlation functions, cluster functions and spacing distributions in random matrices. J. Stat. Phys. **92**, 809–835 (1998)
527. Tracy, C.A., Widom, H.: On orthogonal and symplectic matrix ensembles. Comm. Math. Phys. **177**, 727–754 (1996)
528. Tribe, R., Zaboronski, O.: Pfaffian formulae for one dimensional coalescing and annihilating systems. Electron. J. Probab. **16**, 2080 (2011)
529. Tricomi, F.G.: Asymptotische Eigenschaften der unvollständigen Gammafunktion. Math. Z. **53**, 136–148 (1950)
530. Walters, M., Starr, S.: A note on mixed matrix moments for the complex Ginibre ensemble. J. Math. Phys. **56**, 013301 (2015)
531. Webb, C., Wong, M.D.: On the moments of the characteristic polynomial of a Ginibre random matrix. Proc. Lond. Math. Soc. **118**, 1017–1056 (2019)
532. Whittaker, E.T., Watson, G.N.: A Course of Modern Analysis, 4th edn. Cambridge University Press, Cambridge (1927)
533. Wigner, E.P.: Statistical properties of real symmetric matrices with many dimensions. In: Canadian Mathematical Congress Proceedings, p. 174. University of Toronto Press (1957)
534. Wishart, J.: The generalized product moment distribution in samples from a normal multivariate population. Biometrika **20A**, 32–43 (1928)
535. Witte, N.S., Forrester, P.J.: Loop equation analysis of the circular ensembles. JHEP **2015**, 173 (2015)
536. Xiao, Z., Kawabata, K., Luo, X., Ohtsuki, T., Shindou, R.: Level statistics of real eigenvalues in non-Hermitian systems. Phys. Rev. X **4**, 043196 (2022)
537. Ye, B., Qiu, L., Wang, X., Guhr, T.: Spectral statistics in directed complex networks and universality of the Ginibre ensemble. Commun. Nonlinear Sci. Numer. Simulat. **20**, 1026–1032 (2015)
538. Zabrodin, A.: Matrix models and growth processes: from viscous flows to the quantum Hall effect. Applications of Random Matrices in Physics. NATO Sci. Ser. II Math. Phys. Chem. **221**, 261–318 (2006)
539. Zabrodin, A., Wiegmann, P.: Large-N expansion in the 2D Dyson gas. J. Phys. A **39**, 8933–8963 (2006)
540. Zeng, X.: Eigenvalues distribution for products of independent spherical ensembles. J. Phys. A **49**, 235201 (2016)
541. Zyczkowski, K., Penson, K.A., Nechita, I., Collins, B.: Generating random density matrices. J. Math. Phys. **52**, 06220 (2011)
542. Zyczkowski, K., Sommers, H.-J.: Truncations of random unitary matrices. J. Phys. A **33**, 2045–2057 (2000)